MULTI-CORE
EMBEDDED SYSTEMS

Embedded Multi-Core Systems

Series Editors

Fayez Gebali and Haytham El Miligi
University of Victoria
Victoria, British Columbia

Multi-Core Embedded Systems, *Georgios Kornaros*

MULTI-CORE EMBEDDED SYSTEMS

Edited by
Georgios Kornaros

CRC Press
Taylor & Francis Group
Boca Raton London New York

CRC Press is an imprint of the
Taylor & Francis Group, an **informa** business

Works does not warrant the accuracy of the text of exercises in this book. This book's use or discussion of MATLAB® and Simulink® software or related products does not constitute endorsement or sponsorship by The MathWorks of a particular pedagogical approach or particular use of the MATLAB® and Simulink® software.

CRC Press
Taylor & Francis Group
6000 Broken Sound Parkway NW, Suite 300
Boca Raton, FL 33487-2742

First issued in paperback 2019

© 2010 by Taylor & Francis Group, LLC
CRC Press is an imprint of Taylor & Francis Group, an Informa business

No claim to original U.S. Government works

ISBN-13: 978-1-4398-1161-0 (hbk)
ISBN-13: 978-0-367-38430-2 (pbk)

Library of Congress Cataloging-in-Publication Data

Multi-core embedded systems / editor, Georgios Kornaros.
    p. cm. -- (Embedded multi-core systems)
    "A CRC title."
    Includes bibliographical references and index.
    ISBN 978-1-4398-1161-0 (hard back : alk. paper)
    1. Embedded computer systems. 2. Multiprocessors. 3. Parallel processing
(Electronic computers) I. Kornaros, Georgios. II. Title. III. Series.

    TK7895.E42M848 2010
    004.16--dc22
                                                   2009051515

Visit the Taylor & Francis Web site at
http://www.taylorandfrancis.com

and the CRC Press Web site at
http://www.crcpress.com

Contents

List of Figures   xiii

List of Tables   xxi

Foreword   xxiii

Preface   xxv

1 Multi-Core Architectures for Embedded Systems   1
C.P. Ravikumar
   1.1   Introduction 2
      1.1.1   What Makes Multiprocessor Solutions Attractive? . . 3
   1.2   Architectural Considerations 9
   1.3   Interconnection Networks 11
   1.4   Software Optimizations 13
   1.5   Case Studies 14
      1.5.1   HiBRID-SoC for Multimedia Signal Processing 14
      1.5.2   VIPER Multiprocessor SoC 16
      1.5.3   Defect-Tolerant and Reconfigurable MPSoC 17
      1.5.4   Homogeneous Multiprocessor for Embedded Printer
            Application 18
      1.5.5   General Purpose Multiprocessor DSP 20
      1.5.6   Multiprocessor DSP for Mobile Applications 21
      1.5.7   Multi-Core DSP Platforms 23
   1.6   Conclusions 25
   Review Questions 25
   Bibliography 27

2 Application-Specific Customizable Embedded Systems   31
Georgios Kornaros
   2.1   Introduction 32
   2.2   Challenges and Opportunities 34
      2.2.1   Objectives 35
   2.3   Categorization 37
      2.3.1   Customized Application-Specific Processor Techniques 37

2.3.2   Customized Application-Specific On-Chip Interconnect
Techniques   40
2.4   Configurable Processors and Instruction Set Synthesis   41
    2.4.1   Design Methodology for Processor Customization . . .   43
    2.4.2   Instruction Set Extension Techniques   44
    2.4.3   Application-Specific Memory-Aware Customization . .   48
    2.4.4   Customizing On-Chip Communication Interconnect .   48
    2.4.5   Customization of MPSoCs   49
2.5   Reconfigurable Instruction Set Processors   52
    2.5.1   Warp Processing   53
2.6   Hardware/Software Codesign   54
2.7   Hardware Architecture Description Languages   55
    2.7.1   LISATek Design Platform   57
2.8   Myths and Realities   58
2.9   Case Study: Realizing Customizable Multi-Core Designs . . .   60
2.10 The Future: System Design with Customizable Architectures,
Software, and Tools   62
Review Questions   63
Bibliography   63

3   Power Optimization in Multi-Core System-on-Chip   71
Massimo Conti, Simone Orcioni, Giovanni Vece and Stefano Gigli
3.1   Introduction   72
3.2   Low Power Design   74
    3.2.1   Power Models   75
    3.2.2   Power Analysis Tools   80
3.3   PKtool   82
    3.3.1   Basic Features   82
    3.3.2   Power Models   83
    3.3.3   Augmented Signals   84
    3.3.4   Power States   85
    3.3.5   Application Examples   86
3.4   On-Chip Communication Architectures   87
3.5   NOCEXplore   90
    3.5.1   Analysis   91
3.6   DPM and DVS in Multi-Core Systems   95
3.7   Conclusions   100
Review Questions   101
Bibliography   102

**4   Routing Algorithms for Irregular Mesh-Based Network-on-
Chip   111**
Shu-Yen Lin and An-Yeu (Andy) Wu
4.1   Introduction   112
4.2   An Overview of Irregular Mesh Topology   113

4.2.1   2D Mesh Topology   113
4.2.2   Irregular Mesh Topology   113
4.3   Fault-Tolerant Routing Algorithms for 2D Meshes   115
4.3.1   Fault-Tolerant Routing Using Virtual Channels   116
4.3.2   Fault-Tolerant Routing with Turn Model   117
4.4   Routing Algorithms for Irregular Mesh Topology   126
4.4.1   Traffic-Balanced OAPR Routing Algorithm   127
4.4.2   Application-Specific Routing Algorithm   132
4.5   Placement for Irregular Mesh Topology   136
4.5.1   OIP Placements Based on Chen and Chiu's Algorithm   137
4.5.2   OIP Placements Based on OAPR   140
4.6   Hardware Efficient Routing Algorithms   143
4.6.1   Turns-Table Routing (TT)   146
4.6.2   XY-Deviation Table Routing (XYDT)   147
4.6.3   Source Routing for Deviation Points (SRDP)   147
4.6.4   Degree Priority Routing Algorithm   148
4.7   Conclusions   151
Review Questions   151
Bibliography   151

5   **Debugging Multi-Core Systems-on-Chip**   **155**
Bart Vermeulen and Kees Goossens
5.1   Introduction   156
5.2   Why Debugging Is Difficult   158
5.2.1   Limited Internal Observability   158
5.2.2   Asynchronicity and Consistent Global States   159
5.2.3   Non-Determinism and Multiple Traces   161
5.3   Debugging an SoC   163
5.3.1   Errors   164
5.3.2   Example Erroneous System   165
5.3.3   Debug Process   166
5.4   Debug Methods   169
5.4.1   Properties   169
5.4.2   Comparing Existing Debug Methods   171
5.5   CSAR Debug Approach   174
5.5.1   Communication-Centric Debug   175
5.5.2   Scan-Based Debug   175
5.5.3   Run/Stop-Based Debug   176
5.5.4   Abstraction-Based Debug   176
5.6   On-Chip Debug Infrastructure   178
5.6.1   Overview   178
5.6.2   Monitors   178
5.6.3   Computation-Specific Instrument   180
5.6.4   Protocol-Specific Instrument   181
5.6.5   Event Distribution Interconnect   182

      5.6.6   Debug Control Interconnect 183
      5.6.7   Debug Data Interconnect 183
  5.7 Off-Chip Debug Infrastructure 184
      5.7.1   Overview 184
      5.7.2   Abstractions Used by Debugger Software 184
  5.8 Debug Example 190
  5.9 Conclusions 193
  Review Questions 194
  Bibliography 194

6 System-Level Tools for NoC-Based Multi-Core Design   **201**
Luciano Bononi, Nicola Concer, and Miltos Grammatikakis
  6.1 Introduction 202
      6.1.1   Related Work 204
  6.2 Synthetic Traffic Models 206
  6.3 Graph Theoretical Analysis 207
      6.3.1   Generating Synthetic Graphs Using TGFF 209
  6.4 Task Mapping for SoC Applications 210
      6.4.1   Application Task Embedding and Quality Metrics . . 210
      6.4.2   SCOTCH Partitioning Tool 214
  6.5 OMNeT++ Simulation Framework 216
  6.6 A Case Study 217
      6.6.1   Application Task Graphs 217
      6.6.2   Prospective NoC Topology Models 218
      6.6.3   Spidergon Network on Chip 219
      6.6.4   Task Graph Embedding and Analysis 221
      6.6.5   Simulation Models for Proposed NoC Topologies . . . 223
      6.6.6   Mpeg4: A Realistic Scenario 227
  6.7 Conclusions and Extensions 231
  Review Questions 234
  Bibliography 235

7 Compiler Techniques for Application Level Memory Optimization for MPSoC   **243**
Bruno Girodias, Youcef Bouchebaba, Pierre Paulin, Bruno Lavigueur, Gabriela Nicolescu, and El Mostapha Aboulhamid
  7.1 Introduction 244
  7.2 Loop Transformation for Single and Multiprocessors 245
  7.3 Program Transformation Concepts 246
  7.4 Memory Optimization Techniques 248
      7.4.1   Loop Fusion 249
      7.4.2   Tiling 249
      7.4.3   Buffer Allocation 249
  7.5 MPSoC Memory Optimization Techniques 250
      7.5.1   Loop Fusion 251

	7.5.2	Comparison of Lexicographically Positive and Positive Dependency	252
	7.5.3	Tiling	253
	7.5.4	Buffer Allocation	254
7.6	Technique Impacts		255
	7.6.1	Computation Time	255
	7.6.2	Code Size Increase	256
7.7	Improvement in Optimization Techniques		256
	7.7.1	Parallel Processing Area and Partitioning	256
	7.7.2	Modulo Operator Elimination	259
	7.7.3	Unimodular Transformation	260
7.8	Case Study		261
	7.8.1	Cache Ratio and Memory Space	262
	7.8.2	Processing Time and Code Size	263
7.9	Discussion		263
7.10	Conclusions		264
	Review Questions		265
	Bibliography		266

8 Programming Models for Multi-Core Embedded Software   269
Bijoy A. Jose, Bin Xue, Sandeep K. Shukla and Jean-Pierre Talpin
8.1	Introduction		270
8.2	Thread Libraries for Multi-Threaded Programming		272
8.3	Protections for Data Integrity in a Multi-Threaded Environment		276
	8.3.1	Mutual Exclusion Primitives for Deterministic Output	276
	8.3.2	Transactional Memory	278
8.4	Programming Models for Shared Memory and Distributed Memory		279
	8.4.1	OpenMP	279
	8.4.2	Thread Building Blocks	280
	8.4.3	Message Passing Interface	281
8.5	Parallel Programming on Multiprocessors		282
8.6	Parallel Programming Using Graphic Processors		283
8.7	Model-Driven Code Generation for Multi-Core Systems		284
	8.7.1	StreamIt	285
8.8	Synchronous Programming Languages		286
8.9	Imperative Synchronous Language: Esterel		288
	8.9.1	Basic Concepts	288
	8.9.2	Multi-Core Implementations and Their Compilation Schemes	289
8.10	Declarative Synchronous Language: LUSTRE		290
	8.10.1	Basic Concepts	291
	8.10.2	Multi-Core Implementations from LUSTRE Specifications	291

8.11 Multi-Rate Synchronous Language: SIGNAL 292
   8.11.1 Basic Concepts 292
   8.11.2 Characterization and Compilation of SIGNAL 293
   8.11.3 SIGNAL Implementations on Distributed Systems . . 294
   8.11.4 Multi-Threaded Programming Models for SIGNAL . . 296
8.12 Programming Models for Real-Time Software 299
   8.12.1 Real-Time Extensions to Synchronous Languages . . . 300
8.13 Future Directions for Multi-Core Programming 301
Review Questions 302
Bibliography 305

9 Operating System Support for Multi-Core Systems-on-Chips 309
Xavier Guérin and Frédéric Pétrot
9.1 Introduction 310
9.2 Ideal Software Organization 311
9.3 Programming Challenges 313
9.4 General Approach 314
   9.4.1 Board Support Package 314
   9.4.2 General Purpose Operating System 317
9.5 Real-Time and Component-Based Operating System Models 322
   9.5.1 Automated Application Code Generation and RTOS Modeling 322
   9.5.2 Component-Based Operating System 326
9.6 Pros and Cons 329
9.7 Conclusions 330
Review Questions 332
Bibliography 333

10 Autonomous Power Management in Embedded Multi-Cores 337
Arindam Mukherjee, Arun Ravindran, Bharat Kumar Joshi,
Kushal Datta and Yue Liu
10.1 Introduction 338
   10.1.1 Why Is Autonomous Power Management Necessary? 339
10.2 Survey of Autonomous Power Management Techniques . . . 342
   10.2.1 Clock Gating 342
   10.2.2 Power Gating 343
   10.2.3 Dynamic Voltage and Frequency Scaling 343
   10.2.4 Smart Caching 344
   10.2.5 Scheduling 345
   10.2.6 Commercial Power Management Tools 346
10.3 Power Management and RTOS 347
10.4 Power-Smart RTOS and Processor Simulators 349
   10.4.1 Chip Multi-Threading (CMT) Architecture Simulator 350
10.5 Autonomous Power Saving in Multi-Core Processors 351
   10.5.1 Opportunities to Save Power 353

10.5.2 Strategies to Save Power 354
10.5.3 Case Study: Power Saving in Intel Centrino 356
10.6 Power Saving Algorithms 358
10.6.1 Local PMU Algorithm 358
10.6.2 Global PMU Algorithm 358
10.7 Conclusions 360
Review Questions 362
Bibliography 363

11 Multi-Core System-on-Chip in Real World Products   **369**
Gajinder Panesar, Andrew Duller, Alan H. Gray and Daniel Towner
11.1 Introduction 370
11.2 Overview of picoArray Architecture 371
11.2.1 Basic Processor Architecture 371
11.2.2 Communications Interconnect 373
11.2.3 Peripherals and Hardware Functional Accelerators . . 373
11.3 Tool Flow 375
11.3.1 picoVhdl Parser (Analyzer, Elaborator, Assembler) . . 376
11.3.2 C Compiler 376
11.3.3 Design Simulation 378
11.3.4 Design Partitioning for Multiple Devices 381
11.3.5 Place and Switch 381
11.3.6 Debugging 381
11.4 picoArray Debug and Analysis 381
11.4.1 Language Features 382
11.4.2 Static Analysis 383
11.4.3 Design Browser 383
11.4.4 Scripting 385
11.4.5 Probes 387
11.4.6 FileIO 387
11.5 Hardening Process in Practice 388
11.5.1 Viterbi Decoder Hardening 389
11.6 Design Example 392
11.7 Conclusions 396
Review Questions 396
Bibliography 397

12 Embedded Multi-Core Processing for Networking   **399**
Theofanis Orphanoudakis and Stylianos Perissakis
12.1 Introduction 400
12.2 Overview of Proposed NPU Architectures 403
12.2.1 Multi-Core Embedded Systems for Multi-Service
Broadband Access and Multimedia Home Networks . 403
12.2.2 SoC Integration of Network Components and Examples
of Commercial Access NPUs 405

           12.2.3 NPU Architectures for Core Network Nodes and
                   High-Speed Networking and Switching   407
      12.3 Programmable Packet Processing Engines   412
           12.3.1 Parallelism   413
           12.3.2 Multi-Threading Support   418
           12.3.3 Specialized Instruction Set Architectures   421
      12.4 Address Lookup and Packet Classification Engines   422
           12.4.1 Classification Techniques   424
           12.4.2 Case Studies   426
      12.5 Packet Buffering and Queue Management Engines   431
           12.5.1 Performance Issues   433
           12.5.2 Design of Specialized Core for Implementation of Queue
                   Management in Hardware   435
      12.6 Scheduling Engines   442
           12.6.1 Data Structures in Scheduling Architectures   443
           12.6.2 Task Scheduling   444
           12.6.3 Traffic Scheduling   450
      12.7 Conclusions   453
      Review Questions   455
      Bibliography   459

Index   **465**

List of Figures

1.1 Power/performance over the years. The solid line shows the prediction by Gene Frantz. The dotted line shows the actual value for digital signal processors over the years. The 'star' curve shows the power dissipation for mobile devices over the years. 4

1.2 Performance of multi-core architectures. The x-axis shows the logarithm of the number of processors to the base 2. The y-axis shows the run-time of the multi-core for a benchmark. . 10

1.3 Network-on-Chip architectures for an SoC. 12

1.4 Architecture of HiBRID multiprocessor SoC. 15

1.5 Architecture of VIPER multiprocessor-on-a-chip. 16

1.6 Architecture of a single-chip multiprocessor for video applications with four processor nodes. 18

1.7 Design alternates for MPOC. 19

1.8 Daytona general purpose multiprocessor and its processor architecture. 20

1.9 Chip block diagram of OMAP4430 multi-core platform. . . . 21

1.10 Chip block diagram of C6474 multi-core DSP platform. . . . 24

2.1 Different technologies in the era of designing embedded system-on-chip. Application-specific integrated processors (ASIPs) and reconfigurable ASIPs combine both the flexibility of general purpose computing with the efficiency in performance, power and cost of ASICs. 34

2.2 Optimizing embedded systems-on-chips involves a wide spectrum of techniques. Balancing across often conflicting goals is a challenging task determined mainly by the designer's expertise rather than the properties of the embedded application. 36

2.3 Extensible processor core versus component-based customized SoC. Computation elements are tightly coupled with the base CPU pipeline (a), while (b), in component-based designs, intellectual property (IP) cores are integrated in SoCs using different communication architectures (bus, mesh, NoC, etc.). 41

2.4   Typical methodology for design space exploration of application specific processor customization. Different algorithms and metrics are applied by researchers and industry for each individual step to achieve the most efficient implementation and time to market. 44

2.5   A sample data flow subgraph. Usually each node is annotated with area and timing estimates before passing to a selection algorithm. 46

2.6   A RASIP integrating the general purpose processor with RFUs. 52

2.7   LISATek infrastructure based on LISA architecture specification language. Retargetable software development tools (C compiler, assembler, simulator, debugger, etc.) permit iterative exploration of varying target processor configurations. . 58

2.8   Tensilica customization and extension design flow. Through Xplorer, Tensilica's design environment, the designer has access to the tools needed for development of custom instructions and configuration of the base processor. 61

3.1   Power analysis and optimization at different levels of the design. 76

3.2   Complexity estimation from SystemC source code. 78

3.3   I2C driver instruction set. 79

3.4   Power dissipation model added to the functional model. . . 80

3.5   System level power modeling and analysis. 80

3.6   `power_model` architecture. 83

3.7   Example of association between `sc_module` and `power_model`. 84

3.8   PKtool simulation flux. 84

3.9   NoC performance comparison for a 16-node 2D mesh network: steady-state network average delay for three different traffic scenarios. 92

3.10   NoC performance comparison for a 16-node 2D mesh network: steady-state network throughput for three different traffic scenarios. 92

3.11   Example of probabilistic analysis. The message delay probability density referred to all messages sent and received by a NoC under traffic equally distributed with 50% of messages sent in burst and message generation intensity of 32%; network has 16 nodes, topology is 2D mesh and routing is deterministic. 93

3.12 Example of temporal evolution analysis. The graph shows the number of flits in a router on top side of a 2D mesh network. Each router has globally 120 flit memory of capacity distributed in five input and five out ports. The figure shows that, for this traffic intensity and scenarios, buffer configuration is oversized and the performance is maintained even if the router has a smaller memory. 94

3.13 Example of power graph where power state is indicated over time, router by router. Dark color means high power state. Router power machine has nine power states and follows ACPI standard: values from 1 to 4 are ON states, values from 5 to 8 are SLEEP states and value 9 is the OFF state. 95

3.14 Four ON states, four SLEEP states and OFF state of the ACPI standard. 96

3.15 DPM and communication architectures. 97

3.16 Clock frequency, supply voltage and power dissipation for the different power states of the ACPI standard. 98

3.17 Percentage of the time the three masters and two slaves and the bus are in the different power states during simulation in a low bus traffic test case with local DPM and global DPM. 99

3.18 Energy and bus throughput normalized to the architecture without DPM. 99

3.19 Qualitative results in terms of bus throughput as a function of bus traffic intensity for different DPM architectures and bus arbitration algorithm. 100

3.20 Qualitative results in terms of average energy per transfer as a function of bus traffic intensity for different DPM architectures and bus arbitration algorithm. 101

4.1 (a) A conventional 6 × 6 2D mesh and (b) a 6 × 6 irregular mesh with 1 OIP and 31 normal-sized IPs. 114

4.2 Possible cycles and turns in 2D mesh. 117

4.3 Six turns form a cycle and allow deadlock. 118

4.4 The turns allowed by (a) west-first algorithm, (b) north-last algorithm, and (c) negative-first algorithm. 119

4.5 The six turns allowed in odd-even turn models. 119

4.6 A minimal routing algorithm *ROUTE* that is based on the odd-even turn model. 120

4.7 The localized algorithm to form extended faulty blocks. . . . 121

4.8 Three examples to form extended faulty blocks. 122

4.9 E-XY routing algorithm. 123

4.10 Eight possible cases of the E-XY in normal mode. 123

4.11 Four cases of the E-XY in abnormal mode: (a) south-to-north, (b) north-to-south, (c) west-to-east, and (d) east-to-west direction. 124

4.12   An example to form faulty blocks for Chen and Chiu's algorithm.   125

4.13   Two examples of f-rings and f-chains: (a) one f-ring and one f-chain in a 6 × 6 mesh and (b) one f-ring and eight different types of f-chains in a 10 × 10 mesh.   126

4.14   Pseudo codes of the procedure *Message-Route Modified.* . .   126

4.15   Pseudo codes of the procedure *Normal-Route.*   127

4.16   Pseudo codes of the procedure *Ring-Route.*   128

4.17   Pseudo codes of the procedure *Chain-Route Modified.*   129

4.18   Pseudo codes of the procedure *Overlapped-Ring Chain Route.*   130

4.19   Examples of Chen and Chiu's routing algorithm: (a) the routing paths (RF, CF, and RO) in *Normal-Route*, and (b) Two examples of *Ring-Route* and *Chain-Route.*   131

4.20   Traffic loads around the OIPs by using (a) Chen and Chiu's algorithm   [5] (unbalanced), (b) the extended X-Y routing algorithm   [34] (unbalanced), and (c) the OAPR   [21] (balanced).   131

4.21   The OAPR: (a) eight default routing cases and (b) some cases to detour OIPs.   133

4.22   Restrictions on OIP placements for the OAPR.   133

4.23   The OAPR design flow: (a) the routing logic in the five-port router model, (b) the flowchart of the OAPR design flow, and (c) the flowchart to update LUTs.   134

4.24   Overview of APSRA design methodology.   135

4.25   An example of APSRA methodology: (a) CG, (b) TG, (c) CDG, (d) $ASCDG$, and (e) the concurrency of the two loops.   137

4.26   An example of the routing table in the west input port of node X: (a) original routing table and (b) compressed routing table. 138

4.27   An example of the compressed routing table in node X with loss of adaptivity: (a) the routing table by merging destinations A and B and (b) the routing table by merging regions $R1$ and $R3$.   139

4.28   OIP placement with different sizes and locations.   140

4.29   Effect on latency with central region in NoC.   141

4.30   Latency for horizontal shift of positions.   141

4.31   Latency for vertical shift of positions.   142

4.32   OIP placements with different orientations.   142

4.33   An example of a 12 × 12 distribution graph.   144

4.34   Latencies of one 3 × 3 OIP placed on a 12 × 12 mesh. . . .   144

4.35   Latencies of one four-unit OIP placed on a 12 × 12 mesh: (a) horizontal placements and (b) vertical placements.   145

4.36   (a) Routing paths without turning to destination D and (b) Routing paths with two turns to D.   146

4.37   TT routing algorithm for one destination D.   147

4.38   XYDT routing algorithm for one destination D.   148

4.39 Degree priority routing algorithm. 149
4.40 Examples showing the degrees of the nodes A, B, C, and D. 150
4.41 An example of the degree priority routing algorithm. 150
4.42 Routing tables of nodes 1, 6, 10, C, and X. 150

5.1   Design refinement process. 157
5.2   Safe asynchronous communication using a handshake. 160
5.3   Lack of consistent global state with multiple, asynchronous
      clocks. 161
5.4   Non-determinism in communication between clock domains. 162
5.5   Example of system communication via shared memory. . . . 162
5.6   System traces and permanent intermittent errors. 165
5.7   Scope reduced to include Master 2 only. 166
5.8   Scope reduced to include Master 1 and Master 2 only. . . . 167
5.9   Debug flow charts. 168
5.10 Run/stop debug methods. 175
5.11 Debug abstractions. 177
5.12 Debug hardware architecture. 179
5.13 Example system under debug. 181
5.14 Off-chip debug infrastructure with software architecture. . . 185
5.15 Physical and logical interconnectivity. 189

6.1   Our design space exploration approach for system-level NoC
      selection. 205
6.2   METIS-based NEATO visualization of the Spidergon NoC lay-
      out. 208
6.3   Source file for SCOTCH partitioning tool. 214
6.4   Target file for SCOTCH partitioning tool. 215
6.5   Application models for (a) 2-rooted forest (SRF), (b) 2-rooted
      tree (SRT), (c) 2-node 2-rooted forest(MRF) application task
      graphs. 218
6.6   The Mpeg4 decoder task graph. 218
6.7   The Spidergon topology translates to simple, low-cost VLSI
      implementation. 220
6.8   Edge dilation for (a) 2-rooted and (b) 4-rooted forest, (c) 2
      node-disjoint and (d) 4 node-disjoint trees, (e) 2 node-disjoint
      2-routed and (f) 4 node-disjoint 4-routed forests in function
      of the network size. 222
6.9   Relative edge expansion for 12-node Mpeg4 for different target
      graphs. 223
6.10 Model of the router used in the considered NoC architectures. 225
6.11 Maximum throughput as a function of the network size for (a)
      2-rooted forest, (b) 4-rooted forest (SRF), (c) 2-rooted tree,
      (d) 4-rooted tree (SRT), (e) 2-node 2-rooted forest and (f)
      4-node 2-rooted forest (MRF) and different NoC topologies. 226

6.12 Amount of memory required by each interconnect. 228
6.13 (a) Task execution time and (b) average path length for Mpeg4 traffic on the considered NoC architectures. 228
6.14 Average throughput on router's output port for (a) Spidergon, (b) ring, (c) mesh and (d) unbuffered crossbar architecture. 230
6.15 Network RTT as a function of the initiators' offered load. . 231
6.16 Future work: dynamic scheduling of tasks. 233

7.1 Input code: the depth of each loop nest L_k is n (n loops), A_k is n dimensional. 247
7.2 Code example and its iteration domain. 248
7.3 An example of loop fusion. 249
7.4 An example of tiling. 250
7.5 An example of buffer allocation. 250
7.6 An example of three loop nests. 251
7.7 Partitioning after loop fusion. 252
7.8 Difference between positive and lexicographically positive dependence. 253
7.9 Tiling technique. 254
7.10 Buffer allocation for array B. 255
7.11 Classic partitioning. 257
7.12 Different partitioning. 257
7.13 Buffer allocation for array B with new partitioning. 258
7.14 Sub-division of processor P_1's block. 259
7.15 Elimination of modulo operators. 260
7.16 Execution order (a) without fusion (b) after fusion and (c) after unimodular transformation. 261
7.17 StepNP platform. 262
7.18 DCache hit ratio results for four CPUs. 263
7.19 Processing time results for four CPUs. 264

8.1 Abstraction levels of multi-core software directives, utilities and tools. 272
8.2 Threading structure of fork-join model. 273
8.3 Work distribution model. 274
8.4 Pipeline threading model. 274
8.5 Scheduling threading structure. 277
8.6 Parallel functions in thread building blocks. 281
8.7 Program flow in host and device for NVIDIA CUDA. 283
8.8 Stream structures using *filters*. 285
8.9 OC program in Listing 8.2 distributed into two locations. . . 290
8.10 LUSTRE to TTA implementation flow. 292
8.11 Weakly endochronous program with diamond property. . . . 295
8.12 Process-based threading model. 296
8.13 Fine grained thread structure of polychrony. 297

8.14 SDFG-based multi-threading for SIGNAL. 298
8.15 TAXYS tool structure with event handling and code genera-
      tion [23]. 300
8.16 Task precedence in a multi-rate real time application [37]. . 301

9.1   Example of HMC-SoC. 310
9.2   Ideal software organization. 312
9.3   Parallelization of an application. 313
9.4   BSP-based software organization. 315
9.5   BSP-based application development. 316
9.6   BSP-based boot-up sequence strategies. 316
9.7   Software organization of a GPOS-based application. 318
9.8   GPOS-based application development. 319
9.9   GPOS-based boot-up sequence. 320
9.10 Software organization of a generated application. 323
9.11 Examples of computations models. 324
9.12 Tasks graph with RTOS elements. 325
9.13 Component architecture. 326
9.14 Component-based OS software organization. 327
9.15 Example of a dependency graph. 328

10.1 Pipelined micro-architecture of an embedded variant of Ultra-
      SPARC T1. 352
10.2 Trap logic unit. 352
10.3 Chip block diagram. 353
10.4 Architecture of autonomous hardware power saving logic. . 355
10.5 Global power management unit. 356

11.1 picoBus interconnect structure. 371
11.2 Processor structure. 372
11.3 VLIW and execution unit structure in each processor. . . . 372
11.4 Tool flow. 377
11.5 Behavioral simulation instance 380
11.6 Example of where-defined program analysis. 384
11.7 Design browser display. 385
11.8 Diagnostics output from 802.16 PHY. 386
11.9 Hardening approach. 389
11.10 Software implementation of Viterbi decoder and testbench. . 390
11.11 Partially hardened implementation of Viterbi decoder and
      testbench. 391
11.12 Fully hardened implementation of Viterbi decoder and test-
      bench. 391
11.13 Femtocell system. 393
11.14 Femtocell. 394
11.15 Femtocell reference board. 395

12.1 Taxonomy of network processing functions. 401

12.2 Available clock cycles for processing each packet as a function of clock frequency and link rate in average case (mean packet size of 256 bytes is assumed). 405

12.3 Typical architecture of integrated access devices (IADs) based on discrete components. 406

12.4 Typical architecture of SoC integrated network processor for access devices and residential gateways. 407

12.5 Evolution of switch node architectures: (a) 1st generation (b) 2nd generation (c) 3rd generation. 408

12.6 PDU flow in a distributed switching node architecture. . . 409

12.7 Centralized (a) and distributed (b) NPU-based switch architectures. 409

12.8 Generic NPU architecture. 410

12.9 (a) Parallel RISC NPU architecture (b) pipelined RISC NPU architecture (c) state-machine NPU architecture. 412

12.10 (a) Intel IXP 2800 NPU, (b) Freescale C-5e NPU. 414

12.11 Architecture of PRO3 reprogrammable pipeline module (RPM). 415

12.12 The concept of the EZchip architecture. 416

12.13 Block diagram of the Agere (LSI) APP550. 417

12.14 The PE (microengine) of the Intel IXP2800. 419

12.15 TCAM organization [Source: Netlogic]. 424

12.16 Mapping of rules to a two-dimensional classifier. 426

12.17 iAP organization. 429

12.18 EZchip table lookup architecture. 430

12.19 Packet buffer manager on a system-on-chip architecture. . . 436

12.20 DMM architecture. 437

12.21 Details of internal task scheduler of NPU architecture [25]. . 446

12.22 Load balancing core implementation [25]. 447

12.23 The Porthos NPU interconnection architecture [32]. 448

12.24 Scheduling in context of processing path of network routing/switching nodes. 450

12.25 Weighted scheduling of flows/queues contending for same egress network port. 451

12.26 (a) Architecture extensions for programmable service disciplines. (b) Queuing requirements for multiple port support. 452

List of Tables

1.1   Growth of VLSI Technology over Four Decades   3

4.1   Rules for Positions and Orientations of OIPs   145

6.1   Initiator's Average Injection Rate and Relative Ratio with
       Respect to UPS-AMP Node   229

8.1   SIGNAL Operators and Clock Relations   294

9.1   Solution Pros and Cons   330

10.1   Power Gating Status Register   346
10.2   Power Gating Status Register   356
10.3   Clock Gating Status Register   357
10.4   DVFS Status Register   357

11.1   Viterbi Decoder Transistor Estimates   392

12.1   DDR-DRAM Throughput Loss Using 1 to 16 Banks   434
12.2   Maximum Rate Serviced When Queue Management Runs on
       IXP 1200   435
12.3   Packet Command and Segment Command Pointer Manipula-
       tion Latency   440
12.4   Performance of DMM   441

Foreword

I am delighted to introduce the first book on multi-core embedded systems. My sincere hope is that you will find the following pages valuable and rewarding.

This book is authored to address many challenging topics related to the multi-core embedded systems research area, starting with multi-core architectures and interconnects, embedded design methodologies for multi-core systems, to mapping of applications, programming paradigms and models of computation on multi-core embedded systems.

With the growing complexity of embedded systems and the rapid improvements in process technology the development of systems-on-chip and of embedded systems increasingly is based on integration of multiple cores, either homogeneous (such as processors) or heterogeneous. Modern systems are increasingly utilizing a combination of processors (CPUs, MCUs, DSPs) which are programmed in software, reconfigurable hardware (FPGAs, PLDs), and custom application–specific hardware. It appears likely that the next generation of hardware will be increasingly programmable, blending processors and configurable hardware.

The book discusses the work done regarding the interactions among multicore systems, applications and software views, and processors configuration and extension, which add a new dimension to the problem space. Multiple cores used in concert prove to be a new challenge forming a concurrent architecture with resources for scheduling, with a number of concurrent processes that perform communication, synchronization and input and output tasks. The choice of programming and threading models, whether symmetric or asymmetric, communication APIs, real-time OS services or application development consist of areas increasingly challenging in the realm of modern multi-core embedded systems-on-chip.

Beyond exploration of different architectures of multi-core embedded systems and of the network-on-chip infrastructures that ushered in support of these SoCs in a straightforward manner, the objectives of this book cover also the presentation of a number of interrelated issues. HW/SW development, tools and verification for multi-core systems, programming models, and models of computation for modern embedded systems are also explored.

The book may be used either in a graduate-level course as a part of the subject of embedded systems, computer architecture, and multi-core systems-on-chips, or as a reference book for professionals and researchers. It provides a clear view of the technical challenges ahead and brings more perspectives into the discussion of multi-core embedded systems. This book is particularly useful for engineers and professionals in industry for easy understanding of the subject matter and as an aid in both software and hardware development of their products.

Acknowledgments

I would like to express my sincere gratitude to all the co-authors for their invaluable contributions, for their constructive comments, and essential assistance throughout this project. All deserve special thanks for utilizing their great expertise to make this book exciting.

I also wish to thank Miltos Grammatikakis for his input on chapter organization and his suggestions.

I would also like to mention my publisher, Nora Konopka, Amy Blalock, and Iris Fahrer for their guidance in authoring and organization.

Finally, I am indebted to my family for their enduring support and encouragement thoughout this long and tiring journey.

A windy Sunday morning of February 2010.

Georgios Kornaros

Preface

Multimedia, video and audio content are now part of mobile networks and hand-held mobile Internet devices. Real-time processing of video and audio streams demands computational performance of a few giga-operations per second, which cannot be obtained using a single processor. An embedded system intended for such an application must also support networking and I/O interfaces, which are best handled by dedicated interface processors that are coordinated by a housekeeping processor. Dedicated processors may also be necessary for parsing and processing video/audio stream and, video/graphics rendering.

Chapter 1 provides an overview of multiprocessor architectures that are evolving for such embedded applications. We argue that the VLSI design challenges involved in designing an equivalent uniprocessor solution for the same application may make such a solution prohibitively expensive, making a multiprocessor system-on-chip an attractive alternative. The chapter begins by highlighting the growing demands on computational speed due to the complexity of applications that run on modern-day mobile embedded systems. Next, we point out the challenges of hardware implementation using nanometer CMOS VLSI technology. We show that there are a number of daunting challenges in the VLSI implementation, such as power dissipation and on-chip process variability. Multiprocessor implementations are becoming attractive to VLSI designers since they can help overcome these challenges. In Section 1.2, we provide an introduction to architectural aspects of multiprocessor embedded systems. We also illustrate the importance of efficient interconnect architectures in a multi-core system-on-chip. Software development for embedded devices presents another set of challenges, as illustrated in Section 1.4. Several illustrative case studies are included.

Chapter 2 discusses the recent trend of developing embedded systems using customization which ranges from designing with application-specific integrated processors (ASIPs) to application-specific MPSoCs. There are a number of challenges and open issues that are presented for each category which give an exciting flavor to customization of a system-on-chip. In addition to ASIPs, aspects of memory-aware development or customization of communication interconnect are discussed along with design space exploration techniques. Case studies of successful automated methodologies provide more insight to the essential factors while developing multicore embedded SoCs. This chapter does not seek to cover every methodology and research project in the area of customizable and extendible processors. Instead, it hopes to serve as an introduction to this rapidly evolving field, bringing interested readers quickly up to speed on developments from the last decade.

The design of emerging systems-on-chips with tens or hundred of cores requires new methodologies and the development of a seamless design flow that integrates existing and completely new tools. System-level tools for power and communication analysis are fundamental for a fast and cost-effective design of complex embedded systems. *Chapter 3* presents the aspects related to system-level power analysis of SoC and on-chip communications. The state of the art of system-level power analysis tools and NoC performance analysis tools is discussed. In particular, two SystemC libraries developed by the authors, and available in the sourceforge web site, are presented: *PKtool* for power analysis and *NOCEXplore* for NoC simulation and performance analysis. Chapter 3 also includes an analysis of Dynamic Power Management (DPM) and Dynamic Voltage Scaling (DVS) techniques applied to on-chip communication architectures.

Emerging multi-core systems increasingly integrate hard intellectual property (IP) blocks from various vendors in regular 2D mesh-based network-on-chip (NoC) designs. Different sizes of these hard IPs (Oversized IPs, OIPs) cause irregular mesh topologies and heavy traffic around the OIPs, which also results in hot spots around the OIPs. *Chapter 4* introduces the concept of irregular mesh topology and corresponding traffic-aware routing algorithms. Traditional fault-tolerant routing algorithms in computer networks are firstly reviewed and discussed. The traffic-balanced OIP Avoidance Pre-Routing (OAPR) algorithm is proposed to deal with the problems of heavy traffic loads around the OIP and unbalanced traffic in the mesh-based, network-on-chips. Different placements of OIPs can influence the networks' performance. Different sizes, locations, and orientations of OIPs are discussed. Chapter 4 also introduces the table-reduction routing algorithms for irregular mesh topologies.

Multi-core embedded system design involves an increased integration of multiple heterogeneous programmable cores in a single chip. *Chapter 5* focuses on the debugging of such complex systems-on-chips. It describes the on-chip debug infrastructure that has to be implemented in a chip at design time to support a run-stop communication-centric debug. A multi-core SoC that features on-chip debug support needs to exhibit a higher level interface to the designer than bits and clock cycles. Chapter 5 shows how to provide a debug engineer with a high-level environment for the debugging of SoCs at multiple levels of abstraction and execution granularities. Finally, a method is discussed where the designer can use an iterative refinement and reduction process to zoom in on the location where and to the point in time when an error in the system first manifests itself.

Chapter 6 follows an open approach by extending to NoC domains existing open-source (and free) tools originating from several application domains, such as traffic modeling, graph theory, parallel computing and network simulation. More specifically, this chapter considers theoretical topological metrics, such as NoC embedding quality, for evaluating the performance of different NoC topologies for common application patterns. The chapter considers both conventional NoC topologies, e.g., mesh and torus, and practical, low-cost circulants: a family of graphs offering small network size granularity and good sustained performance for realistic network sizes (usually below 64 nodes). Application performance and embedding quality are also examined by considering bit- and cycle-accurate system-level NoC simulation of synthetic tree based task graphs and a more realistic application consisting of an MPEG4 decoder.

Memory is becoming a key player for significant improvements in multiprocessor embedded systems (power, performance and area). With the emergence of more embedded multimedia applications in the industry, this issue becomes increasingly vital. These applications often use multi-dimensional arrays to store intermediate results during multimedia processing tasks. A couple of key optimization techniques exist and have been demonstrated on SoC architecture. *Chapter 7* focuses on applying loop transformation techniques for MPSoC environment by exploiting techniques and some adaptation for MPSoC characteristics. These techniques allow for optimization of memory space, reduction of the number of cache misses and extensive improvement of processing time extensively.

The recent transition from single-core to multi-core processors has necessitated new programming paradigms, and models of computations, which can capture concurrency in the target application and compile for parallel implementation. Multiprocessor programming models have been attempted as obvious candidates, but the parallelism and communication models differ for multi-cores due to the on-chip communication, shared memory architectures, and other differences. A departure from the conventional von Neumann sequentialization of computation to a highly concurrent strategy requires formulating newer programming models which combine advantages of existing ones with new ideas specific to multi-core target platforms.

Chapter 8 discusses the available programming models spread across different abstraction levels. At a lower level of abstraction, we discuss the different libraries and primitives defined for multi-threaded programming. The mutual exclusion primitives along with transactional memory models for protecting data integrity are discussed as well. Shared memory models such as OpenMP or Thread Building Blocks highlight the use of directives in parallelizing the existing sequential programs, while distributed memory models such as Message Passing Interface draw attention to the importance of communication between execution cores. Current specialized multi-core platforms, whether homogeneous or heterogeneous in their execution core types, leave room for user designed programming models. Graphic processors, the popular specialized multiprocessing platform for a long time, are being converted into a general purpose multi-core execution unit by new programming models such as CUDA. Such customizable multi-core programming models have succeeded in maximizing the efficiency for their target application areas, but have failed to reach consensus for a singular multi-core programming model for the future. In spite of these outstanding issues, discussion of these models may help readers in identifying key aspects of safe multi-threaded implementation such as determinism, reactive response, deadlock freedom etc. Interestingly these aspects were taken into account in the design of synchronous programming languages. A few of the synchronous languages such as Esterel, LUSTRE, and SIGNAL are discussed with their basic constructs and possible multi-processor implementations. The latest research on multi-threaded implementation strategies from synchronous programming languages demonstrates the possibilities and the challenges in this field. The conclusion of this chapter is not in selecting any particular programming model, but rather in posing the question as to whether we are yet to see the right model for effective programming of the emerging multi-core computing platforms.

Designers of embedded appliances rely on multi-core system-on-chip (MC-SoC) to provide the computing power required by modern applications. Due to the inherent complexity of this kind of platform, the development of specific system architectures is not considered as an option to provide low-level services to an application. *Chapter 9* gives an overview of the most widespread industrial and domain-specific solutions. For each of them, the chapter describes their software organization, presents their related programming model, and finally provides several examples of working implementations.

Power management and dynamic task scheduling to meet real-time constraints are key components of embedded system computing. While the industry focus is on putting higher numbers of cores on a single chip, embedded applications with sporadic processing requirements are becoming increasingly complex at the same time. *Chapter 10* discusses techniques for autonomous power management of system-level parameters in multi-core embedded processors. It provides an analysis of complex interdependencies of multiple cores on-chip and their effects on system-level parameters such as memory access delays, interconnect bandwidths, task context switch times and interrupt handling latencies. Chapter 10 describes the latest research and provides links to CASPER, a top-down integrated simulation environment for future multi-core embedded systems.

Chapter 11 presents a real-world product which employs a cutting edge multi-core architecture. In order to address the challenges of the wireless communications domain, picoChip has devised the picoArray^{TM}. The picoArray is a tiled-processor architecture, containing several hundred heterogeneous processors connected through a novel, compile-time scheduled interconnect. This architecture does not suffer from many of the problems faced by conventional, general-purpose parallel processors and provides an alternative to creating an

ASIC. The PC20x is the third generation family of devices from picoChip, containing 250+ processors.

State-of-the-art networking systems require advanced functionality extending to multiple layers of the protocol stack while supporting increased throughput in terms of packets processed per second. *Chapter 12* presents Network Processing Units (NPUs) which are fully programmable chips like CPUs or DSPs but, instead of being optimized for the task of computing or digital signal processing, they have been optimized for the task of processing packets and cells. It describes how the high-speed data path functions can be accelerated by hardwired implementations integrated as processing cores in multi-core embedded system architectures. Chapter 12 shows how each core is optimised either for processing intensive functions so as to alleviate bottlenecks in protocol processing, or intelligent memory management techniques to sustain the throughput for data and control information storage and retrieval. It offers insight on the combination of NPUs' flexibility of CPUs with the performance of ASICs, accelerating the development cycles of system vendors, forcing down cost, and creating opportunities for third-party embedded software developers.

Book Errors

This book covers timely topics related to multi-core embedded systems. It is "probable" that it contains errors or omissions. I welcome error notifications, constructive comments, suggestions and new ideas.

You are encouraged to send your comments and bug reports electronically to kornaros@epp.teiher.gr, or you can fax or mail to:

Georgios Kornaros

Applied Informatics & Multimedia Dept.   Electronic & Computer Engineering Dept.
Tech. Educational Institute of Crete   Technical University of Crete
GR-71004, Heraklion, Crete, Greece   GR-73100, Chania, Crete, Greece
kornaros@epp.teiher.gr   kornaros@mhl.tuc.gr
Tel: +30 2810-379868
Fax: +30 2810-371994

1

Multi-Core Architectures for Embedded Systems

C.P. Ravikumar

Texas Instruments (India)
Bagmane Tech Park, CV Raman Nagar
Bangalore, India
ravikumar@ti.com

CONTENTS

1.1   Introduction   2
     1.1.1   What Makes Multiprocessor Solutions Attractive? . .   3
         1.1.1.1   Power Dissipation   3
         1.1.1.2   Hardware Implementation Issues   6
         1.1.1.3   Systemic Considerations   8
1.2   Architectural Considerations   9
1.3   Interconnection Networks   11
1.4   Software Optimizations   13
1.5   Case Studies   14
     1.5.1   HiBRID-SoC for Multimedia Signal Processing   14
     1.5.2   VIPER Multiprocessor SoC   16
     1.5.3   Defect-Tolerant and Reconfigurable MPSoC   17
     1.5.4   Homogeneous Multiprocessor for Embedded Printer
                Application   18
     1.5.5   General Purpose Multiprocessor DSP   20
     1.5.6   Multiprocessor DSP for Mobile Applications   21
     1.5.7   Multi-Core DSP Platforms   23
1.6   Conclusions   25
Review Questions   25
Bibliography   27

1.1   Introduction

There are many interesting "laws" in the folklore of Information Technology. One of them, attributed to Niklaus Wirth, states that software is slowing faster than hardware is accelerating—a testimonial to the irony of modern-day system design. The "slowing down" in Wirth's law can refer to both the run-time performance as well as software development time. Due to time-to-market pressure, the software designers do not have the luxury of optimizing the code. Software development for modern systems often happens in parallel to the development of the hardware platform, using simulation models of the target hardware. There is increased pressure on software developers to reuse existing IP, which may come from multiple sources, in various degrees of softness. Compilers and software optimization tools either do not exist, have limited capabilities, or are not available during the crucial periods of system development. Due to these reasons, application software development is a slow and daunting task, rarely permitting the use of advanced features supported in hardware due to lack of automated tools. It is quite common for software developers (e.g., video games) to resort to manual assembly-language coding.

Embedded systems for applications such as video streaming require very high MIPS performance, of the order of several giga operations per second, which cannot be obtained through a single on-chip signal processor. As an example, consider broadcast quality video with a specification of 30 frames/second, 720 × 480 pixels per frame, requiring about 400,000 blocks to be processed per second. In telemedicine applications, where the requirement is for 60 frames/second and 1920 × 1152 pixels per frame, about 5 million blocks must be processed per second. Today's wireless mobile Internet devices offer a host of applications, including High-Definition Video playback and recording, Internet browsing, CD-quality audio, and SLR-quality imaging. Some applications require multiple antennas, such as FM, GPS, Bluetooth, and WLAN. For example, if a user who is watching a streamed video presentation on a WLAN network on a mobile device is interrupted by an incoming call, it is desirable that the presentation is paused and the phone switches to the Bluetooth handset. The presentation should resume after the user disconnects the call [6]. The growth of data bandwidth in mobile networks, better video compression techniques, and better camera and display technology have resulted in significant interest in wireless video applications such as video telephony. Set-top boxes can provide access to digital TV and related interactive services, as well as serve as a gateway to the Internet and a hub for a home network [5]. For applications such as these, system architects resort to the use multiprocessor architectures to get the required performance. What has made this decision possible is the power granted by the VLSI system-on-chip technology, which allows the logic of several instruction-set processors and

TABLE 1.1: Growth of VLSI Technology over Four Decades

	1982	**1992**	**2002**	**2012**
Technology (μm)	3	0.8	0.1	0.02
Transistor count	50K	500K	180M	1B
MIPS	5	40	5000	50000
RAM	256B	2KB	3MB	20MB
Power (mW/MIPS)	250	12.5	0.1	0.001
Price/MIPS	$30.00	$0.38	$0.02	$0.003

several megabytes of memory to be integrated in the same package (Table 1.1). Unlike general purpose systems and application-specific servers such as video servers [18], the requirements of an embedded solution are very different; compactness, low-cost, low-power, pin-count, packaging, short time-to-market are among the key considerations.

Historically, multiprocessors were heralded into the scene of computer architecture as early as the 1970s, when Moore's law was not yet in vogue and it was widely believed that uniprocessors cannot provide the kind of performance that future applications will demand. In the 1980s, the notion that we are already very close to the physical limits of the frequency of operation became even more prevalent, and a number of commercial parallel processing machines were built. In a landmark 1991 paper by Stone and Cocke [28], the authors argued that an operating frequency of 250 MHz cannot be achieved due to the challenge metal interconnections will pose in achieving this kind of timing. This prediction, however, was proven false in the same decade, and uniprocessors that worked at speeds over 500 MHz became available. The relentless progress in the speed performance of uniprocessors made parallel processing a less attractive alternative and companies that were making "supercomputers" closed down their operations. Distributed computing on a network of workstations was seen as the right approach to solve computationally difficult problems. We have come full circle, with multiprocessors making a comeback in embedded applications.

1.1.1 What Makes Multiprocessor Solutions Attractive?

1.1.1.1 Power Dissipation

The objectives of system design have changed over the past decade. While performance and cost were the primary considerations in system design until the 1980s, the proliferation of battery-operated mobile devices has shifted the focus to power dissipation and energy dissipation. Figure 1.1 shows the power/performance numbers for mobile devices over the past two decades and extrapolates it for the next few years. The prediction of the power/performance numbers with VLSI technology scaling was made by Gene Frantz and

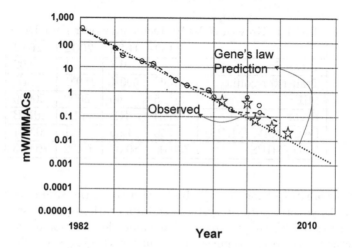

FIGURE 1.1: Power/performance over the years. The solid line shows the prediction by Gene Frantz. The dotted line shows the actual value for digital signal processors over the years. The 'star' curve shows the power dissipation for mobile devices over the years.

has remained mostly true; the deviation from the prediction occurred in the early part of this decade, when leakage power of CMOS circuits became significant in the nanometer technologies. Unless the power dissipation of handheld devices is under check, they will be too hot and demand elaborate cooling mechanisms. Packaging and the associated cost are also related to the peak power dissipation of a device. The distribution of power to the sub-systems gets complex as the average and peak power of a system become larger. In the past decade, we have also seen the concern for "green systems" growing, stemming from the concern about climatic changes, carbon emissions, and e-waste. Energy-efficient system design has therefore gained importance.

Multi-core design is one of the most important solutions for management of system power and the energy efficiency of the system. Systems designed in the 1980s featured a single power supply and a single power domain, allowing the entire system to be powered on or off. As the complexity of the systems has increased, we need an alternate method to power a system, where the system is divided into power domains and power switches are used to cut off power supply to a sub-system which is not required to be active during system operation. In a modern electronic system, there are multiple modes of operation. For example, a user may use his mobile to read e-mail, click a picture or video, listen to music, play a game, or make a phone call. Some sub-systems can be turned off during each of these modes of operation, e.g., when reading mail, the sub-system that is responsible for picture decompression need not be powered on until the user opens an e-mail which has a compressed picture attachment. Similarly, there may be many I/O interfaces in a system, such as USB, credit card, Ethernet, Firewire, etc., not all of which will be necessary in any one mode of operation. Turning off the clock for a sub-system is a way to cut down the dynamic power dissipation in the sub-system. Powering off a sub-system helps us cut down the static as well as dynamic power that would otherwise be wasted.

The traditional way to build high-performance VLSI systems has been to increase the clocking speed. In the late 1980s and the 1990s, we saw the relentless increase in clock speed of personal computers. However, as the VLSI technology used to implement the systems moved from micrometer technology to nanometer technology, a number of challenges intimidated the semiconductor manufacturers. Managing the power and energy dissipation is the most daunting of these challenges. The dynamic power of a VLSI system grows linearly with the frequency of operation and quadratically with the operating voltage. Static power dissipation due to leakage currents in the transistor has different components that increase linearly and as the cube of the operating voltage. Reducing the voltage of operation can result in significant reduction in power, but can also negatively impact the frequency of operation. The selection of operating voltage and frequency of operation must consider both power and performance.

An electronic system is commonly implemented by integrating *IP cores* which operate at different voltages and frequencies. It is also common to use dynamic voltage and frequency scaling (DVFS) in order to manage the power dissipation while constraining the performance. Sub-systems that must provide higher performance can be operated at higher frequency and voltage, while the rest of the system can operate at lower frequency and voltage. An extreme form of frequency scaling is *gated clocking* where the clock signal for a sub-system can be turned off. Similarly, an extreme form of power scaling is *power gating,* where the power supply to a sub-system can be turned off. The OMAP platform for mobile embedded products uses dynamic voltage and frequency scaling to reduce power consumption [10]. Texas Instruments uses its Smart Reflex power management technology and a special 45 nanometer CMOS process for power reduction in the latest OMAP4 series of platforms. Smart Reflex allows the device to adjust the voltage and frequency of operation of sub-blocks based on the activity, mode of operation, and temperature. The OMAP4 processors have two ARM Cortex-A9 processors on-chip and several peripherals (Figure 1.9), but only the core that is required for the target application is activated to minimize power wastage.

Consider a sub-system S that must provide a performance of T time units per operation. Since the switching speed of transistors depends directly on the voltage of operation, building a circuit that implements S may require us to operate the circuit at a higher voltage V, resulting in higher power dissipation. We may be able to use the parallelism in the functionality of the sub-system to break it down into two sub-systems S' and S''. The circuits that implement S' and S'' are roughly half in size and have a critical path that is half of T. As a result, they can be operated at about half the voltage V. This would result in a significant reduction in dynamic and static power dissipation.

Multi-core system design has become attractive from the view point of performance-and-power tradeoff. The tradeoff is between building a "super processor" that can operate at a high frequency (and thereby guzzling power) or building smaller processors that operate at lower frequencies (thereby consuming less power) and yet giving a performance comparable to the super processor.

1.1.1.2 Hardware Implementation Issues

The definition of a *system* in *system-on-a-chip* has expanded to cover multiple processors, embedded DRAM, flash memory, application-specific hardware accelerators and RF components. The cost of designing a multiprocessor system-on-chip, where the processors work at moderate speeds and the system throughput is multiplied by multiplicity of processors, is smaller than designing a single processor which works at a much higher clock speed. This is due to the difficulties in handling the timing closure problem in an automated design flow. The delays due to parasitic resistance, capacitance, and the inductance of the interconnect make it difficult to predict the critical path delays accurately during logic design. Physical designers attempt to optimize the layout subject to the interconnect-related timing constraints posed by the logic design phase. Failure to meet these timing constraints results in costly iterations of logic and physical design. These problems have only aggravated with scaling down of technology, where tall and thin wires run close to one another, resulting in crosstalk. Voltage drop in the resistance of the power supply rails is another potential cause for timing and functional failures in deep submicron integrated circuits. When a number of signals in a CMOS circuit switch state, the current drawn from the power supply causes a drop in the supply voltage that reaches the cells. As a result, the delay of the individual cells will increase. This can potentially result in timing failure on critical paths, unless the power rail is properly designed. Typically, the gates in the center of the chip are most prone to IR drop-induced delays.

Although custom design may be used for some performance-critical portions of the chip, today it is quite common to employ automated logic synthesis flows to reduce the front-end design cycle time. The success of logic synthesis, both in terms of timing closure and optimization, depends critically on the constraints specified during logic synthesis. These constraints include timing constraints, area constraints, load constraints, and so on. Such constraints are easier to provide when a hierarchical approach is followed and smaller partitions are identified. The idea of using multiple processors as opposed to a single processor is more attractive in this scenario.

Another benefit that comes from a divide-and-conquer approach is the concurrency in the design flow. A design that can naturally be partitioned into sub-blocks such as processors, memory, application-specific processors, etc., can be design-managed relatively easily. Different design teams can concurrently address the design tasks associated with the individual sub-blocks of the design.

When a design has multiple instances of a common block such as a processor, the design team can gain significantly in terms of design cycle time. This is possible through the reuse of the following work: (a) insertion of scan chains and BIST circuitry, (b) physical design effort, (c) automatic test pattern generation effort, (d) simulation of test patterns.

In VLSI technologies beyond 90 nm, on-chip variability of process parameters, temperature, and voltage is another challenge that designers have to grapple with. The parameters that determine the performance of transistors and interconnects are known to vary significantly across the die, due to the vagaries of the manufacturing processes. In the past, these variances were known to exist in dies made on different wafers, lots, and foundries. However, due to the small dimension of the circuit components, on-die variation has assumed significance. The exact way in which a transistor or interconnect gets "printed" on the integrated circuit is no longer independent of the surrounding components. Thus, a NAND gate's performance can vary, depending on the physical location of the gate and what logic is in its neighborhood. The temperature of the die varies widely, by as much as 50 degrees Celsius, across the chip. Similarly, due to the impedance drops in the power supply distribution network of the chip, the voltage that reaches the individual gates and flip-flops can vary across the chip.

There are several solutions to combat the problem of on-chip variability. One solution is to apply "optical proximity correction" which subtly transforms the layout geometries so that they print well. Optical proximity correction is a slow and expensive step and is best applied to small blocks. In this context, having regularity and repetitiveness in the system can be an advantage. Homogeneous multiprocessor systems offer this advantage. To alleviate the problem of temperature variability, it would be desirable to migrate computational tasks from hotter regions to cooler portions of the chip. Once again, homogeneous multiprocessors present a natural way of performing task migration. The problem of reducing the variation in power supplies across the power supply network can also be alleviated by building a hierarchical network from smaller, repeatable supply networks. Here again, the use of multiprocessors can be an advantage.

Testing of integrated circuits for manufacturing defects is yet another challenge. Due to the growing complexity and size of integrated circuits, the amount of test data has grown sharply, increasing the cost of testing. Testing of integrated circuits is performed by using an external tester that applies pre-computed test patterns and compares the response of the integrated circuit with the expected results. The test generation software runs very slowly as the size of the circuit grows. A divide-and-conquer approach offers an effective solution to this problem [21]. Multi-core systems have a natural design hierarchy, which lends itself to the divide-and-conquer approach toward test generation, fault simulation, and test pattern validation. When a number of identical cores are present in the integrated circuit, it may be possible to reuse the patterns and reduce the effort in test generation. Similarly, there are interesting "built-in-self-test" approaches where mutual testing can be employed to test a chip. Thus, if we have two processor cores on the same chip, we can apply random patterns to both processor cores and compare their responses to the random tests; a difference in response will indicate an error.

As in the case of design-for-test and test generation, the natural hierarchy imposed by the use of multi-core systems can also pave the way for efficient solutions for other computationally intensive tasks in electronic design, such as design verification, logic synthesis, timing simulation, physical design, and static timing analysis.

1.1.1.3 Systemic Considerations

There are software and system-design issues also that make a multiprocessor solution attractive. There are numerous VLSI design challenges that a design team may find daunting when faced with the problem of designing a high-performance system-on-chip (SoC). These include verification, logic design, physical design, timing analysis, and timing closure.

The way to harness performance in a single processor alternative is to use superscalar computing and very large scale instruction word processors. Compilers written for such processors have a limited scope of extracting the parallelism in applications. To increase the compute power of a processor, architects make use of sophisticated features like out of order execution and speculative execution of instructions. These kinds of processors dynamically extract parallelism from the instruction sequence. However, the cost of extracting parallelism from a single thread is becoming prohibitive, making a *single complex processor* alternative unattractive. With many applications written in languages such as Java or C++ resorting to multithreading, a compiler has more visibility of MIMD-type parallelism (Multiple Instruction Stream, Multiple Data Stream) in the application.

Both homogeneous and heterogeneous multiprocessor architectures have been used in building embedded systems. Heterogeneous multiprocessing is used when there are parts of the embedded software that would need the power of a digital signal processor and other parts need a micro-controller for the housekeeping activity. We shall consider several MPSoC case studies to illustrate the architectures used in modern-day embedded systems. In particular, we shall emphasize the following aspects of MPSoC designs: (a) processor architecture, (b) memory architecture and processor-memory interconnect, and (c) the mapping of applications to MPSoC architectures.

1.2 Architectural Considerations

A wide variety of choice exists for selecting the embedded processor(s) to-day, and the selection is primarily guided by considerations such as overall system cost, performance constraints, power dissipation, system and application software development support which should permit rapid prototyping, and the suitability of the instruction set to the embedded application. The code density, power, and performance are closely related to the instruction set of the embedded processor. Compiler optimizations and application software programming style also play a major role in this. RISC, CISC, and DSP are the three main categories of processors available to a designer. Some design decisions that must be made early in the design cycle of the embedded system are:

- General purpose processors versus application-specific processors for compute-intensive tasks such as video/audio processing

- Granularity of the processor; selecting a small set of powerful CPUs versus selecting a large number of less powerful processors

- Homogeneous or heterogeneous processing

- Reusing an existing CPU core or architecting a new processor

- Security issues

Recently, a simulation study from Sandia National Labs was published [16] after the performance of 8-, 16-, and 32-processor multiprocessor architectures was studied. Refer to Figure 1.2. Memory bandwidth and memory management schemes are reported to be limiting factors in the performance that can be obtained from these multiprocessors. In fact, the study suggests that the performance of the multiprocessors can be expected to degrade as the number of processors is increased beyond 8. For example, a 16-processor machine would behave no better than a 2-processor machine due to memory bandwidth issues. The use of stacked memories (memories stacked in the third dimension over processors) was seen to avert this problem, but the speedup increases only marginally with more processors.

Frantz and Simar point out the multi-core architectures are a blessing in disguise [7]. We have already pointed out that software development can become sloppy due to availability of low-cost, high-performance software and due to short turn-around cycles. Frantz and Simar point out that hardware design can also become sloppy and wasteful since modern VLSI technology allows us to integrate hundreds of millions of transistors on the same chip and the cost of the transistor is falling rapidly; today the cost of an average transistor has dropped to about one hundred nano-dollars. This is encouraging architectures that are wasteful in hardware and wasteful in terms of power. Creating a

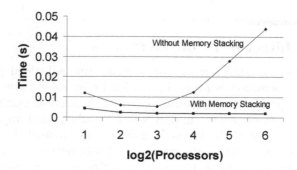

FIGURE 1.2: Performance of multi-core architectures. The x-axis shows the logarithm of the number of processors to the base 2. The y-axis shows the runtime of the multi-core for a benchmark. (Adapted from Moore, S.K. *Spectrum*, 45:5–15, 2008. ©IEEE 2008. With permission.)

large number of identical processors on a single chip may not per se result in a good solution for real problems. The best architecture for the application may require a heterogeneous processor architecture and interconnect architecture evolved through careful analysis. At the same time, it is difficult to always make a custom ASIC for every application since the volumes may not justify the development cost and the turn-around time may be unacceptable. Today, the trend is to create "platforms" for classes of applications. For example, the OMAP platform [10] is intended for multimedia applications; a variety of OMAP chips is available to balance cost and performance. We will further discuss the OMAP platform later in the chapter.

In the examples covered in Section 1.5, we shall see that all the above solutions have their place and considerations such as performance, power, and design cycle time to guide the selection of the processor architecture. This phase in the design is mostly manual, although there is some work on automatic selection [2, 17].

The memory architecture of the MPSoC is equally critical to the performance of the system, since most of the embedded applications are data intensive. In current MPSoC architectures, memory occupies 50 percent of the die area; this number increased to 70 percent by 2005 and is expected to escalate to 92 percent by 2014. Due to numerous choices a system architect has on memory architectures, a systematic approach is necessary for exploring the solution space. Variations in memory architecture come from the choice of sharing mechanism (distributed shared memory or centrally shared memory, or no shared memory at all, as in message-passing architectures), ways to improve the memory bandwidth and latency, type of processor-memory interconnect network, cache coherence protocol, and memory parameters (cache size, type of the memory, number of memory banks, size of the memory banks). Most DSP and multimedia applications require very fast memory close to the

CPU that can provide at minimum two accesses in a processor cycle. Meftali [15] presents a methodology to abstract the memory interfaces through a wrapper for every memory module. The automatic generation of wrappers gives the flexibility to the designers to explore different memory architectures quickly. Cesario [30] addresses the problem of exploring and designing the topology and protocols chosen for communication among processors, memories and peripherals.

As more embedded systems are interconnected over the Internet, with no single "system administrator", there are many security concerns. Embedded systems can control the external environment parameters such as temperature, pressure, voltage, etc. There are even embedded systems that are implanted into the human body. A vulnerable system will permit attacks that can have harmful consequences. To secure an embedded system, there are several solutions, such as the use of public key cryptography with on-chip keys. The booting of the embedded system and flashing of the memory can be secured through a secure password. Access to certain peripherals can be restricted through password protection. Debugging, tracing, and testing must be secure, since the security keys of the embedded system can be read out during scan test. Security solutions for embedded systems can be implemented in hardware and/or software. Software solutions come in the form of programming libraries and toolkits for implementing security features. Security solutions must be cost-effective and energy-efficient. Therefore, many vendors provide security solutions for high-volume products. Texas Instruments provides a solution called "M-shield" for its OMAP platform (Figure 1.9). A multi-core embedded platform is often intended for high-end applications and adding security features to the application would therefore be cost-effective.

1.3 Interconnection Networks

The volume of data that needs to be interchanged between processors in an embedded application intended for video processing is quite high [20]. An efficient interconnection architecture is necessary for interprocessor communication, communication between processors and peripherals, and communication between memories and processors/peripherals. A large number of processor-memory and processor-processor interconnection networks have been explored in the parallel processing literature [8]. The major considerations in designing the interconnection architecture are the propagation delay, testability, layout area, and expandability. Bus-based interconnection schemes continue to remain popular in today's embedded systems, since the number of processors/peripherals in these systems is still quite small. Busses do not scale very well in terms of performance as the number of masters and slave processors connected to the bus increases. Ryu, Shin, and Mooney present a comparison of five different bus architectures for a multiprocessor SoC, taking example applications from wireless communication and video processing to compare the performance [24].

Assuming that Moore's law will continue to hold for several years to come, one can expect a very large number of processors, memories, and peripherals to be integrated on a single SoC in the future. Bus-based interconnection architectures will not be appropriate in such systems. Given the problems that VLSI design engineers already face in closing timing, one can expect that these problems will escalate further in these future systems because the number of connections will be very high. A modular approach to interconnections will therefore be necessary.

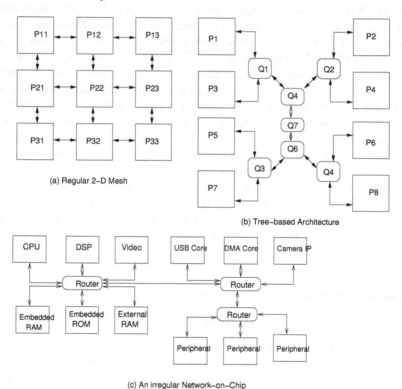

(a) Regular 2–D Mesh

(b) Tree–based Architecture

(c) An irregular Network–on–Chip

FIGURE 1.3: Network-on-Chip architectures for an SoC.

The network-on-chip (NoC) research addresses this problem. Buses on printed circuit boards, such as the PCI bus (peripheral component interconnect) have been implemented as point-to-point high-speed networks (PCI-Express). In the same way, on-chip communications can also benefit from a network-based communication protocol. Such a system-on-chip will have a number of sub-systems (IP cores) that operate on independent clocks and use network protocols for communication of data between IP blocks. These systems are also called Globally Asynchronous, Locally Synchronous or GALS systems since communication within a sub-system may still be based on a synchronous bus.

A number of network-on-chip architectures have been proposed in the literature [12]. Kumar et al. propose a two-dimensional mesh of switches as a scalable interconnection network for SoC [14]. Circuit building blocks such as processors, memory, and peripherals can be placed in the open area of the 2-D mesh. Packet switching is proposed for communication between building blocks. Figure 1.3 shows some possible NoC architectures to connect IP cores on a system-on-chip. The selection of the architecture will be based on power, performance, and area considerations. System integration is a major consideration in the implementation of multi-core SoC. Several efforts toward easing of SoC integration have been reported (see [19] and [29]).

1.4   Software Optimizations

As mentioned in Section 1.2, compilers and other software development support devices play an important role in selecting the processor(s) for an embedded application. Compiler optimizations are important for optimizing the code size, performance, and power [4, 25]. While compiler optimizations are useful in the final phase of software development, a significant difference to the quality of the software comes from the programming style and the software architecture itself. Developing an application for a multiprocessor SoC poses several challenges.

- Partitioning the overall functionality into several parallel tasks

- Allocation of tasks to available processors

- Scheduling of tasks

- Management of inter-processor communication

Identifying the coarse-grain parallelism in the target application is a manual task left for the programmer. For example, in video applications, the image is segmented into multiple macro-blocks (16×16 pixels) and each of the segments is assigned to a processor for computation [23]. Fine-grain parallelism in instruction sequences can be identified by compilers. Vendors of embedded processors often provide software development platforms that help an application programmer develop and optimize the application for the specific processor. An example is the OMAP software development platform by Texas Instruments [10]. The application developer can use a simulator to verify the functional correctness and estimate the run-time of the application on the target processor.

Kadayif [13] presents an integer linear programming approach for optimizing array-intensive applications on multiprocessor SoC. The other key challenge in optimizing an application for a multiprocessor SoC is to limit the number of messages between processors and the number of shared memory accesses. The overall throughput and speed increase obtainable through the multiprocessor solution can be marred by an excess of shared memory accesses and interprocessor communications. Performing worst-case analysis of task run-times and interprocessor communication times, and guaranteeing real-time performance are also challenges in optimizing an application for a multiprocessor SoC. A genetic algorithm for performing task allocation and scheduling is presented in [17]. Chakraverty et al. consider *soft real-time systems* and present a method to predict the deadline miss probability; they also use a genetic algorithm to trade off the deadline miss probability and overall system cost [2].

1.5 Case Studies

In this section, we shall use several examples of multiprocessor system-on-chip designs to illustrate the design choices and challenges involved in these designs.

1.5.1 HiBRID-SoC for Multimedia Signal Processing

The HiBRID system-on-chip solution described by Stolberg et al [27] integrates three CPU cores and several interfaces using the 64-bit AMBA AHB bus. Refer to Figure 1.4. The targeted applications of HiBRID include stationary as well as mobile multimedia applications; as a result, the architecture and design of the SoC focus on programmability. The authors classify multimedia processing into three classes, namely, stream-oriented, block-oriented, and DSP-oriented categories. They see the need for providing all the three types of processing in the same system-on-chip, so that all forms of processing can be done in parallel on the same system.

The following three types of processors are included in HiBRID:

- HiPAR-DSP is intended to provide high throughput for applications that such as the fast Fourier transform and digital filtering. The architecture of this DSP is a 16-way SIMD. Each of the 16 data path units is capable of executing two instructions in parallel. A matrix memory is shared by all the data path units. The DSP operates at 145 MHz and offers a peak performance of 2.3G MAC operations per second.

- A stream processor, which is intended for control-dominated applications. It includes a single five-stage, 32-bit RISC processor that is controlled using a 32-bit instructions.

- A macro-block processor, intended for processing of blocks of images. It consists of a 32-bit scalar data path and a 64-bit vector data path. The vector data path includes a 64 × 64-bit register file and 64-bit data path units. These data path units can execute either two 32-bit, four 16-bit, or eight 8-bit ALU operations in parallel. In addition to these, special functional units capable of executing specialized instructions for video and multimedia applications are provided.

The three processors, host interfaces, and external SDRAM are connected through the AMBA AHB bus. Dual-port memories are used for exchange of data between processors. The underlying philosophy of HiBRID is that one or more of the cores can be removed from the architecture to trade off cost with performance. Note that this architecture can also result in graceful degradation of performance and provide tolerance to faults if a processor core fails during system operation.

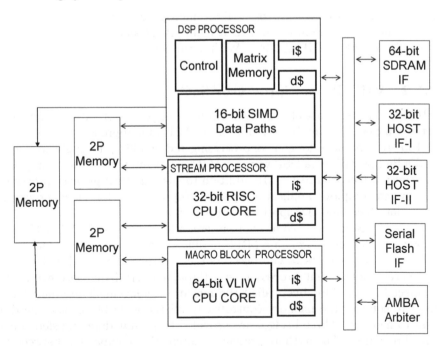

FIGURE 1.4: Architecture of HiBRID multiprocessor SoC. (Adapted from Stolberg, H.-J. et al. HiBRID-SoC: Proceedings of Design Automation and Test in Europe (DATE): *Designer's Forum*, pages 8–13, March 2003.)

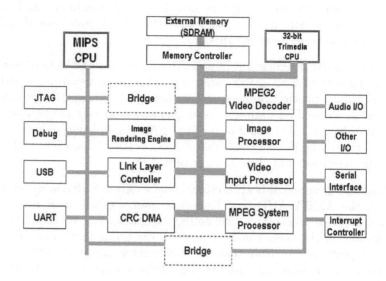

FIGURE 1.5: Architecture of VIPER multiprocessor-on-a-chip. (Adapted from: Dutta, S., Jensen, R., and Rieckmann, A. *Design & Test of Computers*, 18(5):21–31, 2001. ©IEEE 2001. With permission.)

1.5.2   VIPER Multiprocessor SoC

VIPER is an example of a heterogeneous multiprocessor targeted for use in set-top boxes [5]. It makes use of a 32-bit MIPS microprocessor core working at 150 MHz intended for control processing and handling the application layer, and a Philips TriMedia DSP working at 200 MHz intended for handling all the multimedia (see Figure 1.5). In addition to these general purpose processors, the system employs application-specific co-processors for video processing, audio interface, and transport stream processing.

The MIPS processor connects to interface logic such as USB, UART, interrupt controller, etc. through a MIPS peripheral interconnect bus. The TriMedia processor connects to coprocessors such as the MPEG-2 decoder, image composer, video input processor, etc. through a TriMedia peripheral interconnect bus. A third bus, called the memory management interface bus, is used to connect the memory controller to all the logic blocks that need access to memory. Three bridges are provided to permit data transfers among the three buses. The authors present an interesting comparison between two possible implementations of the buses, namely, tri-state buses and point-to-point links. Tri-state buses reduce the number of wires, but present several problems such as poor testability, complicated layout, difficulty in post-layout timing fixes, and poorer performance. In comparison, point-to-point links have a higher number of wires. However, they are simpler to test, simpler to lay out, and lend themselves to post-layout timing adjustments. The scalability and modularity of point-to-point links are not high, since a peripheral that is connected to a bus with n masters must have n interfaces, and adding another master to the bus will necessitate updating the peripheral to include an extra slave interface. The authors report that the area impact of the two schemes is comparable. The choice of whether to select a tri-state bus or point-to-point bus is therefore case-dependent.

In addition to the decision regarding tri-state or point-to-point link topologies, the architect has to also make a decision on the data transfer protocol between external memory and peripherals. There are several choices available:

- High speed peripherals require direct memory access (DMA) protocol

- Combination of programmed I/O and DMA on a common bus

- Combination of programmed I/O and DMA on two different buses

The authors provide guidelines on selecting the appropriate protocol, based on concerns such as expandability, access latency, simplicity, layout considerations, etc.

VIPER was implemented in 180 nm, 6-metal layer process and has about 35 M transistors. Since it is a large design, a partitioned approach was followed for physical design, and the design was divided into nine chiplets, each of which had at most 200 K layout instances. Signals between chiplets get connected through abutment, minimizing the need for top-level routing. The TriMedia CPU core, the MIPS CPU core, and several analog blocks such as phase lock loops were reused in the VIPER design.

1.5.3 Defect-Tolerant and Reconfigurable MPSoC

Rudack et al. describe a homogeneous multiprocessor intended for a satellite-based geographical information system which uses ITU H.263 (video telephony standard) and ISO MPEG-2 (digital TV standard) for image compression [23]. This system has 16 instances of processor nodes, which are based on the AxPe processor. Hardware interfaces are used for DMA, digital video, and satellite communication. The authors state that their design philosophy was not to integrate several different IP cores, but to integrate a few identical cores. The main advantage of this approach, according the authors, is the simplification of the testing and defect tolerance. When one uses multiple IP cores from possibly different vendors, test generation is not easy. The IEEE 1500 standard for testing of core-based SoC designs promotes the use of core wrappers [9]. The authors of [23] decided to use identical processor cores so that they can use the built-in self-test (BIST) as a test methodology and permit parallel testing of cores. Each processor node consists of a bus-based processing unit that uses an AxPe video processor core operating at 120 MHz, a DRAM controller, DRAM frame memory, bus arbiter, and host interface logic. The AxPe processor itself consists of a RISC engine for medium-granularity tasks

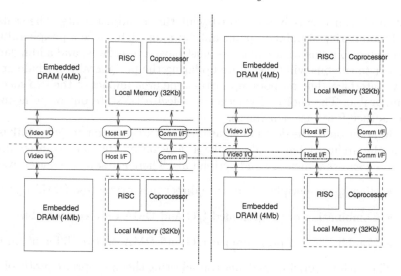

FIGURE 1.6: Architecture of a single-chip multiprocessor for video applications with four processor nodes.

such as Huffman coding and book-keeping; a microprogrammable coprocessor in the AxPe is used for low-granularity tasks such as DCT and motion estimation. Because of its complexity, the AxPe is a large-area integrated circuit, occupying about 2 cm × 2 cm die area. Yield and defect tolerance were therefore major concerns in the design of this system. Since the system consists of 16 identical processor nodes, one can replace the functionality of the other when a failure occurs. The authors describe an interesting manufacturing technique where photocomposition is used to fill a wafer with identical copies of a building block. A building block consists of four copies of the processing node. Since all the processing nodes are identical, it is possible to cut out an arbitrary number of building blocks from the wafer; this improves the yield of the manufacturing process. Reconfiguration techniques described by authors permit one more level of defect tolerance; when a defect is detected in a system, the functionality of the defective block is mapped to a healthy block.

1.5.4 Homogeneous Multiprocessor for Embedded Printer Application

MPOC [22] is an early effort at building a multiprocessor for embedded applications. In this case, the application considered was that of embedded printers. The motivation for using a multiprocessor in this application is turn-around time. In an embedded printer application, high performance is desirable, but a solution based on a state-of-the-art VLIW processor may not be acceptable due to the large turn-around time involved in developing and testing several thousand lines of software for the application on a new processor. A quick fix to such a problem is the use of multiple processors that can offer high performance by exploiting the coarse-grain parallelism in the application. A printer processes images in chunks, called strips. Coarse-grain parallelism refers to the creation of individual tasks for handling individual strips. The software modification to implement the coarse-grain parallelism was quite small, making the solution attractive. Unlike the example of [23], where a single processor was a complex, large-area IC, the processor described in [22] is a simple scalar processor. The following analysis is offered by Richardson to justify the choice of using several simple processors instead of a small number of complex processors. Consider a baseline processor which offers a speed of 1.0 instructions per cycle (IPC) and a die area of 1.0 unit. A possible set of choices for the VLSI architect are:

- Use a die area of 8.0 units on a single complex processor which improves the speed to about 2.0 IPC.

- Use the die area of 8.0 units to implement four processors of medium complexity, each of which offers 1.5 IPC. When parallelism is fully exploited, the speedup will be 6 times.

- Use the die area of 8.0 units to implement eight processors of 0.9 IPC. The effective speedup will be 7.2 times with reference to the baseline processor.

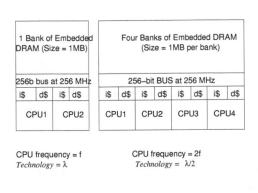

FIGURE 1.7: Design alternates for MPOC. (From Richardson, S. *MPOC: A chip multiprocessor for embedded systems.* Technical report, Hewlett Packard, 2002. With permission.)

The system-on-board prototype described by Richardson uses four MIPS R4700 processors connected to a VME backplane. The design team considered several alternates (Figure 1.7) before deciding on the four-CPU solution, based on cost and performance tradeoffs. It is estimated that a 0.18 micron CMOS logic and DRAM memory process implementation of the system as an SoC would result in a die area of about 55 sq mm.

1.5.5   General Purpose Multiprocessor DSP

Daytona is a multiprocessor DSP described by Ackland et al [1]. The processor can offer a performance of 1.6 billion 16-bit multiply-accumulate operations per second, and is intended for next generation DSP applications such as modem banks, cellular base stations, broad-band access modems, and multimedia processors. One may argue that since these applications have wide variations from one another, an application-specific solution (ASIC) would offer the best price/performance ratio. However, the authors argue that prototyping times are much less, resulting in faster turn-around times, when these applications are implemented on a general purpose processor.

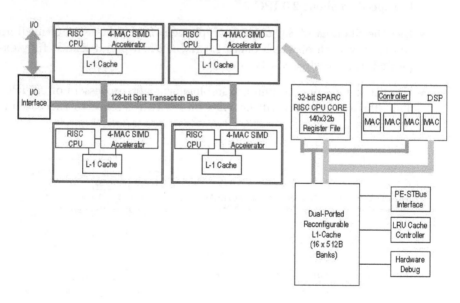

FIGURE 1.8: Daytona general purpose multiprocessor and its processor architecture. (From Ackland, B., et al. *Journal of Solid-State Circuits*, 35(3):412–424, Mar 2000. With permission.)

The goals of Daytona design (Figure 1.8) are to achieve scalable performance, good code density, and programmability. Daytona uses both SIMD and MIMD parallelism to obtain performance. Since it uses a bus-based architecture, the authors argue that adding more processors to scale up performance is easy. However, since a bus is a shared resource for inter-processor communication, it can become a bottleneck for scaling performance. Hence they describe a complex 128-bit bus known as a split transaction bus to mitigate

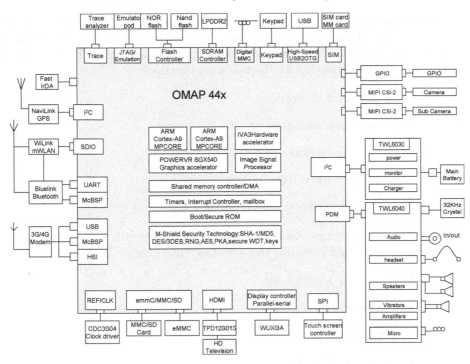

FIGURE 1.9: Chip block diagram of OMAP4430 multi-core platform.

this possibility. Each address transaction has a transaction ID associated with it, which is matched with the transaction ID of the data transactions. Thus multiple transactions can be serviced by the system at one time.

The processing element in Daytona is a SPARC RISC with a vector coprocessor. The overall architecture of Daytona and the PE architecture are both illustrated in Figure 1.8. The 64-bit coprocessor is ideally suited for multimedia and DSP applications which are rich in data parallelism. The coprocessor can operate in 3 modes, namely, 8×8 b, 4×16 b, and 2×32 b. The authors state that video and image processing algorithms can take advantage of the 8 b mode, whereas wireless base-station applications require higher precision. The Daytona processor has four processing elements connected using the split transaction bus.

1.5.6 Multiprocessor DSP for Mobile Applications

OMAP (Open Multimedia Application Platform) is a solution intended primarily for mobile wireless communications and next generation embedded devices [3, 6, 26, 10]. OMAP makes use of an embedded ARM processor core and a Texas Instruments TMS320C55X or TMS320C64X DSP core. OMAP provides support for both 2G and 3G wireless applications. In a 2G wireless architecture, the ARM7 CPU core from Advanced RISC Machines is employed and is intended for the "air interface." More advanced versions of the ARM processor such as ARM Cortex-A8 and ARM9 are used in higher versions of OMAP. ARM is intended for the following functions.

- Modem layer 2/3 protocols

- Radio resource management

- Short message services (SMS)

- Man-machine interface

- Low level operating system functions

The 2G architecture uses the C54X DSP core, which is intended for the "user interface" and performs the following functions.

- Modem layer 1 protocols

- Speech coding/decoding

- Channel coding/decoding

- Channel equalization

- Demodulation

- Encryption

- Applications such as echo cancellation, noise suppression, and speech recognition.

Power is the most important consideration in the design of an SoC intended for wireless application. As per the comparison reported in Chaoui [3], the TMS320C10 offers a reduction of 2 times in terms of power dissipation and an improvement of 3 times in terms of performance when performing applications such as echo cancellation, MPEG4/H.263 encoding or decoding, JPEG decoding, and MP3 decoding. These comparisons were made against a state-of-the-art RISC machine with DSP extension [3]. In the OMAP architecture, multiprocessing is employed in an interesting way to prolong battery life. Had a single RISC processor been used for running a video conferencing application, it would take about three times the time and consume about twice the power, requiring about six times more energy. Employing the TMS320C55X DSP processor reduces the drain on battery, but the DSP is not the best choice for handling control processing and popular OS applications such as word processing and spreadsheets. The ARM processor is used as a "standby" for running such applications. By assigning a task to either of the two processors that gives the best power-performance product, the OMAP prolongs the battery life. Several design techniques are employed to reduce power; for example, unnecessary signal toggling is minimized to reduce switching power and an optimal floorplan is employed to reduce interconnect power. OMAP permits the clock to a particular resource to be turned off when the resource is not required. This clock gating feature can be accessed through application programming as well.

The ARM processor and the DSP communicate with each other through a set of mailboxes [26]. When the ARM processor, which acts as a master, has to dispatch a task to the DSP, it writes a message in the MPU2DSP mailbox. When the DSP completes a task, it places a message in the DSP2MPU mailbox. Since a high-performance graphical display system is a key requirement in 3G wireless applications, OMAP provides a dedicated DMA channel for the LCD controller.

Another advantage of using a multiprocessor platform, namely, hierarchical physical design, is evident in the OMAP design [10]. The physical design and the associated timing closure of the DSP subsystem and the microprocessor subsystem are separated. This permits concurrency in the design flow.

The OMAP platform is available in different versions, depending on the revision of the ARM processor (ARM Cortex-A8, ARM9, etc.), the on-chip DSP core (one of the C64x family of DSP), graphics accelerator and the on-chip peripherals that are included in the SoC. At one extreme is a version of OMAP that supports only the ARM Cortex-A8 processor and peripherals. At the other extreme is an OMAP which supports an ARM Cortex-A8, a C64x DSP, a graphics accelerator and a host of shared peripherals. By creating multiple flavors of the platform, it is possible to offer a cost-effective and power-efficient solution that is right for the target application.

1.5.7 Multi-Core DSP Platforms

The TMPS320C6474 platform from Texas Instruments has three DSP cores, each of which can operate up to 1 GHz speed. This integration is possible due to implementation in 65 nm CMOS VLSI technology. A block diagram of the chip is provided in Figure 1.10 (see [11]). The C6474 platform is suitable for high-performance medical imaging applications such as ultrasound, which are computationally demanding. The measured raw performance of the device is 24,000 million 16-bit multiply-accumulate operations (MMACs). When compared to a solution where designers integrate three discrete DSP devices on a board, the multi-core DSP offers a triple improvement in speed, triple improvement in power, and 1.5 times improvement over cost. The C6474 delivers a performance of 4 MIPS/mW and uses the Smart Reflex technology of Texas Instruments for power management.

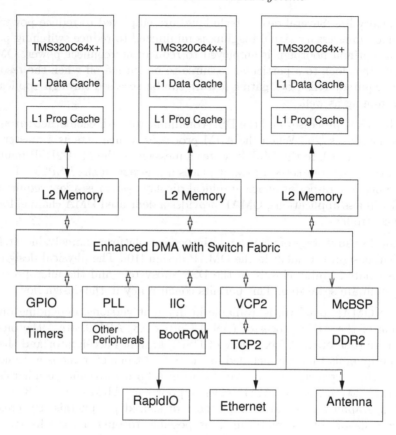

FIGURE 1.10: Chip block diagram of C6474 multi-core DSP platform.

The C6474 integrates several IP cores that are useful in imaging applications. For example, the Viterbi and Turbo accelerators support hardware implementation of Viterbi decoding and Turbo decoding algorithms. To support fast data transfers to/from the chip, C6474 supports several interfaces such as ethernet media access control (EMAC), serial rapidIO, and the antenna interface. The platform supports 32 KB of on-chip L1 cache and 3MB of on-chip L2 memory. High memory bandwidth is made available through DDR2 interfaces that can operate at over 600 MHz. To aid the designers of embedded systems, related products such as analog-to-digital converters, power management, and digital-to-analog converters are available separately. Since the three DSP cores integrated onto the device are code-compatible to single-core TMS320C64 + DSP, migrating to the multi-core platform is expected to be fast.

1.6 Conclusions

Video, audio, and multimedia content are becoming necessary in practically all embedded applications today. The recent growth in interest in telemedicine and medical diagnosis through medical image analysis has created another growth vector for embedded systems. Embedded systems must support access to the Internet and a variety of interfaces to read data, e.g., credit card readers, USB devices, RF antennae, etc. With such requirements, it is natural that multiprocessor architectures are being explored for these embedded systems.

Since multimedia processing requires a lot of computational bandwidth, communication bandwidth, and memory bandwidth, several architectural innovations are necessary to satisfy these demands. Application-specific solutions may be able to deliver the performance demanded by these systems, but since standards are constantly evolving, the flexibility offered by a programmable general purpose multiprocessor solution is attractive. For example, the MPEG-4 standard for video coding was introduced in 1999 and since then, several video profiles have been defined such as Advanced Simple Profile in 2001, and Advanced Video Coding in 2003. Throughputs of the order of 10 giga operations per second are simply not possible using today's uniprocessors. Developing a uniprocessor architecture that can deliver this kind of performance is not easy, and the VLSI design of such a processor would be too expensive to make this endeavor cost-effective. We looked at the VLSI design challenges that a designer of a multiprocessor SoC deals with.

A multiprocessor SoC offers modularity in the design approach, promotes reuse of IP cores, and permits concurrency in the design flow. Issues such as timing closure are easier to tackle with a hierarchical, modular design approach to which multiprocessor SoCs lend themselves. Developing application software and optimizing the application on the multiprocessor platform is, however, more difficult. This is because compilers can only achieve window optimizations within instruction sequences for a uniprocessor. More progress in developing automated solutions for identifying parallelism and performing task partitioning will be needed in the near future. With escalating interest in applications such as digital TV, mobile television, video gaming, etc., we can expect multiprocessor system-on-chip technology to become a focus area in embedded systems R&D.

Review Questions

[Q_1] Assume that you are the system architect for an embedded system for an application such as a medical device. In this chapter, we saw several case studies of suitable multi-core platforms for multimedia applications.

Tabulate the salient features of the platforms under the following headings: performance, power requirements, cost, software availability, peripherals supported. Use relative grades to assess the suitability of the platform to a (fictitious) application that you wish to implement on the platform. Explain your conclusions.

[Q_2] State the following: Amdahl's law, Wirth's law, and Moore's law.

[Q_3] Consider the following statement. "Performance is the only reason why one should consider multiprocessor architectures over uniprocessor architectures." Provide counter-arguments to this statement by enumerating other reasons to move to multiprocessor architectures.

[Q_4] What is meant by a *platform* in the context of embedded systems? What does a platform include? What are the benefits of using a platform for (a) an end user and (b) a provider?

[Q_5] A system architect is considering moving from a PCB with four uniprocessor devices to a system-on-chip with four processors. What are some of the benefits and implementation challenges that the architect will face?

[Q_6] Define the following terms and explain how multi-core architectures are impacted by them:
A. On-chip variability
B. Manufacturing yield

[Q_7] Compare dynamically reconfigurable architectures based on FPGA with programmable media processors for the following applications:
A. A medical imaging application where standards are still evolving
B. A hand-held battery-operated multimedia gaming device

[Q_8] Compare ASIC solutions with programmable media processors on the following counts: cost, performance, power, programmability, extensibility, debugging.

[Q_9] Enumerate some of the opportunities that (a) hardware engineers and (b) software engineers have in optimizing the power dissipation of a system. How do multi-core architectures help in improving power efficiency?

[Q_10] What is the motivation for network-on-chip architectures for implementing on-chip communications in a multi-core platform? What are some good candidates for NoC topologies for a multi-core system with (a) 8 processors and (b) 256 processors?

[Q_11] Explore the Internet and find out what is meant by "cloud computing." Then consider the following statement: "With cloud computing, multi-core platforms may not be required for end-user systems since the computing power is available in the cloud." Debate the statement.

[Q-12] Standards are always evolving in applications such as signal compression, multimedia communication, etc. Explain some of the methods used in product engineering can shield against such a dynamically changing scenario.

[Q-13] An architect is considering two solutions for an embedded SoC platform: (a) integrate four powerful microprocessor cores; (b) integrate 64 moderately powerful processor cores. Assume that the area of the powerful microprocessor core is A_1 and that it provides a performance of I_1 instructions per clock cycle. The moderately powerful processor gives a performance of I_2 instructions per clock cycle and has an area of A_2. If area-delay product is taken as a measure of efficiency, what is the condition under which the second solution is better than the first?

[Q-14] What are the most important design issues that an end user will consider when selecting a multi-core platform for an embedded application?

[Q-15] Consider a medical application such as ultrasound and derive its computational requirements. Also derive the I/O requirements for the application.

Bibliography

[1] B. Ackland et al. A single chip, 1.6 billion, 16-b MAC/s multiprocessor DSP. *Journal of Solid-State Circuits*, 35(3):412–424, Mar 2000.

[2] S. Chakraverty, C.P. Ravikumar, and D. Roy-Choudhuri. An evolutionary scheme for cosynthesis of real-time systems. In *Proceedings of International Conference on VLSI Design*, pages 251–256, 2002.

[3] J. Chaoui. OMAP: Enabling multimedia applications in third generation wireless terminals. *Dedicated Systems Magazine*, pages 34–39, 2001.

[4] V. Dalal and C.P. Ravikumar. Software power optimizations in an embedded system. In *Proceedings of the International Conference on VLSI Design*, pages 254–259, 2001.

[5] S. Dutta, R. Jensen, and A. Rieckmann. VIPER: A multiprocessor soc for advanced set-top box and digital tv systems. *IEEE Design & Test of Computers*, 18(5):21–31, 2001.

[6] S. Eisenhart and R. Tolbert. Designing for the use case: Using the OMAP4 platform to overcome the challenges and integrating multiple applications. Technical report, Texas Insruments, 2008. Available from www.ti.com/omap4.

[7] G. Frantz and R. Simar. Cutting to the core of the problem. *eTech embedded processing e-newsletter*, 2009. Available from www.focus.ti.com/dsp/docs.

[8] J.L. Hennessy and D.A. Patterson. *Computer Architecture: A Quantitative Approach.* Morgan Kaufmann Publishers, San Mateo, CA, 1990.

[9] IEEE. P1500 standard for embedded core test. Technical report, IEEE Standards, grouper.ieee.org/groups/1500/, 1998.

[10] Texas Instruments. OMAP 5910 user's guide. Technical report, Texas Instruments, www.dspvillage.ti.com, 2009.

[11] Texas Instruments. TMS320C6474 multicore digital signal processor. Technical report, Texas Instruments, Available from www.ti.com/docs/prod/folders/print/tms320c6474.html, 2009.

[12] A. Ivanov and De Micheli G. The network-on-chip paradigm in practice and research. *IEEE Design & Test of Computers, Special Issue on Network-on-Chip Architectures*, 22:399–403, 2005.

[13] I. Kadayif et al. An integer linear programming based approach for parallelizing applications in on-chip multiprocessors. In *Proceedings of Design Automation Conference*, 2002.

[14] S. Kumar et al. A network on chip architecture and design methodology. In *Proceedings of the IEEE Computer Society Annual Symposium of VLSI*, 2002.

[15] S. Meftali et al. Automatic generation of embedded memory wrapper for multiprocessor SoC. In *Proceedings of Design Automation Conference*, 2002.

[16] S. K. Moore. Multicore is bad news for supercomputers. *IEEE Spectrum*, 45:5–15, 2008.

[17] V. Nag and C.P. Ravikumar. Synthesis of heterogeneous multiprocessors. Technical report, Electrical Engineering, IIT Delhi, Hauz Khas, New Delhi, India, 1997. Master's Thesis in Computer Technology.

[18] A.L. Narasimha Reddy. Improving the interactive responsiveness in a video server. In *Proceedings of SPIE Multimedia Computing and Networking Conference*, pages 108–112, 1997.

[19] OCPIP. Open core protocol international partnership. Technical report, OCP IP Organization, 2008. www.ocpip.org.

[20] R. Payne Sr. and Wiscombe P. What is the impact of streaming data on SoC architectures? *EE Times*, 2003.

[21] C.P. Ravikumar and Hetherington Graham. A holistic parallel and hierarchical approach towards design-for-test. In *Proceedings of International Test Conference*, pages 345–354, 2004.

[22] S. Richardson. MPOC: A chip multiprocessor for embedded systems. Technical report, Hewlett Packard, 2002.

[23] M. Rudack et al. Large-area integrated multiprocessor system for video applications. *IEEE Design & Test of Computers*, pages 6–17, 2002.

[24] K.K. Ryu, Shin E., and V. J. Mooney. Comparison of five different multiprocessor SoC bus architectures. In *Proceedings of the Euromicro Symposium on Digital Systems*, pages 202–209, 2001.

[25] A. Sharma and C. P. Ravikumar. Efficient implementation of ADPCM Codec. In *Proceedings of the 12th International Conference in VLSI Design*, 2000.

[26] J. Song et al. A low power open multimedia application platform for 3G wireless. Technical report, Synopsys, synopsys.com/sps/techpapers.html, 2004.

[27] H.-J. Stolberg et al. HiBRID-SoC: A multi-core system-on-chip architecture for multimedia signal processing applications. In *Proceedings of Design Automation and Test in Europe Designer's Forum*, pages 8–13, March 2003.

[28] H.S. Stone and J. Cocke. Computer architecture in the 1990s. *IEEE Computer*, 24:30–38, 1991.

[29] VSIA. VSIA - virtual socket interface alliance (1996-2008). Technical report, VSIA, www.vsi.org, 2008.

[30] C. Wander, N. Gabriela, G. Lovic, L. Damien, and A. A. Jerraya. Colif: A design representation for application-specific multiprocessor SOCs. *IEEE Design and Test of Computers*, 18(5):8–20, Sep/Oct 2001.

2

Application-Specific Customizable Embedded Systems

Georgios Kornaros

Applied Informatics & Multimedia Department,
Technological Educational Institute of Crete
Heraklion, Crete, Greece
kornaros@epp.teiher.gr

Electronic & Computer Engineering Department,
Technical University of Crete
Chania, Crete, Greece
kornaros@mhl.tuc.gr

CONTENTS

2.1	Introduction	32
2.2	Challenges and Opportunities	34
	2.2.1 Objectives	35
2.3	Categorization	37
	2.3.1 Customized Application-Specific Processor Techniques	37
	2.3.2 Customized Application-Specific On-Chip Interconnect Techniques	40
2.4	Configurable Processors and Instruction Set Synthesis	41
	2.4.1 Design Methodology for Processor Customization	43
	2.4.2 Instruction Set Extension Techniques	44
	2.4.3 Application-Specific Memory-Aware Customization	48
	2.4.4 Customizing On-Chip Communication Interconnect	48
	2.4.5 Customization of MPSoCs	49
2.5	Reconfigurable Instruction Set Processors	52
	2.5.1 Warp Processing	53
2.6	Hardware/Software Codesign	54
2.7	Hardware Architecture Description Languages	55

2.7.1   LISATek Design Platform   57
2.8   Myths and Realities   58
2.9   Case Study: Realizing Customizable Multi-Core Designs   60
2.10   The Future: System Design with Customizable Architectures,
        Software, and Tools   62
Review Questions   63
Bibliography   63

2.1   Introduction

Embedded system development seeks ever more efficient processors and new automation methodologies to match the increasingly complex requirements of modern embedded applications. Increasing effort is invested to accelerate embedded processor architecture exploration and implementation and optimization of software applications running on the target architecture. Special purpose devices often require application-specific hardware design so as to meet tight cost, performance and power constraints. However, flexibility is equally important to efficiency: it allows embedded system designs to be easily modified or enhanced in response to evolution of standards, market shifts, or even user requirements, and this change may happen during the design cycle and even after production. Hence the various implementation alternatives for a given function, ranging from custom-designed hardware to software running on embedded processors, provide a system designer with differing degrees of efficiency and flexibility. Often, these two are conflicting design goals, and while efficiency is obtained through custom hardwired implementations, flexibility is best provided through programmable implementations.

Unfortunately, even with sophisticated design methodologies and tools, the high cost of hardware design limits the rapid development of application specific solutions and the actual amount of architectural exploration which can be done. Taking new, emerging technologies and putting them on silicon is a great challenge. The complexity is becoming so demanding that the integration and verification of hardware and software components require increasingly more time, thus causing delays to bringing new chips to market.

Recent advances in processor synthesis technology can reduce the time and cost of creating application-specific processing elements. This enables a much more software-centric development approach. A greater percentage of software development can occur up front, and architectures can be better optimized from real software workloads. Application-specific processors can be synthesized to meet the performance needs of functional subsystems while maximizing the programmability of the final system. Essentially, the hardware is adapted to software rather than the other way around.

Configurable processing combines elements from both traditional hardware and software development approaches by incorporating customized and application-specific compute resources into the processor's architecture. These compute resources become additional functional engines or accelerators that are accessible to the designer through custom instructions. Configurable processors offer significant performance gains by exploiting data parallelism through wide paths to memory: operator specialization such as bit width optimization, constant folding and partial evaluation; and temporal parallelism through the use of deep pipelines.

In general, in designing an embedded system-on-chip (SoC) three approaches are historically followed. The first is a purely software-centric approach by mapping of applications to a system-on-chip or multiprocessor SoC (MPSoC) and optimizing them for enhanced performance or for power consumption or real-time response. Using advanced compiler technology often system designers can leverage the knowledge of how to squeeze the ultimate performance out of a specified architecture. Although the C language widely used in developing embedded applications does not support parallelism, parallelizing compilers can give significant advantage to exploit MPSoC architectures. It is even possible for compiler technology to recognize and vectorize data arrays that can be handled through the SIMD (single instruction multiple data) memory-to-memory architectures of certain SoCs.

The second approach is design of application-specific hardware to achieve high-speed embedded systems with varying levels of programmability. Although application-specific integrated circuits (ASICs) have much higher performance and lower power consumption, they are not flexible and involve an expensive and time-consuming design process. Finally, the third recently appeared approach is the development of both the hardware and software architecture of a system in parallel, so as to enhance the flexibility of ASICs for a specific problem domain. Though not as effective as ASICs, custom-instruction processors are emerging as a promisingly effective solution in the hardware/software codesign of embedded systems. The recent emergence of configurable and extensible processors is associated with a favorable trade-off between efficiency and flexibility, while keeping design turn-around times shorter than fully custom designs.

Application-specific integrated processors (ASIPs) fill the architectural spectrum between general-purpose programmable processors and dedicated hardware or ASIC cores (as depicted in Figure 2.1). They allow one to effectively combine the best of both worlds, i.e., high flexibility through software programmability and high performance (high throughput and low power consumption).

The key to customization of an embedded system architecture is the ability to expand the core processor instruction set, and possibly the register files and execution pipelines. Since the application developers in addition to developing the application must also tailor the embedded system and discover the critical processor hotspots for the specific application, it is crucial to use an

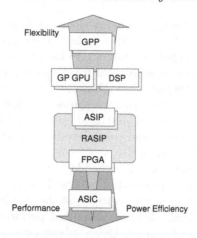

FIGURE 2.1: Different technologies in the era of designing embedded system-on-chip. Application-specific integrated processors (ASIPs) and reconfigurable ASIPs combine both the flexibility of general purpose computing with the efficiency in performance, power and cost of ASICs.

automated framework. Hence, it has become increasingly important to provide also automated software support for extending the processor features. Given a source application, researchers aim at providing a compiler/synthesis tool for a customizable SoC that alone can generate the best cost-efficient processing SoC along with the software tools.

2.2 Challenges and Opportunities: Programmability or Customization

The multi-core revolution has shifted from a hardware challenge (making systems run faster with faster clock cycles) to a software challenge (utilizing the raw computation power provided by the additional cores). Embedded application developers today have more resources at their disposal and have to use concurrent programming techniques to exploit them, making the development and deployment of the applications more challenging. Several parallel programming models do exist: openMP, message passing interface (MPI), POSIX threads, or hybrid combinations of these three. The selection of the most appropriate model in the context of a given embedded application requires expertise and good command of each model, given the complexity imposed by cores competing for network bandwidth or memory resources. Moreover, in one direction, embedded platform providers offer sets of tools and libraries that bring simplicity to multi-core programming and help programmers harness the full potential of their processors. Usually, these involve support for C/C++, standard programming paradigms, and the most advanced multi-core debugging and optimization tools.

Recently, design methodologies for managing exploding complexity consider embedded software from the early stages. Embedded systems are inherently application-specific. While system designers have to traverse the complex path involving different technologies and evolving standards, the success depends on timely reaction to market shifts and minimizing the time to market. Thus, advanced multiprocessor architectures on a single chip are built that mainly rely on programming models (streaming, multi-threading) to support efficiently embedded applications. However, in a different perspective, developing strategies these days employ software in the design and manufacturing process in a different way. Some strategies attempt to tailor the hardware more on the specific domain problem than the other way around.

An embedded system runs one specific application throughout its lifetime. This gives to the designers the opportunity to explore customized architectures for an embedded application. The customization can take many forms: extending the instruction-set architecture (ISA) of the processor with application-specific custom instructions, adding a reconfigurable co-processor to the processor, and configuring various parameters of the underlying microarchitecture such as cache size, register file size, etc. However, given the short time-to-market constraint for embedded systems, this customization should be automatic. Modern techniques face a shift from retargetable compiler technologies to a complete infrastructure for fast and efficient design space exploration of various architectural alternatives.

Although programmability allows changes to the implemented algorithm achieving the requirements of the application, customization allows to specialize the embedded system-on-chip in a way that performance and cost constraints are satisfied for a particular application domain.

2.2.1   Objectives

All research and industrial approaches fundamentally aim to partition an application to core-processor functions and custom functions that are located on the critical execution path. Under certain system constraints (such as area cost, power consumption, schedule length, reliability, etc.) these custom functions are efficiently implemented in hardware to augment the baseline processor. Emerging standards and competitive markets though, stress for more flexible and scalable SoCs than customized hardware solutions.

For embedded SoC developers the objectives are to efficiently explore the huge design space and to combine automatic hardware selection and seamless compiler exploitation of the custom functions. By carefully selecting the customizable functions these can often be generalized to make their use have applicability across a set of applications. This is due to the fact that the computationally intensive portions of applications from the same domain are often similar in structure.

The system designer must be provided with an efficiently automated development environment. This environment can integrate compiler technology and software profiling with a synthesis methodology. Using an analytical approach or benchmark and a simulation methodology significantly enhances an automated environment. The interworking of all these technologies must assist in realistically tuning a multiprocessing SoC to fit a specific application.

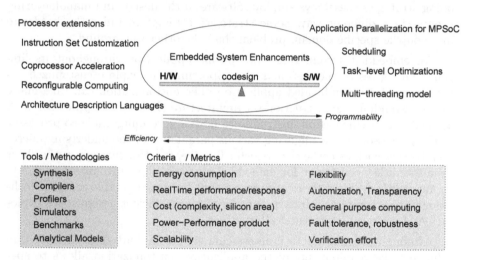

FIGURE 2.2: Optimizing embedded systems-on-chips involves a wide spectrum of techniques. Balancing across often conflicting goals is a challenging task determined mainly by the designer's expertise rather than the properties of the embedded application.

The proliferation of multimillion gate chips and powerful design tools have paved the way for new paradigms. Network-on-chip architecture provides a scalable and more efficient on-chip communication infrastructure for complex systems-on-chips (SoCs). NoC solutions are increasingly used to manage the variety of design elements and intellectual property (IP) blocks required in today's complex SoCs. NoC-based multiprocessor SoCs (MPSoCs) have emerged with a significant impact on the way to develop embedded applications. ASIPs, NoCs, and MPSoCs make the application-specific hardware-software codesign spectrum even wider as discussed in the following sections.

2.3 Categorization

2.3.1 Categorization of Customized Application-Specific Processor Techniques

Different mechanisms to configure and adapt a base system-on-chip (SoC) architecture to specific application requirements have been researched, usually along with a complete design tool and exploration environment. They range from component-based construction of embedded systems, with the aid of architecture description languages or instruction set extensions of a base processor and from design time application specific customization, to runtime system reconfiguration. Extensible processing combines elements from both traditional hardware and software development approaches to provide customized per-application compute resources in the form of additional functional engines or accelerators which are accessible to the designer through custom instructions.

Initial strategic decisions on developing an enhanced embedded SoC (targeting flexibility, i.e., not following an ASIC approach) can be classified as follows.

- *Single processor*, extensible either in the form of its instruction set, or configurable by parameterizing the integrated hardware resources (multipliers, floating-point, DSP units, etc.), or with coprocessors.

- *Symmetric multiprocessor SoC (MPSoC)*. Partitioning and mapping of the embedded application to the processors can be done at compile time at the task or basic block level. Alternatively, the developer can provide hooks to the operating system to schedule tasks on the processors at runtime.

- *Heterogeneous single-chip MPSoC*, or asymmetric multiprocessing that features integration of multiple types of CPUs, irregular memory hierarchies, and irregular communication. Heterogeneous MPSoCs are different from traditional embedded systems due to complexity and heterogeneity of the system that significantly increase the complexity of the HW/SW partitioning problem. Meanwhile, evaluating the performance and verifying its correctness are much more difficult compared to traditional single processor-based embedded systems. Programming a heterogeneous MPSoC is another challenge to be faced. This problem arises simply because there are multiple programmable processing elements. Since these elements are heterogeneous, the software designer needs to have expertise on all of these processing elements and needs to take a lot of care on how to make the software run as a whole.

- *HW/SW codesign* with a combination of the above architectural solutions. Hardware/software partitioning is usually a coarse-grain approach, while custom instruction sets find speedups at finer levels of granularity. Traditionally architecture description languages (ADLs) have been utilized to this direction.

- *Network-on-Chip based multi-core SoCs*. Given the aggregate demands of multi-core architectures, tools are emerging to help chip architects explore new interconnect topologies and perform application-specific analyses. Thus, it is feasible to optimize on-chip communications (bandwidth and latency) between IP cores, along with overall system characteristics such as power, die area, system-level performance, timing closure and time-to-market.

- *Hardware synthesis from high-level languages.* This is a concept that continues to gain momentum in the electronic design automation community. Originating from an academic project (PACT) at Northwestern University a path from the MATLAB® language to an implementation on a heterogeneous embedded computing platform is provided, which later commercialized into the AccelChip MATLAB to RTL VHDL tools targeting FPGAs [6]. Gupta et al. [24] present a framework that treats behavioral descriptions in ANSI-C and generates synthesizable register-transfer level VHDL; emphasis is placed on effectively extracting parallelism for performance. The PACT HDL also is an attempt that converts C programs to synthesizable hardware descriptions targeting both FPGAs and ASICs, optimizing for both power and performance [38]. Catapult C, Handel C and Impulse C are recent products from various EDA companies that synthesize algorithms written in C/C++ directly into hardware descriptions.

Equally important as performance, power and cost is the time-to-market demand, which leads to systems (Tensilica Xtensa [63], ARC 700 [37], MIPS Pro Series [35], Stretch S6000 [36], Altera Nios II [53], Xilinx MicroBlaze [54]) that come with a pre-designed and pre-verified base architecture and an extensible instruction set. The pre-designed and verified base architectures reduce the design effort considerably, and the programmable nature of such processors ensures high flexibility.

The effectiveness of configurable and extensible processors has been demonstrated both for the early single chip processors and for the recent MPSoCs. The main techniques to application-oriented customized processing can be broadly outlined as:

- Extend the instruction-set architecture (ISA) of the processor with application-specific custom instructions

- Configure base processor core with functional engines or attach coprocessor accelerators (maybe using reconfigurable technology)

- Customize memory subsystem (followed with customized load/store semantics)

- Customize various parameters of the resources of the base architecture (cache size, register files, etc.)

- Off-load, use loosely coupled flexible I/O processing.

The methodologies to apply processor configurability and extensibility are various and in principle follow the directions:

- **Processor customization.** *coarse-grain* at block level, by integrating processing units with a CPU, or *fine-grain*, by customizing the instruction set. Customization can be applied on single embedded processor or in the context of homogeneous or heterogeneous multiprocessor architectures.

- **Reconfigurable computing approach.** Use a baseline processor with reconfigurable logic, soft or configurable processors; in addition a few approaches allow run-time reconfiguration.

- **Reverse customization.** Executable code to coprocessor generation: free from source level partitioning and independent from the origin of the source (i.e., multiple source languages can be used), ASIPs are implemented directly from an executable binary targeted at the main processor. The executable code may be translated into a very different application-specific instruction set that is created for each coprocessor. The generated coprocessors range from fixed function hardware accelerators to programmable ASIPs.

- **Hardware Architecture Description Languages (ADL).** ADLs enable embedded processor designers to efficiently explore the design space by modelling their processor using a high level language, and automatically generate instruction set simulators (ISSs) and a complete set of associated software tools including the associated C compiler. Custom processors, such as application-specific instruction processors (ASIPs) for DSP and control applications, are also featured by the automatic generation of synthesizeable register transfer level (RTL) code. Depending on the abstraction level different ADLs have been designed for hardware-software codesign:

  ◇ *High-level ADL*, an attribute grammar-based language is used for processor specification and a synthesis tool next generates structural synthesizable VHDL/Verilog code for the underlying architecture from the specifications. nML ([20], sim-nML([47], [8] and ISDL ([26], FlexWare [51]) belong in this class.

⋄ *Low-level ADL*, MIMOLA [67] hardware specification language enables the designer to write structural specification of a programmable processor at low level, exposing several hardware details.

⋄ *Complete ADL*, both the processor behavior at the instruction level can be described to tailor to the application needs and the architecture design space exploration can be managed via integrated software toolchains and architecture implementation and verification toolchains. LISA [31] is an example of this integrated development environment.

ADL-based methodologies usually offer the maximum flexibility and efficiency at the expense of increased design time and significant effort. Meanwhile, working with pre-designed and pre-verified cores (e.g., Tensilica Xtensa, ARC Tangent, MIPS CorExtend) offers faster timing closure.

The above classification is not very sharp for various reasons. Increasingly, programmable platforms are available with hybrids of the above forms of programmability available in the form of processors and programmable hardware on the same die. Further, the distinction between instruction and hardware programming bits is gradually becoming blurred.

In traditional hardware/software co-synthesis the custom hardware is in the form of predefined hardware computation elements (CEs) that reside in libraries. The outcome of the synthesis flow is principally a processor with a set of CEs permanently bound to it so as to accelerate a fixed assigned task. This is depicted in Figure 2.3 (a) in the shaded part, which may include blocks to assist in DSP computations for example. In a different or complementary approach, design space exploration tools assist in defining the most efficient topology to interconnect an amount of pre-designed and verified computation or interface components with one or more fixed CPUs (Figure 2.3 (b)).

Nowadays, heterogeneous or asymmetric multiprocessing is the most effective and competitive in the cost-conscious embedded SoC market segment. Adoption of embedded SMP is limited mostly because of the immature level of SMP support of embedded OSs and compilation toolchains. For example, general-purpose processors (GPPs) and DSPs have distinctively different characteristics that make them best suited to different application domains. Thus, an embedded application that embraces a mixed workload which demands general purpose and DSP computations, will better be mapped on a heterogeneous SoC in a much more cost-effective way. Using a GPP with SMP is an expensive alternative, while a single-chip DSP is too rigid.

2.3.2   Categorization of Customized Application-Specific On-Chip Interconnect Techniques

Early research on NoC topology design used regular topologies, such as trees, tori, or meshes, like those that have been used in macro-networks for designs

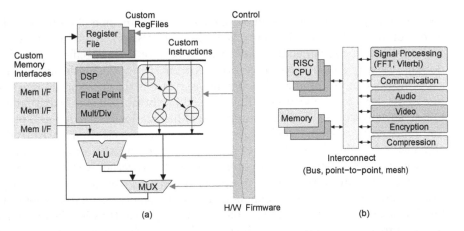

FIGURE 2.3: Extensible processor core versus component-based customized SoC. Computation elements are tightly coupled with the base CPU pipeline (a), while (b), in component-based designs, intellectual property (IP) cores are integrated in SoCs using different communication architectures (bus, mesh, NoC, etc.).

with homogeneous processing cores and memories. However this approach has become rapidly inappropriate for MPSoC designs that are typically composed of heterogeneous cores, since regular topologies result in poor performance, with large power and area overhead. This is due to the fact that the core sizes of the MPSoC are highly non-uniform and the floorplan of the design does not match the regular, tile-based floorplan of standard topologies [10]. Moreover, for most state-of-the-art MPSoCs (like the Cell-Playstation III [12], Philips Nexperia [13] or TI OMAP [3]) the system is designed with static (or semi-static) mapping of tasks to processors and hardware cores, and hence the communication traffic characteristics of the MPSoC can be obtained statically. Thus, an application-specific NoC with a custom topology, which satisfies the design objectives and constraints, must have efficient on-chip interconnects for MPSoCs. Therefore, a lot of research is done in design space exploration of NoC topologies, protocols and automization frameworks.

2.4 Configurable Processors and Instruction Set Synthesis

Application-specific instruction set processor (ASIP) design has long been recognized as an efficient way to meet the growing performance and power demands of embedded applications. Special-purpose hardware, such as coprocessors and special functional units, enables ASIPs to come close to the efficiency of application-specific integrated circuits (ASICs). Using pre-designed and pre-verified components to optimize a processor for a specific application domain creates *configurable* processors. Alternatively, the basic instruction set of ASIPs can be enhanced by custom instructions that use special-purpose hardware. This can be viewed as fine-grained hardware/software partitioning. Often, at more coarse level entire sequence of instructions is treated as a block and replaced by custom circuitry operating in a single cycle. In most ASIP design methodologies the applications are usually represented as directed graphs, and the complete instruction set is generated either together with the microarchitecture or using retargetable compilers based on given hardware descriptions. The search space grows exponentially and globally optimal solutions are hard to achieve. Nowadays, the problem grows significantly with the existence of multiprocessors on chips and when considering both coarse and fine-grained acceleration techniques.

Unlike conventional multiprocessors, where the operating system schedules data-independent processes to different processors, the embedded single-chip multiprocessors and the embedded heterogenous multi-core SoCs usually execute a single or small set of applications (e.g., telecommunications or multimedia). Thus, it is feasible to assign and schedule tasks on different processors or cores in advance, at design time. Tuning a SoC to specific application requirements can additionally integrate design techniques from ASIPs, with simultaneous customization of each processor of the single-chip embedded multiprocessor. At the same time that a uniprocessor can be customized for an application, MPSoCs can exploit parallelism of loops, functions, or coarse-grained tasks, and the application program can be partitioned and assigned to multiple processors.

In addition to the challenges of increasing demand for high performance, low energy consumption and low cost, the success of embedded processors depends equally on time-to-market. Manual selection of instruction set extensions (ISEs) to the base instruction set of the processor for executing the critical portions of the application may achieve the aforementioned blend, using heuristic techniques and matching the expertise of the designer. However, this can be a very time-demanding process and works well for very narrow application fields. Hence it is recognized that automatic identification of ISEs for a given application set is very important.

The following sections investigate the different aspects of processor configurability and extensibility. Various promising automation methodologies and infrastuctures are discussed along with their impact on customized embedded system design.

2.4.1 Design Methodology and Flow for Processor Customization

Within the initial steps in ASIP design is the partitioning of an application into base-processor instructions and custom instructions. Usually the objective is to utilize special purpose functional units tightly coupled to the base processor to perform long operations, or common operations in fewer cycles. At first, the application software is profiled looking for computation intensive segments of the code which, if designed in hardware, increases performance. The processor is then tailored to include the new capabilities.

Design space exploration is often used to include the combined analysis of the application to identify hot spots and the assessment of microarchitecture changes that contribute to the best match of the embedded SoC requirements. To estimate the impact of each customization decision of the microarchitecture (instruction set selection, or datapath optimization, or scratchpad memories insertion, etc.) researchers adopt different methodologies:

- *Analytical-centric.* The advantage of the analytical approach (for example based on the integer linear programming (ILP) model[1], as formulated in [49]) lies in automation, potentially eliminating the need for a manual design process. Because in this case the synthesis problem is basically an ILP problem, existing solvers can be integrated in the design flow to solve an appropriately formulated problem. The challenges of the approach lie in the capturing of all design parameters and constraints of interest, and in ILP tractability.

- *Scheduler-centric.* The instruction set design of an embedded processor can be formulated as a simultaneous scheduling/allocation problem exploiting micro-operation level parallelism. The instructions can be translated into micro-operations ([33]) that are stored in the trace cache for instance, and a resource constrained scheduler optimizes the issuing to the execution units. This methodology is intended to be free from the inefficiency and overhead problems of microarchitecure simulation-based approaches.

- *Simulation-centric.* Several researchers have extensively performed ASIP design exploration using retargetable processor code generation and simulations. Machine description languages such as Expression [46],[27], and LISATek [31] have been developed as the main vehicles to drive the retargeting process.

---

[1]A mathematical programming problem is one in which there is a particular function to be maximized or minimized subject to several constraints. If the function f is linear, then the problem (P) is called a linear programming (LP) problem. If, in addition, the variables x are integer valued, then (P) is called an *integer linear programming* (ILP) problem (A. Schrijver, *Theory of Linear and Integer Programming*, John Wiley & Sons Ltd, [1986].)

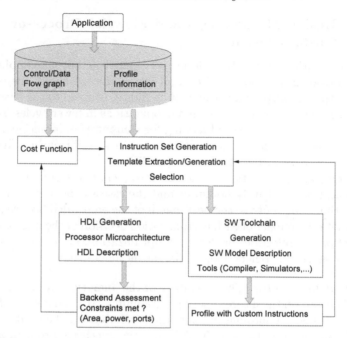

FIGURE 2.4: Typical methodology for design space exploration of application specific processor customization. Different algorithms and metrics are applied by researchers and industry for each individual step to achieve the most efficient implementation and time to market.

An overview of the principal components in a design exploration methodology is shown in Figure 2.4. Simulation-centric approaches are more dependent on the low-level microarchitectural intricacies of the target SoC, while the first strategies attempt to provide a level of abstraction and automation.

2.4.2 Instruction Set Extension Techniques

The customization of a processor instruction set with either a super set of instructions or with new complex instructions can formally be divided into instruction generation and instruction selection. Given the application code, the definition of the instruction set can be classified as follows:

- *Selection based.* Choosing an optimal instruction set for the specific application under the constraints, such as chip area and power consumption, is done by selecting instructions from the fixed super-set of possible instructions. In this context the application is represented as a directed graph and, similarly to the graph isomorphism problem, candidate instructions are identified as subgraphs.

- *Instruction set synthesis based.* The encoding-generation of the entire instruction set architecture usually combines basic operations to create new instructions for specific applications. These new application-specific instructions are called *complex instructions*. The actual synthesis process consists of two phases: complex instruction generation and instruction selection. Complex instructions are generated for each application or a set of applications representing a domain of applications.

- *Scheduling based.* Fitting an instuction set to an application area can be formulated as a modified scheduling problem of micro-operations. In this approach, each micro-operation is represented as a node to be scheduled and a simulated annealing scheme is applied for solving the scheduling problem. This is important in that it triggers the definition, and generation of application-specific complex instructions.

- *Combined selection/synthesis* techniques for processor instruction set extension.

Conceptually, the main concerns of system designers include the constraints imposed by the usage of the system and the stringent limits that it must respect. For instance, power consumption constraints or performance and response time are traditional guide metrics for the system developer. More specifically, one key problem in the hardware/software codesign of embedded systems is the hardware/software partitioning which has been proven to be NP-hard. General-purpose heuristics for HW/SW partitioning include genetic algorithms [55],[58],[17], simulated annealing [52],[29],[19], and greedy algorithms [12],[23].

In the context of extensible processors at the instruction level an application is represented as a directed graph where code transformations, such as loop unrolling and if-conversion are usually applied to selectively eliminate control-flow dependencies.

Most importantly, the decisions of the designer are affected, and at the same time have implications on:

- The area constraints of custom logic

- The throughput between the processor and the custom logic

- Partitioning of a graph to N-input/K-output subgraphs and identification of the optimal (N,K) pair

- Time convergence of algorithms for identification and selection of custom instructions

The design space of area and time tradeoffs grows exponentially with the size of the underlying application. The complete design space cannot be searched with a reasonable time effort. A truly optimal solution could be possible by enumerating all possible subgraphs within the application data flow graphs (DFGs). However, this approach is not computationally feasible, since the number of possible subgraphs grows exponentially with the size of the DFGs. Figure 2.5 shows a sample data flow subgraph. A subgraph discovery mechanism of such subgraphs uses a guide function mostly based on criticality, latency, area, and number of input output operands.

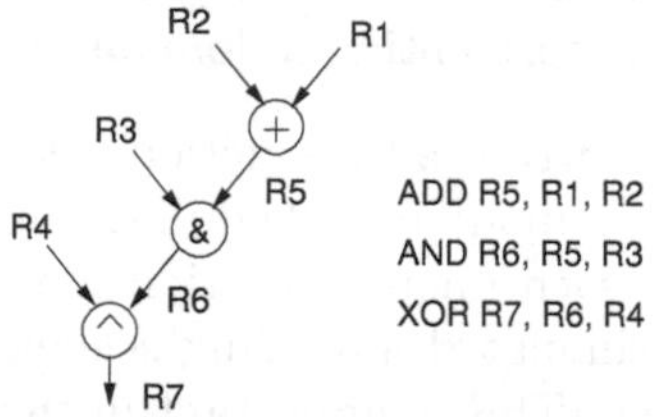

FIGURE 2.5: A sample data flow subgraph. Usually each node is annotated with area and timing estimates before passing to a selection algorithm.

To reduce the complexity different techniques have been devised, as clustering-based approaches, or restricting to single-output operands, or constraint-propagation techniques on the number of input/output operands of the subgraphs. In [70] and [69], Yu and Mitra enumerate only connected subgraphs having up to four input and two output operands and do not allow overlapping between selected subgraphs. Code generation methods traditionally use a tree covering approach (as in [14]) to map the data flow graph (DFG) to an instruction set. The DFG is split into several trees, where each instruction in the instruction set architecture (ISA) covers one or more nodes in the tree. The tree is covered using as few instructions as possible. The purpose behind splitting the DFG into trees is that there are linear time algorithms to optimally cover trees, making the process faster.

Clark et al. [13] enumerates subgraphs in a data flow graph, uses subgraph isomorphism to prune invalid subgraphs, and uses unate covering to select which valid sub-graphs to execute on the targeted accelerators. Thus, their algorithms achieve, on average 10 percent, and as much as 32 percent more speedup than traditional greedy solutions.

The most common technique for instruction generation is mainly based on the concept of *template*. The *template* is a set of program instructions that is a candidate for implementation as a custom instruction. The template is equivalent to a subgraph representing the list of statements selected in the subject graph, where nodes represent the operations and edges represent the data dependencies. Instruction generation can be performed in two non-exclusive ways: using existing templates or creating new templates. A library of templates can be built of identified templates. Usually, the construction and collection of templates is application domain-specific.

One can formulate the problem of matching a library of custom-instruction templates with application DFGs as a subgraph isomorphism problem [42], [14], [10]. In this case instruction generation can be considered as template identification. However, this is not always the case and many researchers develop their own templates [56], [3], [21]. In this case templates are identified inside the graph using a guide function. This function considers a certain number of parameters (often called constraints) and starting from a node taken as a seed, grows a template which respects all the parameters. One such constraint for example that incurs potential complexity is the encoding of multiple input and output operands within a fixed length. Once a certain number of templates is identified the graph is usually re-analyzed to detect recurrences of the built templates.

Improvements of candidate functional unit (FU) identification and selection (or cluster of candidates) can be achieved by restricting the number of port accesses to the register file (bound I/O ports between custom FU and the register file), or serialize them under the actual register file port constraints. This will occur if we allow the algorithm to produce custom FUs which might have more inputs and outputs than available register file ports. Under this formulation additional considerations include the constraint for simultaneous arrival of the operands at the inputs of a subgraph, and pipelining of the candidate subgraph and not the whole graph.

The speedup obtainable by custom instructions is limited by the available data bandwidth to and from the datapaths implementing them. Extending the core register file to support additional read and write ports improves the data bandwidth. However, additional ports result in increased register file size, power consumption, and cycle time. Typical formulation of the instruction-set extension identification problem can have register-port availability as a critical constraint. One way to moderate the problem is to add architecturally visible storage (called AVS in [40]), which intrinsically provides the customized datapath with additional local bandwidth. Architecturally visible storage may simply mean scalar registers to hold local variables mostly used by the customized instruction. It can also mean complete data structures, such as local arrays, whose content is used over and over by the special instruction.

Replication of the register file and use of shadow registers to extend the base processor are indeed strategies to this direction. A complete physical copy (or partial copy) of the core register file allows custom instructions to fetch the encoded operands from the original register file and the extra operands from the replicated register file. Chimaera [28] for instance is capable of performing computations that use up to nine input registers. However, the basic instructions cannot utilize the replicated register file.

Most cost efficient is the use of a small number of shadow registers. Since the shadow registers are mainly used for storing variables with short lifetimes within the basic blocks, the required number of shadow registers is usually much smaller than that of the core register file. Use of shadow registers [15] and exploitation of forwarding paths of the base processor, or custom state registers (Tensilica Xtensa) to explicitly move additional input and output operands between the base processor and custom units are used as efficient architectural approaches.

2.4.3   Application-Specific Memory-Aware Customization

Traditionally, strategies to compensate for memory-latency are *multi-threading, memory hierarchy management,* and *task-specific memories*. Due to the heterogeneity in recent memory organizations and modules, there is a critical need to address the memory-related optimizations simultaneously with the processor architecture and the target application. Through co-exploration of the processor and the memory architecture, it is possible to exploit the heterogeneity in the memory subsystem organizations, and trade off system attributes such as cost, performance, and power. However, such processor memory co-exploration framework requires the capability to explicitly capture, exploit, and refine both the processor as well as the memory architecture [46].

In [9] the authors allow memory instructions to be selected in the set of candidate instructions for acceleration, considering any kind of vector or scalar access. Special instructions were also introduced to perform DMA connection between the local memory inside a FU and the main memory. Open issues remain with pointer accesses and exploitation of the data reuse within the critical section of an application. The architectural model of PICO-NPA [57] also permits the storage of reused memory values in accelerators.

A framework for high-level synthesis and optimization of an application-specific memory accesss network is presented in [65]. A may-dependence flow graph is constructed to represent an ordering dependence at run time. Then, tree-construction heuristics and pruning techniques based on a cost model are applied for efficient design space exploration. They show how to provide a dynamic synchronization mechanism that maintains consistency in the context of memory-ordering dependences that are known only at run time. Optimizations are also explored to identify local regions of memory dependencies and adjust the corresponding memory access network to take advantage of these. However, in a MPSoC the memory access requirements for throughput and synchronization protocols present even more challenges.

2.4.4   Customizing On-Chip Communication Interconnect

Equally significant to processor configurability and extensibility is the interconnect fabric between the processors inside an MPSoC, or between the base processor and its functional units. Different topologies, buffering schemes and protocols and their corresponding user programming models are becoming increasingly essential. Automating retargeting compilers and task mapping tools have adopted a holistic approach to simultaneously consider traffic between the application tasks and instruction-set customization.

Managing Interconnect between Processor and Functional Units

In this case the combinatorial problem consists of selecting specific types of networks for inter-task communications such as buses, rings, meshes, fat trees, hypercubes etc., under given constraints and costs. Different topologies can be mixed. The formulation challenge in this case stems from three aspects: (i) application-dependent dynamic communication patterns, (ii) allowing the mixing of different communication topologies and protocols, and, (iii) allowing the arbitrary sharing of networks.

The trend to embrace heterogeneous processor architectures in modern embedded SoCs often involves these three aspects. This leads to ad hoc interconnection schemes that can complicate the SoC development. However, emphasis is growing on considering interconnect between core processor and accelerating units while solving the overal system optimization problem.

Automated Exploration Infrastructures for On-Chip Interconnect

Application-specific single chip systems increasingly consider the mapping of the embedded application to standard or custom processing resources as a communication-intensive problem. Together with stringent time-to-market requirements and extensive design reuse methodologies network-on-chip (NoC) based multi-core systems ask for automated infrastructures. Hence, NoC design tools focus on exploration of static or dynamic mapping and scheduling of application functionality on NoC platforms. Different frameworks enable user-driven exploration through parameterization under resource constraints trying to optimize performance and power consumption.

Currently available state of the art NoC development tools include the Silistix *ChainWorks* [5], the open-source *On-Chip Communication Network* (OCCN) framework, the *Hermes* [48] NoC design tools and Arteris configurable NoC IP [4]. Additionally *XPipes* [59], a design flow for the generation of synthesizeable and simulatable models for application-specific networks on chip intends to allow designers to explore the design space spanned by various NoC topologies and parameters. *XPipes Lite* is a SystemC library of highly parameterizable, synthesizeable NoC network interface, switch and link modules, optimized for low-latency and high frequency operation. Communication is packet switched, with source routing (based upon street-sign encoding) and wormhole flow control.

Silistix ChainWorks is a set of design tools which offer a graphical way to specify topologies and attributes of asynchronous self-timed interconnects for SoC. It also features adaptation to existing synchronous bus architectures, such as IBM CoreConnect, AMBA AHB and OCP 2.0. The ChainCompiler is a synthesis tool that produces structural Verilog netlist suitable for use by conventional logic synthesis tools.

2.4.5 Customization of MPSoCs

A highly complex multidimensional problem includes a comprehensive integrated framework for ASIP while developing embedded MPSoCs. Complex interdependencies arise while exploring the design space by simultaneously sweeping axes like processing elements, memory hierarchies and chip interconnect fabrics. To this end Angiolini et. al. in [2] combined the use of LISATek processor design platform with MPARM system-level architecture MPSoC platform. At the architecture level they combined exploration of different protocols over shared buses while defining three layers of memory devices: (1) on-tile, strongly coupled to the processor, such as caches and ScratchPad Memories, (2) on-chip, attached to the system interconnect, (3) off-chip, driven by a DRAM memory controller. It is shown that it is hard but necessary to provide a united integrated exploration toolset for MPSoC traditional issues and accurate analysis of the tradeoffs implied by the ASIP/coprocessor paradigm at the system level. Although enhanced infrastructures for exploring extensible processors are very effective, embedded MPSoC applications present even more challenges. For instance, shifting form general-purpose IP cores to ASIPs with a highly parallel task-specific execution engine will doubtless generate more stress for the memory subsystems and interconnection fabric, which may not be able to cope with it. In addition, independent optimization of ASIP instructions may cause unpredictable or decreasing performance, when neglecting cache policies or NoC routing protocols.

In [61] Sun et. al. present an exploration of the interactions between coarse- and fine- grained customizations for application-specific custom heterogeneous single-chip multiprocessors. A methodology is analyzed to simultaneously assign/schedule tasks on single-chip multiprocessors and select custom instructions for each processor, under an area budget for the custom multiprocessor. It is shown that different processors exploit parallelism between tasks that are communication-independent whereas custom instructions try to reduce the execution time of each task.

Jones et al. describe in [39] a multi-core VLIW (very large instruction word) containing several homogeneous execution cores/functional units, which is called SuperCISC MPSoC. By considering the application set at compile time, several SuperCISC hardware functions corresponding to different applications within the set are generated and fabricated into an application-specific MPSoC. After identifying the computationally intensive loops, this information is propagated to a behavioral synthesis flow that consists of a set of compiler transformations, which attempt to convert the loops to the largest data flow graph (DFG) possible for direct implementation in hardware. They combine four homogeneous processor cores within the VLIW with homogeneous asynchronous processor cores to execute the hardware functions. Thus, the system has shown several power and performance improvements, such as cycle compression and efficient control flow execution for performance improvement and power compression, combined with removing the need for clocking via combinational execution.

The multiprocessor SoC design approach (followed by most tools, i.e., XPRESS from Tensilica [63]), assumes that the application can be decomposed into a set of communicating tasks, and that the functionality of each task can be defined in software using a high level programming language. The processors in the system are then tailored for specific tasks, enhancing the performance, area, and power efficiency. The software-based MPSoC approach is expected to reduce the SoC development effort and allows adaptation of the design to changes in the system specification that occur late in the design process, even after the chip fabrication. The development of a MPSoC involves multiple steps: (1) decomposition of an application into a set of tasks; (2) mapping of the tasks to a set of customizable processors; (3) optimization of each processor for the tasks assigned to it; (4) optimization of the communication between the processors; (5) optimization of the memory subsystem.

Key problems for developers of application specific MPSoCs are:

- The number, type (symmetric or heterogeneous, general purpose, DSP or VLIW) and configuration of processors required for the application

- Interprocessor communications choosing the right mix of standard buses, point to point communications, shared memory, and emerging network on chip approaches

- Concurrency, synchronization, control and programming models or mixed strategies

- Memory hierarchy, types, and access methods; instruction set extension techniques are hard to include operations that access memory

- Application partitioning, use of appropriate APIs and communications models, and associated design space exploration

Design space exploration for multiprocessor architectures is presented by Zivkovic [72]. This work focuses on the comparison of fast estimations against accurate estimations generated by simulation traces. Trace driven (TD) co-simulation exploration and executable control data-flow graph (CDFG) are the two most common exploration methodologies. Together with symbolic programs as application workload (in Zivkovic) they offer a few conceptual levels for accurate and fast exploration methodologies. System optimisation and exploration with respect to power consumption are presented within the work of Henkel [30]. In their work effects of certain system parameters like cache size and main memory size are considered.

Although tuning the instruction set of a processor to match the performance or power and cost of an embedded application are the primary objectives in ASIP design, generation of custom instructions to replace complex ones has two noticeable advantages. First, replacing multi-cycle with single-cycle instructions can reduce the program memory size, which might be crucial in embedded systems. In addition, it can reduce the number of required code fetches, thus speeding up the execution, especially if the code is stored in external memory that is much slower than the ASIP. In addition, the fewer memory accesses lead to a reduction in power consumption since fetching codes from external memory consumes much power.

2.5   Reconfigurable Instruction Set Processors

Similar to application-specific instruction set processors (ASIPs) and extensible processors, reconfigurable instruction set processors (RASIP or RISPs)
introduce a cost-efficient approach for implementing embedded systems by
taking advantage of reconfigurable technology [7], [50]. A RASIP consists of a
base processor for executing the non-critical parts of an application and custom instructions (CIs) which are generated and added after chip fabrication.
CIs are the instruction set extensions which are extracted from hot portions
of target applications. CIs are mapped onto the reconfigurable fabric forming
the custom functional units (RFUs) and a configuration bitstream is generated for each CI and stored in the configuration memory prior to application
execution. Figure 2.6 shows an outline of a reconfigurable ASIP paradigm.

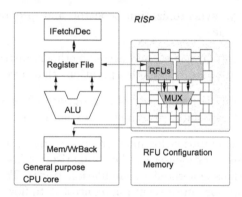

FIGURE 2.6: A RASIP integrating the general purpose processor with RFUs.

The baseline CPU actually has an instruction set that is fixed during
the entire application. The process of selecting which instructions are to be
used is the same in both types of processors, ASIP and RASIP. The achieved
speedup depends on the proper selection of instructions. This selection process
is constrained by the number of instructions that can be implemented. In an
ASIP, there is an area limit, and with RASIP, the limit comes from the size
of the RFU.

RASIPs in which reconfiguration takes place at run-time offer an additional opportunity. The flexibility increases as the type and number of RFUs
increases; and in consequence, the more execution on the reconfigurable fabric,
the higher speedup is achievable. However, the instruction selection process
is more complex and the impact on area and energy consumption is not usually appealing. One major issue is configuration overhead; various techniques
such as compression of configuration data may be applied, or scheduling by
predicting the required configurations and loading them in advance.

Commercial customizable *soft* processors, (processors built on an FPGA programmable fabric) are available (for example, NiosII [53] and MicroBlaze [54]), which allow designers to tune the processor with additional hardware functional units, either at processor configuration time, or custom designed and tightly coupled to the processor, so as to better match their application requirements. Whereas these solutions facilitate a limited number of configuration parameters, researchers have exploited reconfigurable technology to automate RASIP design flow using soft processors [68], [50]. CUSTARD [18] is a flexible customizable multi-threaded soft-processor representing an FPGA implementation of a parameterizable core supporting the following options: different number of hardware threads and types, custom instructions, branch delay slot, load delay slot, forwarding, and register file size. The CUSTARD compiler generates custom instructions using a technique called *similar sub-instructions*. The principle is to find instruction datapaths that can be re-used across similar pieces of code. These datapaths are added to the parameterizable processor and then the decoding logic is updated to map the new instructions to unused portions of the opcode space.

A different approach is followed by *Molen*, a polymorphic processor paradigm which incorporates both general purpose and custom computing processing [64]. The Molen machine consists of two main components, namely the core processor, which is a general-purpose processor (GPP), and the reconfigurable processor (RP). Instructions are issued to either processor by the arbiter and data are fetched (stored) by the data fetch unit. The memory MUX unit is responsible for distributing (collecting) data. This scheme allows instructions, entire pieces of code, or their combination to execute on microcoded reconfigurable units. The reconfigurable processor is further subdivided into the $\rho\mu$-code unit and the custom configured unit (CCU). The CCU consists of reconfigurable hardware and memory. The $\rho\mu$-code unit comprises of the control store which is used as storage for the microcodes and the sequencer which determines the microinstruction execution sequence. All code runs on the GPP except pieces of (application) code implemented on the CCU in order to speed up program execution. The envisioned support of operations by the reconfigurable processor can be initially divided into two distinct phases: set and execute. In the set phase, the CCU is configured to perform the supported operations. Subsequently, in the execute phase, the actual execution of the operations is performed. This decoupling allows the set phase to be scheduled well ahead of the execute phase, thereby hiding the reconfiguration latency. As no actual execution is performed in the set phase, it can even be scheduled upward across the code boundary in the code preceding the RP targeted code.

2.5.1   Warp Processing

A paradigm has been proposed for multiprocessing systems, in which one processor performs optimizations that benefit other processors [44],[16],[32], [71].

Such optimizations might include detecting, just-in-time compiling critical regions with optimizations, scheduling threads, scaling voltages, etc.

Warp processing uses an on-chip processor to dynamically remap critical code regions from processor instructions to FPGA circuits [45] using run time synthesis. Warp processing dynamically detects critical regions of a running program and dynamically reimplements code regions on an FPGA, requiring partitioning, decompilation, synthesis, placement, and routing tools, all having to execute with minimal computation time and data memory so as to coexist on a chip with the main processor.

2.6 Hardware/Software Codesign

Assigning and scheduling an application to a set of heterogeneous processing elements (PEs) has been studied in the area of hardware/software codesign. The problem consists of selecting the number and type of PEs, and then assigning or scheduling the tasks to those PEs. PEs can include different programmable processors or custom hardware implementations of specific application tasks. The current target architectures for codesign mainly focus on integrating CPU and custom hardware coprocessors at a coarse-grained level.

Traditionally the codesign approach assumes a processor and a coprocessor integrated via a general purpose bus interface [24], [49]. Hardware/software partitioning is done at the task or basic block level. The system usually is represented as a graph, where the nodes represent tasks or basic blocks, and the edges are weighted based on the amount of communication between the nodes. An approach is to initially allocate all nodes in hardware. Area cost is reduced by iterative movements from hardware to software while trying not to exceed a constraint on the schedule length. Henkel and Ernst [29] propose a simulated annealing-based methodology. Niemann et al. [49] formulate the hardware/ software partitioning problem under area and schedule length constraints as an ILP problem. However, hardware/software partitioning under area and schedule length constraints is an NP-hard problem. The partitioning algorithms need a description of the system often in languages like C. In the recent years SystemC (www.systemc.org) and SpecC (www.specc.gr.jp/eng/index.html) have emerged as system-level design languages. In addition to the system modeling languages, hardware/software codesign is essentially influenced by promising new architectures in embedded systems. Reconfigurable computing and VLIW-based architectures have rapidly been adopted to codesign as designers can now more efficiently develop embedded multimedia, networking and signal processing applications.

Codesign methodologies are often implemented as a set of design tools to aid the rapid development of systems. *POLIS*, for instance, was developed from the Hardware/Software Codesign Group at Berkeley. It is an infrastructure specifically created to support the concurrent design of both hardware and software, effectively reducing multiple iterations and major redesigns. Design is then done in a unified design model, with a unified view of how the hardware/software partition can be built in practice, so as to prejudice neither hardware nor software implementation. This model is maintained throughout the design process, in order to preserve the design and ensure both hardware and software build is optimized for peak performance of both.

Chinook from University of Washington, is a hardware/software co-synthesis CAD tool for embedded systems. It is designed for control-dominated, reactive systems under timing constraints, with a new emphasis on distributed architectures. The partitioning is performed by the designer, while Chinook works at the mapping, thus enabling designers to make informed design decisions at the high level early in the design cycle, rather than reiterate after having worked out all the low level details.

2.7   Hardware Architecture Description Languages

Hardware architecture description languages (ADLs) are principally concerned with describing the hardware components. This is often the case when dealing with application-specific instruction-set processor (ASIPs) within a design process. Therefore, the languages describe the processors in terms of their instruction sets. Hence, they are sometimes called machine description languages. ADLs concentrate on representation of components trying in principle to provide a level of abstraction found in traditional programming languages and at the same time hardware features such as synchronization or parallelism. In practice ADLs are a blend of programming languages, modeling languages, and hardware description languages.

Increasingly, even relatively simple consumer devices must now implement a wide range of functions. Hence, realizing that balancing generality with efficiency is a key goal in new products, companies are deciding to create their own programmable processors, typically embedded processors or ASIPs, because these devices provide the necessary flexibility for performing algorithmic acceleration, with the added benefit of easier re-use for derivatives or other projects. ADLs in different forms of formalism try to offer fast design exploration through high degree of automation. The designer can optimize the instruction set of a processor to fit the target application requirements through simulation profiling, to understand and remove any performance bottlenecks and achieve the optimum architecture.

In this direction different hardware ADLs appear, both in research and in commercial use. The challenging issues of each ADL are to provide compiler tools, simulation environment, synthesis and validation methodologies. Combined with an automation infrastructure each ADL presents various features in an effort to cover this wide spectrum. However, few tools (such as nML or ISDL) may make decisions about the structure of the architecture that are not under the control of the designer.

- nML [20] is a formalism that supports both structural or behavioral descriptions. The language describes the architecture at the register-transfer level. The nML description is obtained from analysing the instruction set of the target machine. The CHESS/CHECKERS environment [62], which incorporates nML, is used for automatic and efficient software compilation and instruction-set simulation. CHESS/CHECK-ERS is a retargetable tool suite that supports the different phases of designing application-specific processor cores, developing application software for these cores, and verifying the correctness of the design.

- The machine-independent microprogramming language MIMOLA [67] is an early ADL which is structure oriented and thus is suitable for hardware synthesis. The features supported by MIMOLA are: the behavioral and register-transfer level description of hardware modules, hierarchical hardware specifications, a simple timing model, and an overloading mechanism. MIMOLA can be seen as a high-level programming language, a register-transfer level language or a hardware description language. Actually, the same description can be used for compilation, synthesis, simulation and test generation.

- The instruction set description language, ISDL [26] primarily describes the instruction set of processor architectures. ISDL can specify a variety of architectures, supports constraints on instructions for grouping operations, and generates code generator, assembler, and instruction set simulator automatically. It also contains an optimization information section that can be used to provide certain architecture-specific hints for the compiler to make better machine-dependent code optimizations. ISDL accepts input in the form of the processor description (from a CAD tool) and a source program in C or C++. The program is parsed into SUIF 6.2 which, together with the ISDL description, is used generate the assembly code. An assembler is also generated and used to translate the binary code which becomes the input to the ISDL. ISDL is mainly targeted toward VLIW processors. In fact ISDL is an enhanced version of the nML formalism and allows the generation of a complete tool suite consisting of high level language (HLL) compiler, assembler, linker and simulator.

- The PEAS-III system [41] is an ASIP development environment based on a micro-operation description of instructions that allows the generation of a complete tool suite consisting of HLL compiler, assembler, linker and simulator including HDL code. This system works with a set of predefined components and thus limits the resulting flexibility in modeling arbitrary processor architectures.

- The language HMDES [25] is part of the Trimaran tool set [11]. The Trimaran system is an integrated compilation and performance monitoring infrastructure, which uses HPL-PD as base processor. The HPL-PD machine supports predication, control and data speculation, and compiler controlled management of the memory hierarchy. HMDES is essentially used to target HPL-PD processors. The target processor is described using a relational database description language. The machine database reads the low level files and supplies information for the compiler back end through a predefined query interface.

- The LISA [31] ADL is oriented to ASIP development, offering a high degree of automation so as to achieve design efficiency. LISA is a language designed for the formalized description of ASIP architectures, their peripherals, and interfaces. It supports different description styles and models at various abstraction and hierarchical levels.

2.7.1 LISATek Design Platform

The LISATek processor design platform is built around the LISA 2.0 ADL [43], [31]. The LISATek platform provides a set of processor development tools such as instruction-set simulator, C compiler, assembler, and linker, which are automatically generated to support architecture exploration. A graphical user front end is also available for software debugging and profiling purposes. Moreover, RTL hardware models in the most popular hardware description languages, VHDL, SystemC and Verilog, can also be generated from the LISA model for hardware implementation.

LISATek provides a library of sample models which contains processors for different architecture categories like VLIW (very large instruction word), SIMD (single instruction multiple data), RISC (reduced instruction set computer) and superscalar architectures of real products currently on the market.

The user is provided with powerful profiling tools to identify hotspots in his application and modify the LISA model of the architecture and the corresponding software tools. The objective is a fully automated closed loop through a rapid modeling and retargetable simulation and code generation. Taking sample models as basis processor has a major advantage to directly have compiler support for the architecture due to the existence of an instruction set.

The features of LISA include also strong orientation to C, support for instruction aliasing and complex instruction coding schemes, and support of cycle-accurate processor models, including constructs to specify pipelines and their mechanisms.

LISA descriptions are composed of both resource declarations and operations. The declared resources represent the storage objects of the hardware architecture (registers, memories, pipelines) which capture the state of the system and which can be used to model the limited availability of resources

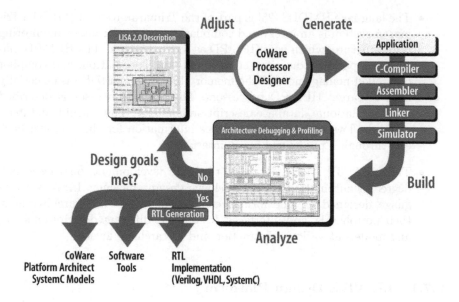

FIGURE 2.7: LISATek infrastructure based on LISA architecture specification language. Retargetable software development tools (C compiler, assembler, simulator, debugger, etc.) permit iterative exploration of varying target processor configurations. (From CoWare Inc. LISATek. http://www.coware.com With permission.)

for operation access. Operations are the basic objects in LISA. They represent the designer's view of the behavior, the structure, and the instruction set of the programmable architecture. Operation definitions collect the description of different properties of the system, operation behavior, instruction set information, and timing.

LISATek and similar state-of-the-art infrastructures, as the Tensilica XPRES, or the instruction set generator at the EPFL [66] greatly increase the design efficiency, enabling the automatic exploration of a large number of alternatives. Nevertheless, the optimal application specific embedded SoC still depends on the expertise of designers, since tools cannot explore all types of architecture customization and parallelization (instruction level, data level, fused operations), combined with a complete application parallelization and optimization.

2.8 Myths and Realities

State-of-the-art ASIP toolchains and modern CAD tool methodologies have enabled SoC designers to effectively investigate the large configuration space and interactions of IP cores, memory hierarchies and interconnects and the impact on embedded applications. However, configurability and extensibility of multi-core SoCs deals with numerous tradeoffs implied by the various forms of the ASIP paradigm. In brief, different issues and challenges are raised by researchers:

⋄ Highly automated infrastructures versus manual expert optimizations of hot-spots

⋄ Efficient exploration of the huge design space and effort/cost versus benefit

⋄ Competitive technologies, compiler technology

⋄ Limitations of customization, automation methodologies

Even with more attention to architecture, the high cost of hardware design limits the actual amount of architectural exploration which can be done. Recent advances in processor synthesis technology dramatically reduce the time and cost of creating application-specific processing elements. However, purely software approaches are far more rapid and less costly compared to even the most automated customized methodology.

Moreover, ADLs and toolchains offer a promising possibility to increase designers' productivity by automation; abstractions make it hard to model some features, since architects for example, can create unusual pipelines with varying numbers of register files and memory ports for better data-level concurrency.

Notwithstanding their technological advantages, it is sometimes argued that the introduction of ASIPs is risky. Perceived risks include the extra time needed to design the architecture and the RTL implementation, potential reliability issues due to the introduction of new hardware, and the difficulty of programming ASIPs due to a lack of software development tools.

Architectures can be better optimized from real software workloads. Synthetic benchmarks may sometimes produce conclusions that deviate from the real application characteristics. Fine-grain optimizations can benefit a specific subset of an application domain but can be inefficient when similar applications have varying run-time behavior.

Compilers need a high level model of the target machine, whereas other tools like simulators or synthesis require detailed information about the accurate cycle and bit behavior of machine operations. ASIP design automated environments promise to bind these technologies harmonically; robust cross-checking tools seem hard to develop and run.

ILP-based custom instruction selection can provide solutions in a systematic way, but may become computationally expensive for large number of custom instruction instances. Heuristics are used therefore after defining weight functions.

2.9   Case Study: Realizing Customizable Multi-Core Designs – Commercial ASIP

Several commercial products in the customizable processor domain offer integrated toolchains for design space exploration, implementation, and verification. Promising time closure can be realized by building automatically both the processor hardware and the matching software tools. Such products include Xtensa from Tensilica [63], [22], ARCtangent from ARC [37], Jazz from Improv Systems [34], SP-5flex from 3DSP [1], and LISATek products from CoWare [43].

Tensilica developed the *Xtensa* Series of *configurable and extensible* processors. They offer designers a set of predefined parameters which they can configure in order to tailor the processor to the intended application. Additionally, the designer can invent custom instructions and execution units and integrate them directly into the processor core. For this purpose the Xtensa processor is extended using the proprietary *Tensilica Instruction Extension* (TIE) language, which is a Verilog-like language that can be used to describe custom instructions. The user can analyze and carefully profile its application, and consequently determine candidate kernels for instruction set extension. Such kernels are then described in TIE. Designers can also write TIE code manually and compile it using the TIE Compiler, or they can use the XPRES (Xtensa Processor Extension Synthesis) Compiler to automatically create TIE descriptions of processor extensions. The XPRES Compiler can analyze a given algorithm written in C/C++ and automatically configure and extend the Xtensa processor so that it is optimized to run that particular algorithm. Optimizations can be a combination of performance improvement, area minimization, and energy reduction that best meet users' design objectives.

Tensilica's objective is to provide a complete user abstraction to the automatic TIE generation process. Using the TIE language and Xtensa Xplorer toolkit, the generation and verification of the instructions used to extend the processor ISA are automated. Such automation, outlined in Figure 2.8, helps to reduce the hardware verification time that typically consumes a large percentage of the project duration of a typical hardware developed for the same functionality. The Xtensa Processor Generator can be used to generate HDL descriptions of the customized processor, as well as a set of electronic design automation (EDA) scripts and a full suite of software development tools specifically suited for that processor design. In sequence, customization includes levels of validation and testing required verifying the functionality. Software testing, after integration of TIE code with user C code testing of the software running on the Xtensa core is performed with an instruction set simulator. Hardware verification is achieved with a hardware/software co-simulation environment.

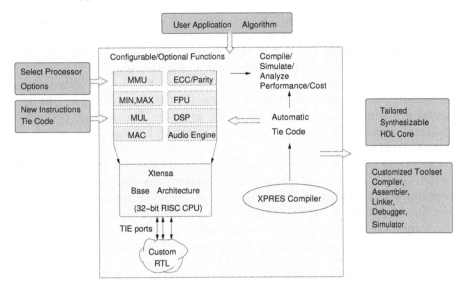

FIGURE 2.8: Tensilica customization and extension design flow. Through Xplorer, Tensilica's design environment, the designer has access to the tools needed for development of custom instructions and configuration of the base processor.

Real-life applications have been mapped on Xtensa platforms and even more importantly heterogeneous multiprocessor systems-on-chips (MPSoCs) have been designed, in which different processors are customized for specific tasks. In general, MPSoCs can provide high levels of efficiency in performance and power consumption, while maintaining programmability. However, in order to best exploit processor heterogeneity, designers are still required to manually customize each processor, while mapping the application tasks to them, so that the overall performance and/or power requirements are satisfied.

In [60] and [61] designers propose a methodology to automatically synthesize a custom heterogeneous architecture, consisting of multiple extensible processors, to evaluate multimedia (MPEG2, and MediaBench applications) and encryption applications (AES, RSA, PGPENC). Their methodology simultaneously customizes the instruction set and task assignment to each processor of the MPSoC. The need for such an integrated approach is motivated by demonstrating that custom instruction selection has complex interdependencies with task assignment and scheduling, and performing these steps independently may result in significant degradation in the quality of the synthesized multiprocessor architecture. Their methodology uses an iterative improvement algorithm to assign and schedule tasks on processors and select custom instructions along the critical path in an interleaved manner. It utilizes the concept of expected execution time to better integrate these two steps. It not only considers the currently selected custom instructions for the current task assignment and schedule, but also the possibility of better custom instructions selected in future iterations. Authors also enhance their methodology to integrate task-level software pipelining to further increase the parallelism and provide opportunities for multiprocessing.

Their results, while using their methodology for custom instruction extension on the Xtensa platform, indicate that the processors in the multiprocessor system can achieve significant speedup. The average performance improvements of 2.0 times, to 2.9 times relate to homogeneous multiprocessor systems with well-optimized task assignment and scheduling. Promising conclusions bring to light that the impact of the area budget and number of processors on completion times is nearly orthogonal. Different processors can exploit parallelism between tasks that are independent and thus not connected by any edge in the application task graph. Meanwhile, custom instructions try to reduce the total execution time of tasks connected by edges (i.e., those on the critical path). Designers can first obtain the task-graph completion time on a single processor under different custom instruction area budgets. Then, task-graph completion times can be obtained on multiple processors, assuming no custom instruction is used and task assignment and scheduling on the heterogeneous MPSoC.

2.10   The Future: System Design with Customizable Architectures, Software, and Tools

Concurrency modeling

New models of concurrency are required, in order to move from the multi-thread paradigm, useful for uniprocessor systems, toward the multiprocessor approaches. Such models must span all over the system hierarchy, characterized by possibly different models of computation at each level.

A modern programming model that is capable of exporting critical features of ASIPs that will enable exploitation of their specific features is necessary. Even if separate language features must be devised for different architecture classes, it is critical to ensure consistency among the architecture tools, the compiler, the simulator and the software environment.

Interconnect architectures, arbitration, synchronization, routing and repeating schemes

Synergetic behavior of heterogeneous components is a must, to be achieved both through intelligent interfacing and through middleware development. A large scale integrability of IP blocks is necessary for speeding up the time-to-market directives. Complex and fragmented natures of diverse components inside customizable multi-core architectures become barriers to their rapid deployment, which system architects, chip vendors and software experts help to gradually overcome.

What characterizes this new breed of ASIPs in the embedded world is that unlike their predecessors, these ASIPs are created not just to provide flexibility through programmability, but in a large part, also to provide an easier implementation alternative to ASICs for their respective application domains. This trend is expected to grow significantly into other domains (and sub-domains as evidenced by the networking and communication spaces) in the near future.

Review Questions

[Q-1] What are the different ways to customize an embedded system?

[Q-2] Describe the methodologies to customize a single embedded CPU.

[Q-3] How do the CPU extension methodologies vary compared to instruction-set customization?

[Q-4] What are the principles of the template-based custom instruction generation?

[Q-5] Describe the techniques to manage and reduce the complexity of the design space for custom instruction generation.

[Q-6] Today, extending the base processor with custom units and generating complex instructions from primitive ones are handled efficiently by research or commercial methodologies. What are the additional challenges in the MPSoC era and which are the issues that are more acute for customizing heterogeneous multi-core systems?

[Q-7] Given the ASIP categorization of methodologies and techniques describe the Tensilica's design environment.

Bibliography

[1] 3DSP. http://www.3dsp.com.

[2] Federico Angiolini, Jianjiang Ceng, Rainer Leupers, Federico Ferrari, Cesare Ferri, and Luca Benini. An integrated open framework for heterogeneous MPSoC design space exploration. In *DATE'06: Proceedings of the conference on Design, Automation and Test in Europe*, pages 1145–1150, 2006.

[3] Jeffrey M. Arnold. The architecture and development flow of the S5 software configurable processor. *J. VLSI Signal Process. Syst.*, 47(1):3–14, 2007.

[4] Arteris. http://www.arteris.com.

[5] John Bainbridge and Steve Furber. Chain: A delay-insensitive chip area interconnect. *IEEE Micro*, 22(5):16–23, 2002.

[6] P. Banerjee, M. Haldar, A. Nayak, V. Kim, V. Saxena, S. Parkes, D. Bagchi, S. Pal, N. Tripathi, D. Zaretsky, R. Anderson, and J.R. Uribe. Overview of a compiler for synthesizing MATLAB programs onto FPGAs. *Trans. on VLSI*, 12(3):312–324, 2004.

[7] Francisco Barat, Rudy Lauwereins, and Geert Deconinck. Reconfigurable instruction set processors from a hardware/software perspective. *IEEE Trans. Softw. Eng.*, 28(9):847–862, 2002.

[8] Souvik Basu and Rajat Moona. High level synthesis from Sim-nML processor models. In *VLSID'03: Proceedings of the 16th International Conference on VLSI Design*, pages 255–260. IEEE Computer Society, 2003.

[9] Partha Biswas, Nikil Dutt, Paolo Ienne, and Laura Pozzi. Automatic identification of application-specific functional units with architecturally visible storage. In *DATE'06: Proceedings of the conference on Design, Automation and Test in Europe*, pages 212–217, 2006.

[10] P. Bonzini and L. Pozzi. A retargetable framework for automated discovery of custom instructions. In *ASAP'07: Application Specific Systems, Architectures and Processors*, pages 334–341. IEEE, 2007.

[11] Lakshmi N. Chakrapani, John Gyllenhaal, Wen-mei W. Hwu, Scott A. Mahlke, Krishna V. Palem, and Rodric M. Rabbah. Trimaran: An infrastructure for research. In *Instruction-Level Parallelism. Lecture Notes in Computer Science*, 2004.

[12] Karam S. Chatha and Ranga Vemuri. MAGELLAN: multiway hardware-software partitioning and scheduling for latency minimization of hierarchical control-dataflow task graphs. In *CODES'01: Proceedings of the Ninth International Symposium on Hardware/Software Codesign*, pages 42–47. ACM, 2001.

[13] Nathan Clark, Amir Hormati, Scott Mahlke, and Sami Yehia. Scalable subgraph mapping for acyclic computation accelerators. In *CASES '06: Proceedings of the 2006 International Conference on Compilers, Architecture and Synthesis for Embedded Systems*, pages 147–157. ACM, 2006.

[14] Nathan Clark and Hongtao Zhong. Automated custom instruction generation for domain-specific processor acceleration. *IEEE Trans. Comput.*, 54(10):1258–1270, 2005.

[15] Jason Cong, Yiping Fan, Guoling Han, Ashok Jagannathan, Glenn Rein-man, and Zhiru Zhang. Instruction set extension with shadow regis-ters for configurable processors. In *FPGA'05: Proceedings of the 2005 ACM/SIGDA 13th International Symposium on Field-programmable Gate Arrays*, pages 99–106. ACM, 2005.

[16] Abhinav Das, Jiwei Lu, and Wei-Chung Hsu. Region monitoring for local phase detection in dynamic optimization systems. In *CGO'06: Proceed-ings of the International Symposium on Code Generation and Optimiza-tion*, pages 124–134. IEEE Computer Society, 2006.

[17] Robert P. Dick and Niraj K. Jha. MOGAC: A multiobjective genetic al-gorithm for hardware-software cosynthesis of distributed embedded sys-tems. *IEEE Transactions on Computer-Aided Design of Integrated Cir-cuits and Systems*, 17:920–935, 1998.

[18] R. Dimond, O. Mencer, and Wayne Luk. CUSTARD: a customisable threaded FPGA soft processor and tools. *International Conference on Field Programmable Logic and Applications*, 0:1–6, 2005.

[19] P. Eles, Zebo Peng, K. Kuchcinski, and A. Doboli. System level hard-ware/software partitioning based on simulated annealing and tabu search. *Des. Automat. Embedd. Syst.*, 2(1):5–32, 1997.

[20] A. Fauth, J. Van Praet, and M. Freericks. Describing instruction set processors using nML. In *Proceedings on the European Design and Test Conference*, pages 503–507, 1995.

[21] Carlo Galuzzi, Koen Bertels, and Stamatis Vassiliadis. A linear complex-ity algorithm for the generation of multiple input single output instruc-tions of variable size. *LNCS, Embedded Computer Systems: Architectures, Modeling, and Simulation*, 4599/2007:283–293, 2007.

[22] David Goodwin and Darin Petkov. Automatic generation of application specific processors. In *CASES'03: Proceedings of the 2003 International Conference on Compilers, Architecture and Synthesis for Embedded Sys-tems*, pages 137–147. ACM, 2003.

[23] J. Grode, P. V. Knudsen, and J. Madsen. Hardware resource allocation for hardware/software partitioning in the LYCOS system. In *DATE'98: Proceedings of the Conference on Design, Automation and Test in Europe*, pages 22–27. IEEE Computer Society, 1998.

[24] Sumit Gupta, Rajesh Kumar Gupta, Nikil D. Dutt, and Alexandru Nico-lau. Coordinated parallelizing compiler optimizations and high-level syn-thesis. *ACM Trans. Des. Autom. Electron. Syst.*, 9(4):441–470, 2004.

[25] C. Gyllenhaal, B.R. Rau, and W.W. Hwu. Hmdes version 2.0 specifica-tion. In *Technical Report, IMPACT-96-3, The IMPACT Research Group*. Springer-Verlag, 1996.

[26] George Hadjiyiannis, Silvina Hanono, and Srinivas Devadas. ISDL: an instruction set description language for retargetability. In *DAC'97: Proceedings of the 34th Annual Conference on Design Automation*, pages 299–302. ACM, 1997.

[27] Ashok Halambi, Peter Grun, Vijay Ganesh, Asheesh Khare, Nikil Dutt, and Alex Nicolau. EXPRESSION: a language for architecture exploration through compiler/simulator retargetability. In *DATE'99: Proceedings of the Conference on Design, Automation and Test in Europe*, pages 485–490. ACM, 1999.

[28] Scott Hauck, Thomas W. Fry, Matthew M. Hosler, and Jeffrey P. Kao. The Chimaera reconfigurable functional unit. *IEEE Trans. Very Large Scale Integr. Syst.*, 12(2):206–217, 2004.

[29] Jörg Henkel and Rolf Ernst. An approach to automated hardware/software partitioning using a flexible granularity that is driven by high-level estimation techniques. *Trans. on Very Large Scale Integration (VLSI) Systems*, 9(2):273–289, 2001.

[30] Jörg Henkel and Yanbing Li. Avalanche: an environment for design space exploration and optimization of low-power embedded systems. *IEEE Trans. Very Large Scale Integr. Syst.*, 10(4):454–468, 2002.

[31] Andreas Hoffmann, Tim Kogel, Achim Nohl, Braun Gunnar, Schliebusch Oliver, Wahlen Oliver, Wieferink Andreas, and Meyr Heinrich. A novel methodology for the design of application-specific instruction-set processors (ASIPs) using a machine description language. *IEEE Transactions on Computer-Aided Design of Integrated Circuits and Systems*, 20:1338–1354, 2001.

[32] Shiwen Hu, Madhavi Valluri, and Lizy Kurian John. Effective management of multiple configurable units using dynamic optimization. *ACM Trans. Archit. Code Optim.*, 3(4):477–501, 2006.

[33] Ing-Jer Huang and Ping-Huei Xie. Application of instruction analysis/scheduling techniques to resource allocation of superscalar processors. *IEEE Trans. Very Large Scale Integr. Syst.*, 10(1):44–54, 2002.

[34] Improv Systems Inc. http://www.improvsys.com.

[35] MIPS Technologies Inc. http://www.mips.com.

[36] Stretch Inc. http://www.stretchinc.com.

[37] ARC International. http://www.arc.com.

[38] Alex Jones, Debabrata Bagchi, Sartajit Pal, Prith Banerjee, and Alok Choudhary. *PACT HDL: a compiler targeting ASICs and FPGAs with power and performance optimizations.* Kluwer Academic Publishers, Norwell, MA, 2002.

[39] Alex Jones, Raymond Hoare, Dara Kusic, Gayatri Mehta, Josh Fazekas, and John Foster. Reducing power while increasing performance with SuperCISC. *Trans. on Embedded Computing Sys.*, 5(3):658–686, 2006.

[40] Theo Kluter, Philip Brisk, Paolo Ienne, and Edoardo Charbon. Speculative DMA for architecturally visible storage in instruction set extensions. In *CODES/ISSS '08: Proceedings of the 6th IEEE/ACM/IFIP International Conference on Hardware/Software Codesign and System Synthesis*, pages 243–248. ACM, 2008.

[41] Shinsuke Kobayashi, Yoshinori Takeuchi, Akira Kitajima, and Masaharu Imai. Compiler generation in PEAS-III: an ASIP development system. In *SCOPES'01: Workshop on Software and Compilers for Embedded Systems*, 2001.

[42] C. Liem, T. May, and P. Paulin. Instruction-set matching and selection for DSP and ASIP codegeneration. In *European Design and Test Conference, EDAC, European Conference on Design Automation, ETC European Test Conference*, pages 31–37. IEEE Computer Society, 1994.

[43] CoWare Inc. LISATek. http://www.coware.com.

[44] Jiwei Lu, Howard Chen, Pen-chung Yew, and Wei-chung Hsu. Design and implementation of a lightweight dynamic optimization system. *Journal of Instruction-Level Parallelism*, 6:2004, 2004.

[45] Roman Lysecky, Greg Stitt, and Frank Vahid. Warp processors. *ACM Trans. Des. Autom. Electron. Syst.*, 11(3):659–681, 2006.

[46] Prabhat Mishra, Mahesh Mamidipaka, and Nikil Dutt. Processor-memory coexploration using an architecture description language. *Trans. on Embedded Computing Sys.*, 3(1):140–162, 2004.

[47] Rajat Moona. Processor models for retargetable tools. In *Proceedings of Eleventh IEEE International Workshop on Rapid Systems Prototyping*, pages 34–39, 2000.

[48] Fernando Moraes, Ney Calazans, Aline Mello, Leandro Möller, and Luciano Ost. HERMES: an infrastructure for low area overhead packet-switching networks on chip. *Integr. VLSI J.*, 38(1):69–93, 2004.

[49] Ralf Niemann and Peter Marwedel. An algorithm for hardware/software partitioning using mixed integer linear programming. In *Proceedings of the Design Automation for Embedded Systems*, pages 165–193. Kluwer Academic Publishers, 1997.

[50] Hamid Noori, Farhad Mehdipour, Kazuaki Murakami, Koji Inoue, and Morteza Saheb Zamani. An architecture framework for an adaptive extensible processor. *The Journal of Supercomputing*, 45(3):313–340, Sep. 2008.

[51] Pierre G. Paulin and Miguel Santana. FlexWare: A retargetable embedded-software development environment. *IEEE Des. Test*, 19(4):59–69, 2002.

[52] Zebo Peng and Krzysztof Kuchcinski. An algorithm for partitioning of application specific systems. In *Proceedings of the European Conference on Design Automation (EDAC)*, pages 316–321, 1993.

[53] Altera Nios II Processor. http://www.altera.com/products/ip/processors/nios2/ni2-index.html.

[54] Xilinx MicroBlaze Processor. http://www.xilinx.com/products/design_resources/proc_central/microblaze.htm.

[55] G. Quan, X. Hu, and G. Greenwood. Preference-driven hierarchical hardware/software partitioning. In *Proceedings of the IEEE/ACM International Conference on Computer Design*, pages 652–658, 1999.

[56] Rahul Razdan, Karl S. Brace, and Michael D. Smith. PRISC software acceleration techniques. In *ICCS'94: Proceedings of the 1994 IEEE International Conference on Computer Design: VLSI in Computer & Processors*, pages 145–149. IEEE Computer Society, 1994.

[57] Robert Schreiber, Shail Aditya, Scott Mahlke, Vinod Kathail, B. Ramakrishna Rau, Darren Cronquist, and Mukund Sivaraman. PICO-NPA: High-level synthesis of nonprogrammable hardware accelerators. *J. VLSI Signal Process. Syst.*, 31(2):127–142, 2002.

[58] Vinoo Srinivasan, Shankar Radhakrishnan, and Ranga Vemuri. Hardware software partitioning with integrated hardware design space exploration. In *DATE'07: Proceedings of the Conference on Design, Automation and Test in Europe*, pages 28–35, 1998.

[59] S. Stergiou, F. Angiolini, S. Carta, L. Raffo, D. Bertozzi, and G. De Micheli. XPipes Lite: a synthesis oriented design library for networks on chips. In *Design, Automation and Test in Europe, 2005*, volume 2, pages 1188–1193, 2005.

[60] Fei Sun, Srivaths Ravi, Anand Raghunathan, and Niraj K. Jha. Synthesis of application-specific heterogeneous multiprocessor architectures using extensible processors. In *VLSID'05: Proceedings of the 18th International Conference on VLSI Design held jointly with 4th International Conference on Embedded Systems Design*, pages 551–556. IEEE Computer Society, 2005.

[61] Fei Sun, Srivaths Ravi, Anand Raghunathan, and Niraj K. Jha. Application-specific heterogeneous multiprocessor synthesis using extensible processors. *IEEE Trans. Comput.*, 25(9):1589–1602, 2006.

[62] Target Compiler Technologies. http://www.retarget.com.

[63] Tensilica. http://www.tensilica.com.

[64] Stamatis Vassiliadis, Stephan Wong, and Sorin Cotofana. The MOLEN rho-mu-coded processor. In *FPL'01: Proceedings of the 11th International Conference on Field-Programmable Logic and Applications*, pages 275–285. Springer-Verlag, 2001.

[65] Girish Venkataramani, Tobias Bjerregaard, Tiberiu Chelcea, and Seth C. Goldstein. Hardware compilation of application-specific memory access interconnect. *IEEE Transactions on Computer Aided Design of Integrated Circuits and Systems*, 25(5):756–771, 2006.

[66] Scott J. Weber, Matthew W. Moskewicz, Matthias Gries, Christian Sauer, and Kurt Keutzer. Fast cycle-accurate simulation and instruction set generation for constraint-based descriptions of programmable architectures. In *CODES+ISSS'04: Proceedings of the 2nd IEEE/ACM/IFIP International Conference on Hardware/Software Codesign and System Synthesis*, pages 18–23. ACM, 2004.

[67] Lehrstuhl Informatik Xii, Steven Bashford, Ulrich Bieker, Berthold Harking, Rainer Leupers, Peter Marwedel, Andreas Neumann, and Dietmar Voggenauer. The MIMOLA language, version 4.1, 1994.

[68] Peter Yiannacouras, J. Gregory Steffan, and Jonathan Rose. Application-specific customization of soft processor microarchitecture. In *FPGA'06: Proceedings of the 2006 ACM/SIGDA 14th International Symposium on Field Programmable Gate Arrays*, pages 201–210. ACM, 2006.

[69] Pan Yu and Tulika Mitra. Characterizing embedded applications for instruction-set extensible processors. In *DAC'04: Proceedings of the 41st Annual Conference on Design Automation*, pages 723–728. ACM, 2004.

[70] Pan Yu and Tulika Mitra. Disjoint pattern enumeration for custom instructions identification. In *FPL'07: Field Programmable Logic and Applications*, pages 273–278. IEEE, 2007.

[71] Weifeng Zhang, Brad Calder, and Dean M. Tullsen. An event-driven multithreaded dynamic optimization framework. In *PACT'05: Proceedings of the 14th International Conference on Parallel Architectures and Compilation Techniques*, pages 87–98. IEEE Computer Society, 2005.

[72] Vladimir D. Zivkovic, Erwin de Kock, Pieter van der Wolf, and Ed Deprettere. Fast and accurate multiprocessor architecture exploration with symbolic programs. In *DATE'03: Proceedings of the Conference on Design, Automation and Test in Europe*, page 10656. IEEE Computer Society, 2003.

3

Power Optimization in Multi-Core System-on-Chip

Massimo Conti, Simone Orcioni, Giovanni Vece and Stefano Gigli

Università Politecnica delle Marche
Ancona, Italy
{m.conti, s.orcioni, g.vece, s.gigli}@univpm.it

CONTENTS

3.1	Introduction	. . . .	72
3.2	Low Power Design	. . . .	74
	3.2.1	Power Models	75
	3.2.2	Power Analysis Tools	80
3.3	PKtool	. . . .	82
	3.3.1	Basic Features	82
	3.3.2	Power Models	83
	3.3.3	Augmented Signals	84
	3.3.4	Power States	85
	3.3.5	Application Examples	86
3.4	On-Chip Communication Architectures	. . . .	87
3.5	NOCEXplore	. . . .	90
	3.5.1	Analysis	91
3.6	DPM and DVS in Multi-Core Systems	. . . .	95
3.7	Conclusions	. . . .	100
	Review Questions	. . . .	101
	Bibliography	. . . .	102

3.1   Introduction

In recent years, due to the continuous development in the field of silicon technology, it is possible to implement complex electronic systems in a single integrated circuit. Systems-on-chips (SoCs) have favored the explosion of the market of electronic appliances: small mobile devices, which provide communications and information capabilities for consumer electronics and industrial automation. These devices require complex electronic and high levels of system integration and need to be delivered in a very short time in order to meet their market window.

The design complexity of these systems requires new design methodologies and the development of a seamless design flow that integrates existing and emerging tools. The International Technology Roadmap for Semiconductors (ITRS) and MEDEA+ Roadmap evidence some key points that electronic design automation companies must consider in order to deal with such design complexity, among them:

- **Intellectual Property Reuse**

  Intellectual property (IP) reuse is becoming critical for an efficient system development; the need to shorten the time to market is stimulating reusability of both hardware and software. A good way to keep design costs under control is to minimize the number of new designs that are required each time a new SoC is developed: reuse existing design components where possible.

  The development of reusable IPs requires:

    - The development of standards, including general constraints and guidelines, as well as executable specifications for intra- and inter-company IP exchange, such as SystemC, XML and UML
    - The creation of parameterizable, qualified and validated IPs
    - The use of hierarchical reuse methodology, allowing the reuse of the IPs and of the testbenches at different levels of abstraction

  Furthermore, the IP reuse methodology is indispensable when the design of a system is developed in cooperation between different companies, or when the design center is distributed all over the world and consequently the project management is distributed.

  A lot of work has been done on the development of standards for IP qualification. The SPIRIT Consortium developed the IP-XACT specification to enable rapid, reliable deployment of IPs into advanced design environments. The Virtual Socket Interface Alliance (VSIA) developed the international standard QIP (Quality Intellectual Property) for measuring IP quality. OpenCores is the world's largest community for development of open source hardware IPs.

- **Low Power Design**

  The continuous progress of micro and nano technologies led to a grow-ing integration and clock frequency increment in electronics systems. These combined effects led to an increase both in power density and energy dissipation, with important consequences above all in portable systems. Some design and technology issues related to power efficiency are becoming crucial, in particular for power optimized cell libraries, clock gating and clock trees optimization, and dynamic power manage-ment. Emphasis is now moving to architectural level (software energy optimization), optimum memory hierarchy organization and run time system management.

- **System Level Design Methodologies and On-Chip Communi-cation**

  The design of complex systems-on-chips and multi-core systems requires the exploration of a large solution space. Current design approaches start with low level models of components and interconnect them when most architectural decisions have been fixed. Multi-core system design methodologies perform architecture exploration at high level, taking into account constraints at this level. Multi-core system design methodologies must select:

  - The global communication architecture, which may be multi-level bus architecture, network-on-chip (NoC) architecture or mixed-bus NoC
  - Synchronous or asynchronous architectures for local and global communication
  - The partitioning of system specification and the allocation of com-ponents, such as software (real time operating system) or hardware IPs to execute them

  Transaction level modeling (TLM) [39] has been widely used to explore the space solution at system level in a fast and efficient way.

- **Design for Testability and Manufacturability**

  When the complexity increases the time spent in the verification and validation increases much more than the time spent in the design, a designer must consider, among other specifications, the simplification of the test phase in prototyping and in production. Design methodologies that take these aspects into account are:

  - Formal verification
  - Hierarchical specification and verification and reuse of test benches at different levels of abstraction
  - HW/SW co-verification

– Reuse of qualified IPs

– Virtual prototyping

The chapter is organized as follows. In Section 3.2, system level power models and the state of the art of power analysis tools are presented. Section 3.3 presents a SystemC library, called PKtool, for system level power analysis. Some design considerations and existing analysis tools for network-on-chip are reported in Section 3.4. Section 3.5 presents a SystemC library, called NOC-EXplore, for network-on-chip performance analysis. Section 3.6 presents the application of dynamic voltage scaling techniques in different on-chip communication architectures. Finally Section 3.7 reports the conclusions.

3.2   Low Power Design

The mean energy dissipated during a time period T in a CMOS circuit can be modeled by the following equation

$$E_M(T) = E_{\text{dyn}} + E_{\text{leak}} + E_{\text{sc}} =$$
$$= \sum_{i=1}^{N} C_i V_{\text{DD}}^2 D_i + V_{\text{DD}} I_{\text{sc},i} \tau_i D_i + V_{\text{DD}} I_{\text{leak},i} T \quad (3.1)$$

where the first term represents the capacitive switching energy, the second the energy dissipated due to leakage currents, the third term represents the short circuit energy, N represents the number of nodes of the circuit, C_i is the capacitance associated to the i-th node, D_i is the number of commutations of the i-th node during the period T, $I_{\text{sc},i}\tau_i$ is the charge lost during commutation of the i-th node due to short circuit effect, $I_{\text{leak},i}$ is the mean leakage current of the i-th node, and V_{DD} is the supply voltage.

The different techniques, applied at different levels of the design to reduce the power dissipation, have the objective of reducing one or more terms of Equation (3.1). A resume of some design techniques for low power is the following.

- **Leakage Current Reduction**

  The feature size reduction gives, as a drawback, the increment of the sub-threshold current, the bulk leakage current and the leakage current through the gate oxide. As a consequence the leakage power is no more negligible with respect to the other terms and it can be reduced and controlled using techniques such as multi-threshold MOS transistors, silicon on insulator technologies, back biasing, or switching off the complete block when it is inactive.

- **Short Circuit Current Reduction**

  Short circuit current flows in a CMOS gate when both the pMOSFET and nMOSFET are on. The increment of clock frequency makes the commutation period of the logic devices comparable with the clock period, increasing the short circuit effect. A reduction of short circuit current is obtained using low level design techniques, trying to reduce the period of time in which both the pMOSFET and nMOSFET are on.

- **Capacitance Reduction**

  From low level design to high level design the objective is the reduction of the complexity and therefore the area required to implement the desired functionality, with the additional objective of the reduction of cost of the silicon and the increment of clock frequency.

- **Switching Activity Reduction**

  With the increment of the number of devices implemented in a single chip, the interconnections increase more than linearly. A great part of the power is actually dissipated by the interconnections with respect to the logic part and the delay due to the interconnections is more relevant with respect to the delay of the logic gates. Placement and routing algorithms should optimize not only the delay, but the power dissipation too. This means that the algorithms should reduce the length of the interconnections of the signals whose switching activity is higher for the particular application for which the hardware will be used.

  The clock gating technique is used to stop the clock in parts of the circuit where no active computation is required. Some conditions for stopping the clock signal can be found directly from the state machine specification of the circuit [10, 12].

3.2.1 Power Models

System level design and IP modeling is the key to fast SoC innovation with the capability to quickly examine different alternatives early in the design process, to establish the best possible architecture, taking into account HW/SW partitioning, cost, performance and power consumption trade-offs.

The first necessary step to make toward low-power design is the dissipated power estimation of the system under development. This kind of analysis should be performed in the early phases of the design when some good ideas on optimizing power dissipation can drive the choice between different architectures.

Power analysis at system level is less accurate than at lower levels since the details of the real implementation of the functionality are not defined yet, but conversely the simulation time is much faster, due to the absence of

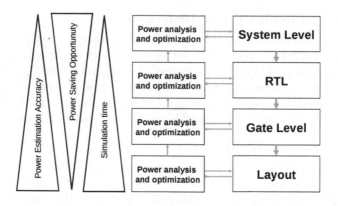

FIGURE 3.1: Power analysis and optimization at different levels of the design.

these details, and the power saving opportunity with an optimization is much higher. This concept is summarized in Figure 3.1.

Essentially two methodologies exist for estimating the power dissipation at different levels of abstraction: simulation-based methods and probabilistic methods.

- **Simulation-based methods.** The power dissipation is obtained applying specific input patterns to the circuit, see for example [46]. Therefore the estimation depends not only on the accuracy of the model description, but on the input patterns too. The input patterns should be strictly related to the real application in which the circuit will be applied. Simulation-based methods are widely used, since they are strictly related to the timing and functional simulation and test of the system.

- **Probabilistic methods.** These methods require the specification of the typical behavior of the input patterns through their probabilities; in this way it is possible to cover a large number of patterns with limited computational effort [25]. The switching activity, necessary to perform power estimation, is computed from the signal probabilities of the circuit nodes. Approaches to such methods are represented by probabilistic simulation [65, 80], symbolic simulation [40] and simulation of transition densities [63, 64].

Many consolidated and accurate tools estimate power dissipation from RTL to circuit level, but at higher levels there is still a lot of research to be done. Power models are classified on the basis of the level of abstraction of the description of the system and are reviewed in the following.

- **Transistor Level Power Estimation**

  An accurate estimate of power consumption can be carried out at transistor level, simulating the analog behavior of the circuit, analyzing the supply current, using SPICE-like simulators. The CPU time requested for the simulation is extremely high, making the simulation possible only for circuits with hundreds of transistors and few input patterns.

- **Gate Level Power Estimation**

  At gate level it is possible to analyze the behavior of the circuit using digital simulators if one has the details of the single logic gate. The estimation of power consumption is obtained by using switching activity and single node capacity using the relationship reported in Equation(3.1). At this level the results of the power estimation strongly depend on the delay model used, that may correctly estimate the presence or absence of glitches. In a "zero delay" model all transitions happen simultaneously, glitches are not considered, so power estimation is very optimistic.

- **RT Level Power Estimation**

  At register transfer level (RTL) power can be estimated using more complex blocks like multiplexers, adders, multipliers and registers. The source of inaccuracy at this level depends on the poor modeling of dynamic effect (e.g., glitches), causing an inaccurate estimation of the switching activity, and on the poor description details of the functional blocks and interconnections with a consequent inaccurate estimation of the capacitances.

  The improvement in the automatic synthesis tools from RTL description allows us to estimate the power dissipation using a fast synthesis with a mapping into a technology and a library defined by the user.

  Some analytical methods at RTL use complexity, or an equivalent gate count, as a capacitance estimate [62, 54]. In this way the power dissipated by a block can be roughly estimated as the number of equivalent gates multiplied by the power consumption of a single reference gate; a fixed activity factor is assumed.

  Some methods are based on analytical macromodels (linear, piecewise linear, spline, ...) of the power dissipation of each block. The model fits the experimental data obtained from numerical simulations at lower levels or experimental data. The model is affected by an error intrinsic in the model, by an estimation error due to the limited number of experiments and by an error due to the dependence of the measurements on the input patterns. The model can be represented as an equation [6, 84] or as a multi-dimensional look-up table (LUT) [58, 45, 72].

- **System Level Power Estimation**

  System level power estimation relies upon the power analysis of the hardware and software parts of the system. The components in a system level description are microprocessors, DSPs, buses, peripherals, whose internal architecture is, in general, not defined. Battery, thermal dissipation and cooling system modeling should also be considered at this level. Because the complete architecture of the system is not defined, power estimation is highly inaccurate; conversely, design exploration opportunity is high and so is power optimization.

At this level of abstraction power estimation usually is performed for the evaluation of different system architectures, in order to choose the best one in terms of power consumption too.

To enable power estimation, a model of the power dissipated by each block is created and the coefficients of the model are estimated from the information derived from the lower levels. The system level power model can be derived from the power dissipation of the single CMOS device, as reported in Equation (3.1), and can be represented by the following relationship

$$E = N \left(CV_{\mathrm{DD}}^2 D + Q_{\mathrm{sc}} V_{\mathrm{DD}} D + I_{\mathrm{leak}} V_{\mathrm{DD}} T \right) \qquad (3.2)$$

where V_{DD} is the supply voltage, D is the average number of commutations of the gates of the block, N is the number of gates, C is the average capacitance of the gates, Q_{sc} is the average charge lost due to short-circuit current during commutation, I_{leak} is the average leakage current of the block.

The average number of commutations D must be calculated during the system level simulation and therefore depends on the specific application and test vector. The coefficients C, Q_{sc}, I_{leak} are related to the specific technology chosen, N is the number of equivalent gates necessary to implement the block described at system level. If the block described at system level has already been implemented, these coefficients can be obtained from the low level implementation. If the block has not yet been implemented, the complexity of the block, that is, an estimation of the number of gates required for its implementation should be given. Of course, if the detailed architecture of the system is not yet defined, only a rough estimation can be given. An example of this procedure is given in Figure 3.2. From the SystemC code of each module the number of equivalent gates required for the implementation of the module is estimated.

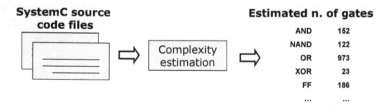

FIGURE 3.2: Complexity estimation from SystemC source code.

The mathematical operations on different SystemC types (`sc_int`, `sc_uint`, `sc_bigint`, `sc_biguint`, `sc_fixed`, `sc_ufixed`, `sc_fix`, `sc_ufix`), the bitwise and comparison operators, the assignments and the C++ control instructions (`if else`, `switch case`, `for` and `while`) are recognized and a module from a library of a reference technology is associated to each operator. A software has been developed to give these results in an automatic way [83].

Instruction-based power analysis has been presented in [79, 41] and applied in many other works [34, 22]. The term "instruction" is used to indicate an action that, together with others, covers the entire set of core behaviors. At system level a core can be seen as a functional unit executing a sequence of instructions or processes without any information on their hardware or software implementations. Instruction-based power analysis associates an energy model to each instruction, for example, the one reported in Equation (3.2). The power model should be parametric in order to allow the reuse not only of the IP functional description, but of the power model too.

An example is the power model of an I2C driver reported in [22]; in this case two power models have been used: a model that associates a constant value to each block and instruction independently on the data transmitted, and a model with a linear dependence on switching activity, and clock frequency obtained during high level functional simulations. The instruction set of an I2C driver is reported in Figure 3.3.

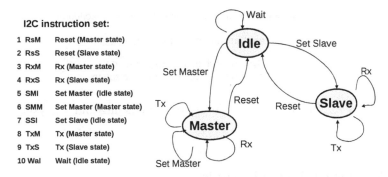

I2C instruction set:

1 RsM   Reset (Master state)
2 RsS   Reset (Slave state)
3 RxM   Rx (Master state)
4 RxS   Rx (Slave state)
5 SMI   Set Master (Idle state)
6 SMM   Set Master (Master state)
7 SSI   Set Slave (Idle state)
8 TxM   Tx (Master state)
9 TxS   Tx (Slave state)
10 Wal   Wait (Idle state)

FIGURE 3.3: I2C driver instruction set.

The second step of the instruction based power analysis, is the association of the power model to the functional model, as shown in Figure 3.4. Functional and power models are described in the same language (VHDL, SystemC ...).

The simulation of a complete SoC, that uses system level IP models, can be several hundreds times faster than an RTL simulation, so in a short time it is possible to evaluate hundreds of different configurations

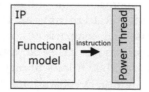

FIGURE 3.4: Power dissipation model added to the functional model.

and architectures in order to reach the desired trade-offs in terms of different parameters like speed, throughput and power consumption. The complete steps for instruction-based power modeling and analysis are reported in Figure 3.5.

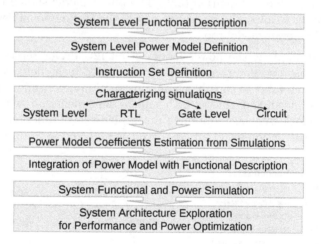

FIGURE 3.5: System level power modeling and analysis.

3.2.2   Power Analysis Tools

A great effort has been put forth in the development of tools for a complete design flow that can implement a top-down design methodology from high level modeling languages, i.e., C/C++, to silicon, see for example [20]. Some EDA companies started developing design tools with the goal of an automatic or semiautomatic synthesis from a subset of system level languages, for example RT level descriptions generated by SystemC co-simulation and synthesis tools. In recent years low level synthesis has been replaced by behavioral synthesis, as proposed for example in CoCentric SystemC Compiler and Behavioral Compiler by Synopsys, PACIFIC by Alternative System Concepts (ASC) and

Cynthesizer by Forte Design Systems. Cadence recently developed Palladium Dynamic Power Analysis at pre-RT level. Palladium Dynamic Power Analysis helps in full-system power analysis of designs, including both hardware and software.

There are also some emerging tools and methodologies that perform power estimation without the need for synthesis, often working at high levels of abstraction. PowerChecker, by BullDAST, avoids synthesis; it performs power estimation by working on a mixed RT/gate level description obtained through source HDL analysis, elaboration and hardware inferencing.

In ORINOCO [78], by ChipVision, the analysis of the power consumption is based on a compiler which extracts the control flow and the execution of the binary to collect profiling data. The expected circuit architecture is derived from a control data flow graph without carrying out a complete synthesis. The control data flow graph and the collected data statistics build the foundation for the calculation of the power dissipation.

ChipVision recently developed PowerOpt a low-power system synthesis tool. PowerOpt analyzes power consumption at system level. It automatically optimizes for low power, while synthesizing ANSI C and SystemC code into Verilog RTL designs, producing the lowest-power RTL architecture. Chipvision states that the tool automatically achieves power savings of up to 75% compared to RTL designed by hand and it is up to 60 times faster than lower level power analysis methods.

JouleTrack [75] is a tool for software energy estimation. It is instruction-based and computes the energy consumption of a given software. The model of power dissipation has been derived from experimental measurements of the supply current of the processor while executing different instructions. It has been applied to StrongARM SA-1100 and Hitachi SH-4 microprocessors.

Wattch [18] is an architectural level framework for power analysis. The authors created parameterized power models of common structures present in modern superscalar microprocessors. The models have been integrated into the Simplescalar [19] architectural simulator to obtain functional and power simulations. Recently the Wattch power simulator has been integrated in a complete simulation framework called SimWattch [23].

SimplePower [85] is an execution-driven, cycle-accurate, RT level power estimation tool. The framework evaluates the effect of high level algorithmic, architectural, and compilation trade-offs on energy. The simulation flow converts the C source benchmarks to SimplePower [19] executables. Simplepower provides cycle-by-cycle energy estimates for processor datapath, memory and on-chip buses.

Recently, since transaction level modeling (TLM) in SystemC is becoming an emerging architectural modeling standard, many works apply power estimation in SystemC-TLM environment. Many tools for power estimation from SystemC description have been recently presented [5, 66, 34, 4, 51].

3.3 PKtool

This section presents the Power Kernel Tool (PKtool) [1], developed for system level power estimation. PKtool is a simulation environment dedicated to power analysis of digital systems modeled in SystemC language. The main result provided is the estimation of power dissipation under specific operative conditions and power models. Its application needs the same efforts necessary for creating and simulating an ordinary SystemC description, except for some additional steps.

Like SystemC, PKtool is based on C++ class libraries and distributed as an open source software framework [1]. In comparison with typical commercial tools, the design capabilities provided by PKtool show both strength and weakness points. Among the formers, commercial tools usually represent more optimized and user-friendly environments as concerns both graphical-interfacing aspects and analysis means. Considering PKtool design potentialities, the strict embedding with SystemC framework gives PKtool a high and natural integration in a SystemC design flux. In particular, it is possible to reach a strong merging in the simulation phases, with a very limited intrusion in the original workflow. As a further consequence, the whole power analysis does not need ad hoc execution tools, but relies on the same simulation means required by SystemC applications. Moreover, the open source nature leads to great flexibility with regard to user interaction and evolution opportunities.

3.3.1 Basic Features

PKtool can be directly applied to each module constituting a system described in SystemC language. While the module abstraction is realized in SystemC through a suitable entity called `sc_module`, in PKtool it is realized through the definition of a new component called `power_module`. A `power_module` allows to extend the internal behavior of a traditional `sc_module` for PKtool analysis. This enhancement mainly consists in the linkage to a power model and in additional functionalities related to power estimation tasks. From an external point of view (in particular as regards the I/O port layout) a `power_module` retains the original `sc_module` structure, as can be seen in Figure 3.6.

In order to select an `sc_module` for a PKtool analysis, it is necessary to replace the original `sc_module` instance with a corresponding `power_module` instance. This operation can be made selectively, considering only some `sc_modules`, as shown in Figure 3.7.

A PKtool simulation is handled by a customized simulation engine, called Power Kernel, which deals with all the execution and synchronization tasks. Power Kernel acts simultaneously with the SystemC kernel in a hidden and non-intrusive way. The main tasks constituting a PKtool simulation concern the handling of the power models and the linkage to the required data, the

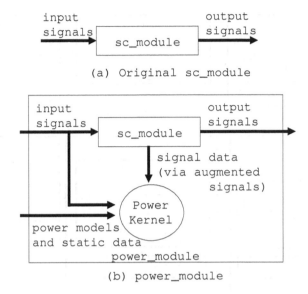

(a) Original sc_module

(b) power_module

FIGURE 3.6: **power_model** architecture.

computation of the power estimations, and the printing of the results, as shown in Figure 3.8.

3.3.2 Power Models

A power model gives an estimate of the power dissipated by a digital system, commonly by means of an analytical/algorithmic formulation. The PKtool environment is not related to a particular power model, but is linked to a library that makes available several power models. During a PKtool simulation, each monitored **sc_module** has to be associated to a specific power model, that will be applied for computing the related power estimation. This association must be carried out at the beginning of the simulation by the user.

The application of a power model is usually based on specific data required in its formulation (model data). We can subdivide model data into two distinct categories:

- *Static data*: data known a priori, available before the beginning of a simulation, for example technology parameters
- *Dynamic data*: data available only during simulation, on the basis of the run-time evolution of the module, for example switching activity

PKtool implements different solutions for the acquisition and the handling of static and dynamic data. Static data are communicated by the user at the beginning of a PKtool simulation, while dynamic data are handled at simulation time by ad hoc components called *augmented signals*.

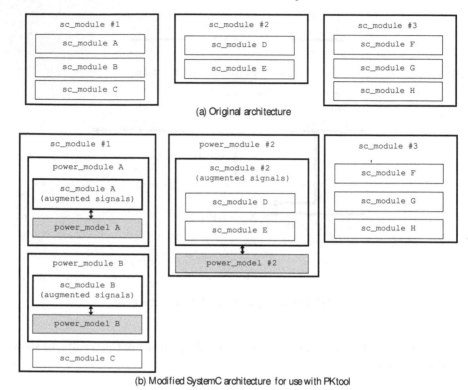

(a) Original architecture

(b) Modified SystemC architecture for use with PKtool

FIGURE 3.7: Example of association between sc_module and power_model.

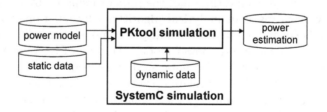

FIGURE 3.8: PKtool simulation flux.

3.3.3 Augmented Signals

An augmented signal is a smart signal, able to show a traditional behavior with the additional capabilities of computing and making available to the power model signal information such as commutations and probabilities. The class implementations of augmented signals are already incorporated inside the PKtool class library, constituting a framework of augmented signal types. The augmented types currently available cover many of the possible types which can be used for modeling signals in a SystemC description. From the user's point of view, the application of augmented signals consists of simple modifications in the code of the sc_module selected for PKtool analysis. As an example, let us consider the following code, which represents the class definition of an sc_module called example_mod:

```
SC_MODULE(example_mod)
{ sc_in<sc_uint<32> >    in_1, in_2;
  sc_in<bool>            reset;
  sc_in_clk              clk;
  sc_out<unsigned>       out;
  sc_uint<3>             ctr_1;
  sc_uint<2>             ctr_2;
  ...      // rest of the code, not shown
}
```

If, for example, we want to monitor the input ports in_1 and in_2, we have to cite the corresponding augmented signals, as shown in the following code.

```
SC_MODULE( example_mod)
{ sc_in_aug<sc_uint<32>> in_1, in_2;
  sc_in<bool>            reset;
  sc_in_clk              clk;
  sc_out<unsigned>       out;
  sc_uint<3>             ctr_1;
  sc_uint<2>             ctr_2;
  ...      // rest of the code, not shown
}
```

During a PKtool simulation the augmented signals will retain their original behavior and, in addition, will be able to provide their run-time commutations for the output power estimations. The instance of augmented signals represents the only modification to be made on the original code of an sc_module for PKtool analysis.

3.3.4 Power States

PKtool provides some functionalities for enhancing and refining the related power analysis. The most important one is the power state characterization, which allows a configurable control over the temporal evolution of a PKtool simulation. Power states are utility entities that can be optionally introduced in the configuration of an sc_module for PKtool simulations. Their main function is to distinguish distinct working states of the sc_modules behavior, on the basis of operative conditions specified by the user. Each of these working states is associated to a power state and can be handled in distinct way as regards power estimation tasks. The realization of a power state approach requires the definition of the power state objects, the association between power states and sc_module working states, and the definition of the rules for updating the power states.

3.3.5   Application Examples

The PKtool simulator has been used in different applications to estimate the power dissipation of systems described in SystemC. In some of these applications, the design has been implemented and simulated in VHDL too, at gate level, observing a CPU time increment of about two orders of magnitude with respect to SystemC. This result shows that, even if a lower level power simulation gives more accurate results, system level simulations must be used in the case of complex systems, such as a complete H.264/AVC codec or a Bluetooth network.

In [83] many simulations of the power dissipated by the Bluetooth baseband layer during the life of the piconet have been performed. Noise has been inserted in the channel in order to verify the performances in terms of power dissipated by the baseband during the creation of the piconet as a function of the noise. Another result shown is the mean value and the standard deviation of the energy dissipated by the baseband of the master during the transmission of data of different sizes and with different packet types (DH1, DH3, DH5, DM1, DM3, DM5).

In [21] a system level power analysis has been applied to the AMBA AHB bus, described in SystemC, to get information about the power dissipated during a system level simulation.

In [28] the application of the sum of absolute transformed differences (SATD) function in the motion estimation of the H.264/AVC codec have been studied. The developed SystemC models allowed a comparison of the architectures in terms of latency, area occupancy of the hardware, SNR and power dissipation. The simulation that uses system level IP models, can be several hundreds times faster than an RTL simulation, so it is possible to evaluate different configurations and architectures.

The discrete cosine transform (DCT) and the inverse discrete cosine transform (IDCT) are widely used techniques in processing of static images (JPEG) and video sequences (H.261, H.263, MPEG 1-4, and with some modification in H.264) with the aim of data stream compression. The diffusion of video processing in portable devices makes the power constraint extremely relevant. Different DCT/IDCT architectures have been modeled in SystemC for the system level power analysis in [82].

In [22] the system level power analysis methodology has been applied to the design of an I2C bus driver. The power dissipated by the I2C driver during the execution of each instruction has been derived from gate level VHDL simulations. In Section 3.6 examples of the application of PKtool to different communication architectures will be shown.

3.4 On-Chip Communication Architectures: Power, Performances and Reliability

The canonical multi-core embedded system view consists of various processing elements (PEs) responsible for the computation of the desired functions, including embedded DRAM, FLASH, FPGA and application-specific IP, programmable components, such as general purpose processor cores, digital signal processing (DSP) cores and VLIW cores, as well as analog front-end, peripheral I/O devices and MEMS.

A global on-chip communication architecture (OCCA) interconnects these devices, using a bus system, a crossbar, a multistage interconnection network, or a point-to-point static topology. Crossbars are attractive for very high speed communications. The crossbar maps incoming packets to output links, avoiding bottlenecks associated with shared bus lines and centralized shared memory switches.

OCCA provides communication mechanisms that allow distributed computation among different processing elements. Currently there are two common types of communication architectures: bus and network-on-chip (NoC). Bus networks, such as AMBA or STBus, are usually synchronous and offer several variants. Buses may be reconfigurable, partitionable into smaller sub-systems, provide multicasting or broadcasting facilities, etc.

NoC, such as the Spidergon by STMicroelectronics, uses a point-to-point topology. It can be visualized as a ring of communication nodes with several middle links. Each communication node is directly connected to its adjacent neighbors. One or more processors may be connected to each communication node. The network provides high concurrency, low latency on-chip communication architecture.

SoC design requires the exploration of a large solution space to select the global communication architecture, the partitioning of system functionalities, the allocation of components to execute them, and the local communication architectures to interconnect components to the global communication architectures.

Many issues arise in communication architecture when the number of IPs to be connected increases, for example: large bandwidth requirements, additional services associated to the communication protocol, clock domain partitioning. The more traditional communication architecture, the bus, has an intrinsic limit on bandwidth; the NoC paradigm [60] tries to overcome this limit. NoC is composed by three types of modules: routers, links and interfaces. The messages are sent from the source IP to a router and forwarded to other routers until they arrive to the router connected to the destination IP. Routers are connected to each other by links forming a net of chosen topology, size and connection degree.

A NoC architecture has many degrees of freedom. The topology of regular networks can be chosen from a wide variety of topologies: the most common ones are two-dimensional mesh and torus, but examples of other topologies are hypercubes, Spidergon [30, 17], hexagonal [35, 86], binary tree and variants [43, 47, 67], butterfly and benes network [61]. The topology affects performance factors such as cost (router and link number), communication throughput, maximum and average distance between nodes and fault-tolerance through alternative paths.

A network can use circuit switching and/or packet switching techniques and can support different *quality-of-service* (QoS) levels [16]. The links are characterized by the communication protocol (synchronization between sender and receiver), width (number of bits per transmission), presence or absence of error detection/correction scheme and dynamic voltage scaling [76, 73]. In general NoC links are unidirectional.

The router architecture has a strong impact on network performance. The router has input ports and output ports, where messages enter and go to and from the router; each *flit* (FLow control digIT) that represents the information quantum circulating in the network, is stored in internal *buffers* close to the input ports and/or to the output ports; the *routing* module indicates to the *switch* module how flits advance from input stage to output stage; contentions are resolved by specific *arbitering* rules and the *DPM* (*dynamic power management*) module implements power saving policies by slowing down or speeding up or turning off the whole router or some parts of it; the *flow control* indicates how the router resources are coordinated.

Buffer dimensions, structure (shift register or inserting register [15]) and parallelism degree must be chosen. The *switch* structure could be a complete or incomplete crossbar between input and output ports and can have some additional ports for delayed contention resolution [52, 53]. The *routing* algorithm and *DPM* policy should be implemented in a cheap and efficient way. Most common *flow control techniques* used in NoCs are virtual channel [20], virtual cut-through [50], wormhole [32], and flit-reservation flow control [69].

Compared to a bus, NoC has the following advantages:
- The bandwidth increases because message transaction takes place at the same time, but in different part of the network
- The arbitering is distributed and it is less complex, therefore the router is simpler and faster
- Regular topologies make NoC scalable and the use of the same blocks (routers and links) allows a high degree of reuse
- NoC, using GALS (globally asynchronous locally synchronous) synchronization paradigm, allows communication between modules with different clock domains
- The network, as a distributed architecture, can be more robust to faults, because messages can be redirected in areas not damaged or busy
- NoC can dynamically adjust power consumption depending on current communication requirements

On the other hand, NoC design is more complex than bus design. New problems and trade-offs arise:
- Routing algorithms should verify deadlock and livelock conditions [33, 42, 36, 37]
- More complicated and fault-tolerant routing schemes improve performances and reliability, but they need more complex, more expensive and slower routers
- Complex and efficient power management schemes need additional circuitry
- Routers and interfaces must implement appropriate arbitration schemes and must have suitable hardware in order to manage different QoS

NoC configuration parameters must be carefully tuned in order to improve throughput, cost and power performances. System level tools allowing solution space exploration, pruning non-optimal solution of network and router architectures, help the designers to reduce time-to-market.

The following part of the section will discuss the recent research toward the design of efficient NoC architectures and some existing tools, used to compare optimize cost, performance, reliability and power dissipation.

LUNA [38] is a system level NoC power analysis tool; LUNA extracts power consumption based on network architecture, routing and traffic application and link bandwidth calculated by sums of message flows routed in links and router. Power consumption estimation is directly proportional to recalculated flows.

Garnet [2] is a router model for the GEMS [57] simulator. The network can be simulated with different topologies, static routing, virtual channel number, flit and buffer size. Router architecture has no buffer at input ports and does not allow adaptive routing.

Xpipes [14] serves as a library of components for NoC. The modules are implemented in hardware macros and SystemC modules. Components written in the library are links, switch and interfaces OCP-compliant. Xpipes comprise a compiler and a simulator.

Nostrum [55] can simulate networks with two-dimensional topologies, wormhole flow control and deflection routing [56]. In [70] a power model for links and switches of the Nostrum NoC validated with Synopsys Power Compiler was integrated in the NoC SystemC-based simulator.

Other works are directly related to power modeling in NoCs. In all the works presented the power model is relative only to the routers and not to the IP connected. The power models of the routers are derived from a detailed low level description, and in some case applied to a SystemC NoC description. In [3] a VHDL-based cycle-accurate RTL model of the routers of a NoC is presented and used to evaluate latency, throughput, dynamic and leakage power consumption of NoC interconnection architecture.

In [59] and in [44] a power modeling methodology for NoC is proposed. The model coefficients are derived by a fitting with data obtained by the synthesis of several configurations of the switch architecture with Synopsys Design Compiler and PrimePower. The model of power consumption of a NoC switch takes traffic conditions into account.

The *PIRATE* [68] framework is mainly composed of the following modules: (1) generator of Verilog RTL models for the configurable NoC that can be automatically synthesized; (2) automatic power characterization; (3) cycle-based SystemC simulation model for dynamic profiling of power and performance. The parametric power model depends on the NoC architecture and on a traffic factor that is the activity of the router. The power characterization is based on a standard gate level power estimation using Synopsys Design Power, the results show an accuracy of the model of about 5 percent with respect to gate level simulations.

3.5 NOCEXplore

In this section the SystemC class library for modeling and simulation NoCs, recently proposed by the Universitá Politecnica delle Marche, is presented. The library has been integrated with tools allowing a statistical analysis of NoC performances and the investigation of communication bottlenecks. The integration between NOCEXplore and PKtool allows a deep analysis of the power dissipation of the IP and of the router. The simulation environment allows the exploration of the best communication architecture, the best routing algorithm and the placement of the IPs in the network.

Networks are configurable by a set of nineteen parameters that represent the network configuration and can be divided in two main categories: network architecture and router architecture. The traffic description involves three additional parameters. Globally, the configuration space has 22 dimensions. Each dimension could have a physical value, a numeric value or an identification. The list of the 22 parameters is reported in the following.

1) Network *quality of service* is an identification and describes global network services and main router architecture. At the moment packet switching with services of best effort delivery and no priority scheduling is implemented; the router main architecture has buffers on input and output ports.

2) *Network size* is a numeric value indicating how many modules are connected to the network.

3) *Topology* is an identification related to how routers are connected by links.

4-7) *Link type, link width, link delay* and the number of *physical links per topological arc* are four parameters that describe links. Link type identifies link protocol and communication scheme. The link delay can be a constant or data-dependent. The flit dimension depends on link width parameter.

8-9) *Flit-per-packet* and *packet-per-message* define how many flits correspond to a packet and, in communications with bursts, how many packets are in a message. Generally, flits of the same packet go through the same path; different packet, even if of the same message, can be routed in different ways.

10) Each router, if it is a synchronous machine, has a local clock generator of

a certain *frequency*; each generator has own starting delay, independent from the others.

11-19) *Routing algorithm, arbitering scheme, switch structure, DPM policy, flow control* and four other parameters that describe buffer length and parallelism.

20) *Traffic intensity* indicates the amount of messages injected in the network and it is normalized to the maximum value of one flit per clock cycle per node connected.

21) *Traffic scenario* describes the spatial distribution of message flows, that is, the flow $\lambda_{i,j}$ between each source node i and sink node j of the network.

22) *Burstyness* is a normalized value of traffic with burst over total traffic emitted by each source node.

The set of the 22 parameter values is defined as *network configuration*. At the moment the nodes attached to network are traffic generators and they are source and sink at the same time. The platform has been created to be easily expandable: to add a new numeric or physical value, for example, a new network size, simply insert the new value in the list of this parameter; to add a new behavior, for example, a new topology or a new routing algorithm, designers must create a new topology class, derived from the topology base class and overload one or few virtual methods that describe topology or routing algorithm. For example, a new traffic scenario with a certain value of locality requires about 30 lines of code. It is also possible to add new parameters in an easy and fast way. The great number of possible solutions creates some managing issues. A simulation manager coordinates all the actions for performing simulations and data postprocessing.

3.5.1 Analysis

NOCEXplore performs a statistical analysis of communication performances. All message delays are collected and global statistical parameters such as mean value and standard deviation are calculated. Moreover, the throughput, that is the number of delivered flit per source node per clock cycle, is computed on the basis of steady-state messages generated. Figures 3.9 and 3.10 show NoC communication performances at different traffic intensity and percentage of burst in the traffic.

Each emitted flit has a unique identifier and can be recognized in each part of the network. Source and sink nodes record identification and respectively creation and arrival time of the messages. Based on these records, both overall and source/sink pair communication performances (statistics on delays and throughput) can be calculated: this feature can be seen as a table where the i-th/j-th position is related to the source i and sink j pair.

A probabilistic analysis can be done on these records. Post-process is able to produce delay density probability tables and graphs of sets of messages records: all messages, messages emitted by a specific source node, messages collected by a specific sink node and messages of a particular message flow

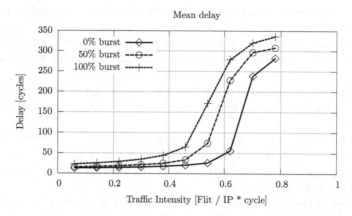

FIGURE 3.9: NoC performance comparison for a 16-node 2D mesh network: steady-state network average delay for three different traffic scenarios.

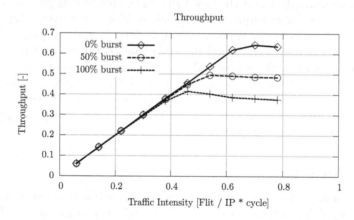

FIGURE 3.10: NoC performance comparison for a 16-node 2D mesh network: steady-state network throughput for three different traffic scenarios.

starting from a specific source node i delivered to a specific sink node j. Figure 3.11 shows an example.

Furthermore, source/sink pair statistics return information about *which* message flows are in greater delay with respect to the others, about congestion and location of the congestion. NOCEXplore gives information about transaction time specifications margins and *how many* messages fail to respect the limits. The tool allows other investigations, mainly concentrated on where congestions occur for bottleneck performance determination.

Each link and router, source node and sink node records information about its activities. Link activity and link switching activity can be monitored.

pdf of messages from all nodes to all nodes:mean=941 std=659; on 4637 messages;

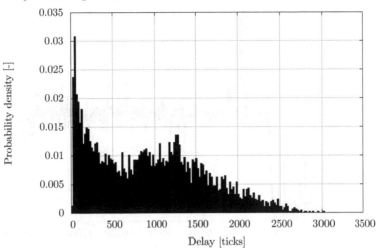

FIGURE 3.11: Example of probabilistic analysis. The message delay probability density referred to all messages sent and received by a NoC under traffic equally distributed with 50% of messages sent in burst and message generation intensity of 32%; network has 16 nodes, topology is 2D mesh and routing is deterministic.

Routers, seen as black boxes, record information about when each flit enters and exits.

Internal buffers record their own utilization level, switches and line controllers record which flit transversal per clock cycle has been performed. Routing modules record information about routing function calls: when each packet called the function and the corresponding result. Dynamic power modules record information about the router internal variables, such as router traffic rate and buffer utilization and utilization of neighbor routers, and actual power state.

Our tool processes previously mentioned activities and events in two ways:

- A statistical processing: for example mean value and standard deviation are calculated

- Temporal evolution quantities: for example n cycle moving average of link or switch activities and buffer or memory router utilization (see Figure 3.12)

Power analysis can be performed associating a power model to each router. This power model depends on the router activities such as link data commutations, incoming to and outgoing from router of a flit, routing function calls and flit crossings in the switch.

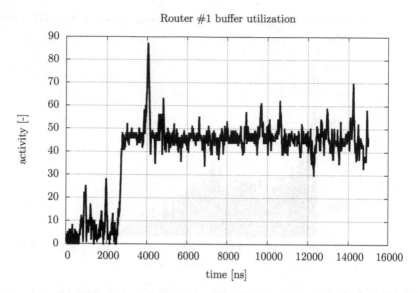

FIGURE 3.12: Example of temporal evolution analysis. The graph shows the number of flits in a router on top side of a 2D mesh network. Each router has globally 120 flit memory of capacity distributed in five input and five out ports. The figure shows that, for this traffic intensity and scenarios, buffer configuration is oversized and the performance is maintained even if the router has a smaller memory.

An interesting analysis that can be performed by NOCEXplore is the adoption of dynamic power management to each router of the NoC. This analysis can highlight repercussions of some router state on the neighboring routers and can be used to modify the combination of topology, routing, traffic scenario and DPM policy in order optimize communication performances and power dissipation.

Figure 3.13 shows a graph where the power state of each router is reported: on the x-axis the time is reported and on the y-axis the router identification number is reported; the color indicates the state and the legend on the right side reports the relationship between colors and states.

Some improvements can be made on the NOCEXplore tool, in order to reduce CPU time of simulations, that strongly depends on network size and traffic intensity. At the moment, simulating and post-processing a 16 node and 16 router NoC at maximum traffic intensity requires about 8 minutes on a commercial PC. We consider that this computation performance is quite good, since simulations are cycle accurate and a user can access many of the event details for investigation.

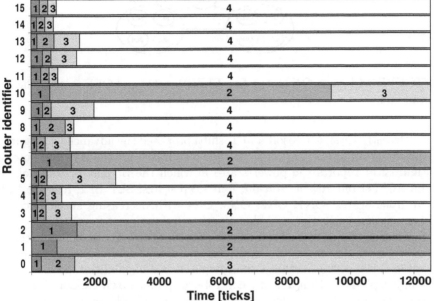

FIGURE 3.13: Example of power graph where power state is indicated over time, router by router. Dark color means high power state. Router power machine has nine power states and follows ACPI standard: values from 1 to 4 are ON states, values from 5 to 8 are SLEEP states and value 9 is the OFF state.

3.6 DPM and DVS in Multi-Core Systems

Energy consumption is extremely important for portable devices such as new generation of mobile phones, laptops, MP3 players, wireless sensor networks. The workload conditions in which these devices operate usually change over time. The techniques that have been recently adopted to reduce power dissipation of the device are dynamic power management (DPM) and dynamic voltage scaling (DVS). DPM is a technique that dynamically reduces the performances of the system by placing the components in low-power states in order to reduce power consumption. Many DPM algorithms have been introduced to force sleep or standby states when a device is idle. Dynamic voltage scaling is a technique that reduces supply voltage and frequency to reduce power consumption. The DVS is usually implemented in software: the processor spends part of the time to apply DVS when it is required.

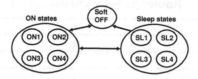

FIGURE 3.14: Four ON states, four SLEEP states and OFF state of the ACPI standard.

Recently Intel, Microsoft and Toshiba proposed the advanced configuration and power interface (ACPI) to provide a standard for the HW/SW interface. Figure 3.14 reports the power states in which the IP may operate following the ACPI standard: the soft-off, four sleep states: SL1, SL2, SL3, SL4 , four execution states: ON1, ON2, ON3, ON4 with decreasing speed and power consumption using the variable-voltage technique.

Shutting down some components increases the latency of the system and consequently decreases its overall performance [11]. DPM requires the observation of the system activity, the computational capabilities for management policy implementation and the control over power down capabilities of hardware resources [9]. An efficient power manager should measure inter-arrival and service times and at the same time should provide a low impact on resource usage and idle times [7, 77, 13, 71, 74, 8].

Many companies introduced DVS strategies in their processors: Intel introduced *SpeedStep* in 1999, Transmedia developed *LongRun* in 2000, AMD introduced *PowerNow!* in 2000, and National Semiconductor adopted *Power-Wise*.

Some ARM, AMD, Hitachi and Intel microprocessor-based systems and multi-core-systems [49, 48, 81] support frequency scaling and voltage scaling, with a significant energy reduction. Intel is applying DVS in multicore systems.

The application of DVS and DPM in multicore systems is essential for reducing power dissipation, but the interaction between power management architecture and communication architecture is very complex. Therefore, power reduction can have an unacceptable effect of communication throughput, if the interaction between these design variables is not considered.

Some DPM and communication architectures for a multi-core system are shown in Figure 3.15. The architecture (a), a bus-based communication and global DPM, is not efficient in terms of power dissipation when some of the cores are inactive. In the architecture (b), a bus-based communication and local DPM proposed in [26, 24], the communication throughput is strongly reduced when one of the cores involved in the bus communication is in sleep or low power and low clock frequency state. This effect is emphasized by the fact that usually buses are synchronous. The NoC communication is more suitable for local DPM, as for example, in architecture (c) in Figure 3.15. In fact, the delay occurring when a core must wake up due to a communication, may not cause a delay of the communications between the other cores, if

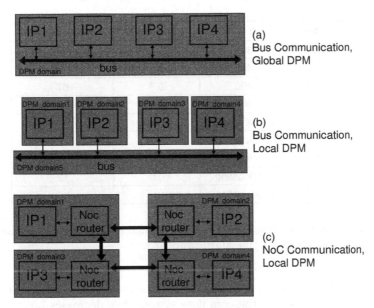

FIGURE 3.15: DPM and communication architectures.

routing algorithm and NoC architecture are properly chosen. Furthermore, GALS (global asynchronous and local synchronous) architectures are suitable for local DPM and NoC.

In [26], [29] and [27], DVS and DPM with different arbitration algorithm have been applied to architectures of the types (a) and (b) in Figure 3.15 of a system-on-chip based on the AMBA AHB bus. In the SystemC developed IP, the four ON states of the ACPI standard differ from the supply voltage and clock frequency, and the clock gating technique is applied in the sleep mode. Figure 3.16 reports the clock frequency, supply voltage and power dissipation for the different power states of the ACPI standard for the IPs used. The values have been derived from the data of the Intel Xscale processor.

The architectures (a) and (b), reported in Figure 3.15, have been modeled in SystemC and simulated with different situations of traffic in the AMBA AHB bus, and different bus arbitration algorithms:

1) No DPM: the complete system is always in ON1 state, operating at maximum frequency. The results of the next architectures have been normalized to the results of this architecture.

2) Global power management: a central power manager applies the DPM and DVS techniques to all the masters and slaves and bus at the same time. Therefore the power state is the same for all the blocks.

3) Local power management: each master and slave has its local energy manager that establishes the power state on the basis of battery status, chip temperature, predicted time for which the block will remain in idle state, whether the block is used for the bus or not.

State	Freq. (MHz)	Vdd (V)	Power (mW)
ON1	800	1.65	955
ON2	400	1.1	228
ON3	200	0.7	57
ON4	100	0.6	28
SL1	Clock gating	1.65	34
SL2	Clock gating	1.1	22
SL3	Clock gating	0.7	14
SL4	Clock gating	0.6	12
OFF	Clock gating	0	0.2

FIGURE 3.16: Clock frequency, supply voltage and power dissipation for the different power states of the ACPI standard.

A local DPM applied separately to the bus, masters and slaves can decrease the power dissipation of the complete system, but decreases bus throughput because the bus AMBA AHB is synchronous. In fact, the master that wants to use the bus must be at the same frequency of the bus before sending the bus request and must wait for the instant the slave is awake and working at the same frequency of the bus before sending data. Therefore the performance of the system may be extremely degraded if the local power managers are not coordinated with each other and with the bus arbitration policy.

The three DPM architectures (no DPM, global DPM and local DPM) have been tested in different bus traffic conditions. Some results of the simulation are reported in Figures 3.17 and 3.18. The results have been normalized to the corresponding results of the no DPM architecture.

Figure 3.17 reports the percentage of time each component is in the different states, for all the architectures in low bus traffic condition. The results in terms of energy dissipation are related to the percentage of time the IPs are in the different states. An energy reduction can be achieved when the IP is in sleep mode. When the IP is changing state (state transition in Figure 3.17) and it is executing a bus transfer task, the time and energy are wasted. It can be seen that the time spent in changing state is low. In global DPM case the system cannot go into sleep mode since the bus is always used by some master, and the masters or slaves not involved cannot go into sleep mode, as they can do with a local DPM.

Figure 3.18 reports the normalized energy dissipation, the normalized bus throughput, and the ratio between energy and throughput of the DPM architectures for different conditions of bus traffic (high, low).

Some conclusions can be briefly drawn: in critical conditions, when the battery is low, all the proposed DPM architectures have a strong reduction in power dissipation with a decrement factor of 4 of the bus throughput. Local

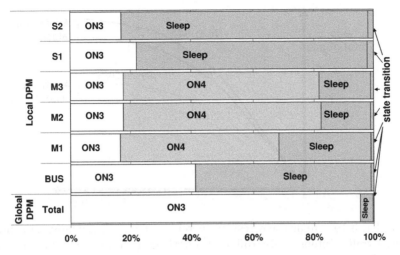

FIGURE 3.17: Percentage of the time the three masters and two slaves and the bus are in the different power states during simulation in a low bus traffic test case with local DPM and global DPM.

Energy	High Bus Traffic	Low Bus Traffic
Global DPM	101%	12%
Local DPM	98%	18%

Throughput	High Bus Traffic	Low Bus Traffic
Global DPM	96%	63%
Local DPM	39%	28%

Energy/Throughput	High Bus Traffic	Low Bus Traffic
Global DPM	105%	20%
Local DPM	250%	64%

FIGURE 3.18: Energy and bus throughput normalized to the architecture without DPM.

management gives a strong decrement on power dissipation at the cost of a worst throughput.

Figure 3.19 and Figure 3.20 summarize the comments on DPM and global and local DPM on a bus-based communication.

When the bus use is high, all the power management techniques are inefficient, as expected. The inefficiency in terms of communication throughput is more relevant with respect to the energetic inefficiency. This is due to the relevant waste of time required to resynchronize master and slave to the bus. The energy gained in sleep mode is wasted during synchronization. Conversely, when the bus use is low, a strong energy reduction is obtained with local and global power management architectures. The energetic gain is reached with

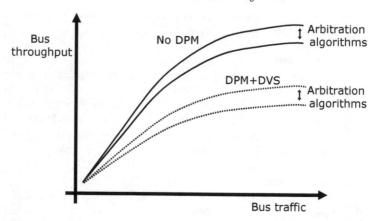

FIGURE 3.19: Qualitative results in terms of bus throughput as a function of bus traffic intensity for different DPM architectures and bus arbitration algorithm.

an increment of the time required to complete the tasks with respect to the time required without energy management.

The arbitration algorithm affects the energy dissipation, but bus efficiency dependence with DPM is stronger with respect to the arbitration algorithm. Energy reduction with DPM is stronger for low bus traffic. Energy efficiency depends on the type of traffic in the bus: Local power management is very efficient when some masters do not use the bus.

The DPM architecture and algorithm and NoC topology and routing algorithms should be selected considering that they both affect in a complex and complementary way the network throughput, power dissipation and system reliability.

3.7 Conclusions

Today's design methodologies must consider power dissipation constraint in the first phases of the design of a complex system on chip. The improvement of silicon technology allows the implementation of many cores in the same system, therefore the design of the communication architecture is fundamental to reach acceptable system performance. System level techniques for power reduction, communication architectures and routing algorithms have strong interaction and exert a strong effect both on power dissipation and communication throughput.

System level tools for power and communication analysis are fundamental for a fast and cost effective design of complex systems. This chapter presented

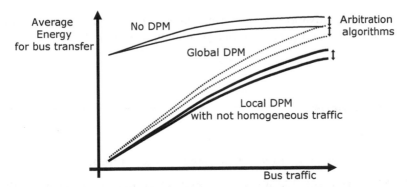

FIGURE 3.20: Qualitative results in terms of average energy per transfer as a function of bus traffic intensity for different DPM architectures and bus arbitration algorithm.

general aspects related to system level power analysis of SoC and on-chip communications. The state of the art of system level power analysis tools and NoC performance analysis tools was reported. In particular two SystemC libraries developed by the authors, and available in the sourceforge web site, have been presented: PKtool for power analysis and NOCEXplore for NoC simulation and performance analysis.

Finally, the application of dynamic voltage scaling techniques in on-chip communication architectures has been presented and general considerations are reported.

Review Questions

[Q_1] Summarize the power estimation methodologies at different levels of abstraction.

[Q_2] What are the main characteristic of instruction-based power models?

[Q_3] Indicate the basic features of the software PKtool.

[Q_4] Compare bus and network-on-chip communication architectures and indicate their advantages and disadvantages.

[Q_5] Indicate the characteristics of a typical router architecture of a network-on-chip.

[Q_6] Summarize the main characteristics of the dynamic voltage scaling technique.

[Q_7] Indicate advantages and disadvantages of dynamic voltage scaling in bus-based and network-on-chip communication architectures.

Bibliography

[1] PKtool documentation. http://sourceforge.net/projects/pktool/.

[2] Niket Agarwal, Li-Shiuan Peh, and Niraj Jha. Garnet: A detailed interconnection network model inside a full-system simulation framework. Technical report, Princeton University, 2008.

[3] N. Banerjee, P. Vellanki, and K.S. Chatha. A power and performance model for network-on-chip architectures. In *Proceedings of Design, Automation and Test in Europe Conference and Exhibition, 2004*, volume 2, pages 1250–1255, February 2004.

[4] N. Bansal, K. Lahiri, and A. Raghunathan. Automatic power modeling of infrastructure IP for system-on-chip power analysis. In *20th International Conference on VLSI Design, 2007*. Held jointly with 6th International Conference on Embedded Systems, pages 513–520, January 2007.

[5] Giovanni Beltrame, Donatella Sciuto, and Cristina Silvano. Multi-accuracy power and performance transaction-level modeling. *IEEE Trans. Comput.-Aided Design Integr. Circuits Syst.*, 26(10):1830–1842, October 2007.

[6] L. Benini, A. Bogliolo, M. Favalli, and G. De Micheli. Regression models for behavioral power estimation. *Integr. Comput.-Aided Eng.*, 5(2):95–106, 1998.

[7] L. Benini, A. Bogliolo, and G. De Micheli. A survey of design techniques for system-level dynamic power management. *IEEE Trans. VLSI Syst.*, 8(3):299–316, June 2000.

[8] L. Benini, G. Castelli, A. Macii, and R. Scarsi. Battery-driven dynamic power management. *IEEE Des. Test. Comput.*, 18(2):53–60, April 2001.

[9] L. Benini, R. Hodgson, and P. Siegel. System-level power estimation and optimization. In *Proc. of ACM/IEEE International Symposium on Low Power Electronics and Design(ISLPED'98)*, pages 173–178, Monterey, CA, August 1998.

[10] L. Benini and G. De Micheli. Transformation and synthesis of FSMs for low power gated clock implementation. *IEEE Trans. Comput.-Aided Design Integr. Circuits Syst.*, 15(6):630–646, June 1996.

[11] L. Benini and G. De Micheli. *Dynamic Power Management of Circuits and Systems: Design Techniques and CAD Tools.* Kluwer Academic Publishers, 1997.

[12] L. Benini, G. De Micheli, A. Lioy, E. Macii, G. Odasso, and M. Poncino. Synthesis of power-managed sequential components based on computational kernel extraction. *IEEE Trans. Comput.-Aided Design Integr. Circuits Syst.*, 20(9):1118–1131, September 2001.

[13] L. Benini, G. Paleologo, A. Bogliolo, and G. De Micheli. Policy optimization for dynamic power management. *IEEE Trans. Comput.-Aided Design Integr. Circuits Syst.*, 18(6):813–833, June 1999.

[14] Davide Bertozzi and Luca Benini. Xpipes: A network-on-chip architecture for gigascale systems-on-chip. *IEEE Circuits and Systems Magazine*, 4, 2004.

[15] Shubha Bhat. *Energy Models for Network on Chip Components.* PhD thesis, Technische Universiteit Eindhoven, 2005.

[16] E. Bolotin, I. Cidon, R. Ginosar, and A. Kolodny. QNoC: QoS architecture and design process for network on chip. *Journal of Systems Architecture*, 50:105–128, February 2004.

[17] L. Bononi and N. Concer. Simulation and analysis of network on chip architectures: ring, spidergon and 2D mesh. In *Proc. Design, Automation and Test in Europe (DATE)*, March 2006.

[18] D. Brooks, V. Tiwari, and M. Martonosi. Wattch: a framework for architectural-level power analysis and optimizations. In *Proc. of the 27th International Symposium on Computer Architecture*, pages 83–94, 2000.

[19] D. Burger, T. M. Austin, and S. Bennett. Evaluating future microprocessors: The simplescalar tool set, 1996. University of Wisconsin, Madison, Technical Report, CS-TR-1996-1308.

[20] L. Cai, P. Kritzinger, M. Olivares, and D. Gajski. Top-down system level design methodology using SpecC, VCC and SystemC. In *Proc. of Design, Automation and Test in Europe Conference and Exhibition*, page 1137, Paris, France, March 2002.

[21] Marco Caldari, Massimo Conti, Paolo Crippa, Simone Orcioni, Lorenzo Pieralisi, and Claudio Turchetti. System-level power analysis methodology applied to the AMBA AHB bus. In *Proc. of Design Automation and Test in Europe, (DATE'03)*, pages 32–37, Munchen, Germany, March 2003.

[22] Marco Caldari, Massimo Conti, Paolo Crippa, Simone Orcioni, and Claudio Turchetti. Design and power analysis in SystemC of an I2C bus

driver. In *Proc. of Forum on Specifications & Design Languages. FDL '03*, Frankfurt, Germany, September 2003.

[23] Jianwei Chen, M. Dubois, and P. Stenstrom. Simwattch and learn. *Potentials, IEEE*, 28(1):17–23, January-February 2009.

[24] C. W. Choi, J. K. Wee, and G. S. Yeon. The proposed on-chip bus system with GALDS topology. In *International SoC Design Conference, 2008. ISOCC '08*, volume 1, pages 292–295, November 2008.

[25] M. A. Cirit. Estimating dynamic power consumption of CMOS circuits. In *Dig. of IEEE Int. Conf. on Computer-Aided Design. ICCAD-87*, pages 534–537, Santa Clara, CA, November 1987.

[26] M. Conti and S. Marinelli. Dynamic power management of an AMBA AHB system on chip. In *Proc. of SPIE'07, Int. Conference VLSI Circuits and Systems 2007*, Maspalomas, Gran Canaria, Spain, 2007.

[27] Massimo Conti, Marco Caldari, Giovanni B. Vece, Simone Orcioni, and Claudio Turchetti. Performance analysis of different arbitration algorithms of the AMBA AHB Bus. In *Design Automation Conference. DAC '04*, pages 618–621, San Diego, CA, June 2004.

[28] Massimo Conti, Francesco Coppari, Simone Orcioni, and Giovanni B. Vece. System level design and power analysis of architectures for SATD calculus in the H.264/AVC. In *SPIE Int. Conference on VLSI Circuits and Systems II 2005*, volume 5837, pages 795–805, Seville, Spain, 2005.

[29] Massimo Conti, S. Marinelli, Giovanni B. Vece, and Simone Orcioni. SystemC modeling of a dynamic power management architecture. In *Proc. of Forum on Specifications & Design Languages. FDL '06*, pages 229–234, Darmstadt, Germany, September 2006.

[30] M. Coppola, M. Grammatikakis, R. Locatelli, G. Maruccia, and L. Pieralisi. *Design of Cost-efficient Interconnect Processing Units: Spidergon STNoC*. CRC Press, 2009.

[31] William J. Dally. Virtual-Channel Flow Control. In *Proc. of the 17th Annual International Symposium on Computer Architecture (ISCA)*, pages 60–68, Seattle, Washington, May 1990.

[32] William J. Dally and Charles L. Seitz. The torus routing chip. *Journal of Parallel and Distributed Computing*, 1986.

[33] W.J. Dally and C.L. Seitz. Deadlock-free message routing in multiprocessor interconnection networks. In *IEEE Transactions on Computers*, 1987.

[34] Nagu Dhanwada, Ing-Chao Lin, and Vijay Narayanan. A power estimation methodology for SystemC transaction level models. In *CODES+ISSS '05: Proceedings of the 3rd IEEE/ACM/IFIP Int. Conf. on Hardware/Software Codesign and System Synthesis*, pages 142–147, 2005.

[35] James W. Dolter, P. Ramanathan, and Kang G. Shin. Performance analysis of virtual cut-through switching in HARTS: A hexagonal mesh multicomputer. *IEEE Transactions on Multicomputers*, 1991.

[36] J. Duato. A necessary and sufficient condition for deadlock-free adaptive routing in wormhole networks. *IEEE Transactions on Parallel and Distributed Processing*, 1995.

[37] J. Duato. A necessary and sufficient condition for deadlock-free routing in cut-through and store-and-forward networks. *IEEE Transactions on Parallel and Distributed Processing*, 1996.

[38] Noel Eisley and Li-Shiuan Peh. High-level power analysis for on-chip networks. In *Proceedings of CASES*, pages 104–115. ACM Press, 2004.

[39] Frank Ghenassia. *Transaction Level Modeling with SystemC*. Springer, 2005.

[40] A. Ghosh, S. Devadas, K. Keutzer, and J. White. Estimation of average switching activity in combinational and sequential circuits. In *Proc. of 29th ACM/IEEE Design Automation Conf.*, pages 253–259, June 1992.

[41] Tony Givargis, Frank Vahid, and Jörg Henkel. A hybrid approach for core-based system-level power modeling. In *Proc. of Conf. on Asia South Pacific Design Automation. ASP-DAC '00*, pages 141–146. ACM, 2000.

[42] C. J. Glass and L. M. Ni. *The turn model for adaptive routing. ACM*, 1994.

[43] Pierre Guerrier and Alain Greiner. A generic architecture for on-chip packet-switched interconnections. In *Proc. of DATE*, pages 250–256. ACM Press, 2000.

[44] G. Guindani, C. Reinbrecht, T. Raupp, N. Calazans, and F. G. Moraes. NoC power estimation at the RTL abstraction level. In *IEEE Computer Society Annual Symposium on VLSI, 2008. ISVLSI '08*, pages 475–478, April 2008.

[45] S. Gupta and F. N. Najm. Power macromodeling for high level power estimation. In *Proc. of 34th ACM/IEEE Design Automation Conf.*, pages 365–370, June 1997.

[46] S. M. Kang. Accurate simulation of power dissipation in VLSI circuits. *IEEE Trans. Syst. Sci. Cybern.*, 21(5):889–891, October 1986.

[47] Heikki Kariniemi and Jari Nurmi. New adaptive routing algorithm for extended generalized fat trees on-chip. In *Proc. International Symposium on System-on-Chip*, pages 113–188, Tampere, Finland, 2003.

[48] H. Kawaguchi, Y. Shin, and T. Sakurai. uITRON-LP: Power-conscious real-time OS based on cooperative voltage scaling for multimedia applications. *IEEE Transactions on Multimedia*, 7(1), February 2005.

[49] H. Kawaguchi, Y. Shin, and T. Sakurai. Case study of a low power MTCMOS based ARM926 SoC: Design, analysis and test challenges. In *IEEE International Test Conference*, 2007.

[50] P. Kermani and L. Kleinrock. Virtual cut-through: a new computer communication switching technique. *Computer Networks*, 1979.

[51] F. Klein, G. Araujo, R. Azevedo, R. Leao, and L.C.V. dos Santos. An efficient framework for high-level power exploration. In *50th Midwest Symposium on Circuits and Systems, 2007. MWSCAS 2007*, pages 1046–1049, August 2007.

[52] Andrew Laffely, Jian Liang, Prashant Jain, Ning Weng, Wayne Burleson, and Russell Tessier. Adaptive system on a chip (aSoC) for low-power signal processing. In *Thirty-Fifth Asilomar Conference on Signals, Systems, and Computers*, November 2001.

[53] Jian Liang, S. Swaminathan, and R. Tessier. aSOC: A scalable, single-chip communications architecture. In *IEEE International Conference on Parallel Architectures and Compilation Techniques*, pages 524–529, October 2000.

[54] Dake Liu and C. Svensson. Power consumption estimation in CMOS VLSI chips. *IEEE Trans. Syst. Sci. Cybern.*, 29(6):663–670, June 1994.

[55] Zhonghai Lu. *A User Introduction to NNSE: Nostrum Network-on-Chip Simulation Environment*. Royal Institute of Technology, Stockholm, November 2005.

[56] Zhonghai Lu, Mingchen Zhong, and Axel Jantsch. Evaluation of onchip networks using deflection routing. In *Proceedings of GLSVLSI*, 2006.

[57] M. Martin, D. Sorin, B. Beckmann, M. Marty, M. Xu, A. Almadeen, K. Moore, M. Hill, and D. Wood. Multifacet's general execution-driven multiprocessor simulator (GEMS) toolset. *Computer Architecture News*, 2005.

[58] H. Mehta, R.M. Owens, and M. J. Irwin. Energy characterization based on clustering. In *Proc. of 33rd ACM/IEEE Design Automation Conf.*, pages 702–707, June 1996.

[59] P. Melonit, S. Carta, R. Argiolas, L. Raffo, and F. Angiolini. Area and power modeling methodologies for networks-on-chip. In *1st International Conference on Nano-Networks and Workshops, 2006. NanoNet '06*, pages 1–7, September 2006.

[60] G. De Micheli and L. Benini. Networks on chip: A new paradigm for systems on chip design. In *DATE '02: Proceedings of the Conference on Design, Automation and Test in Europe*, page 418, 2002.

[61] H. Moussa, O. Muller, A. Baghdadi, and M. Jezequel. Butterfly and benes-based on-chip communication networks for multiprocessor turbo decoding. In *Design Automation and Test in Europe Conference*, 2007.

[62] K. D. Muller-Glaser, K. Kirsch, and K. Neusinger. Estimating essential design characteristics to support project planning for ASIC design management. In *Dig. of IEEE Int. Conf. on Computer-Aided Design. ICCAD-91*, pages 148–151, November 1991.

[63] F. N. Najm. Transition density: a new measure of activity in digital circuits. *IEEE Trans. Comput.-Aided Design Integr. Circuits Syst.*, 12(2):310–323, February 1993.

[64] F. N. Najm. Low-pass filter for computing the transition density in digital circuits. *IEEE Trans. Comput.-Aided Design Integr. Circuits Syst.*, 13(9):1123–1131, September 1994.

[65] F. N. Najm, R. Burch, P. Yang, and I. N. Hajj. Probabilistic simulation for reliability analysis of CMOS VLSI circuits. *IEEE Trans. Comput.-Aided Design Integr. Circuits Syst.*, 9(4):439–450, Apr 1990.

[66] V. Narayanan, Ing-Chao Lin, and N. Dhanwada. A power estimation methodology for SystemC transaction level models. In *Third IEEE/ACM/IFIP International Conference on Hardware/Software Codesign and System Synthesis, 2005. CODES+ISSS '05*, pages 142–147, September 2005.

[67] S. R. Ohring, M. Ibel, S. K. Das, and M. J. Kumar. On generalized fat trees. In *Proceedings of 9th International Parallel Processing Symposium*, 1995.

[68] G. Palermo and C. Silvano. *PIRATE: A Framework for Power/Performance Exploration of Network-on-Chip Architectures*. Book Series Lecture Notes in Computer Science, Publisher Springer Berlin / Heidelberg, 2004.

[69] Li-Shiuan Peh and William J. Dally. Flit-reservation flow control. In *Proc. of the 6th Int. Symp. on High-Performance Computer Architecture (HPCA)*, pages 73–84, January 2000.

[70] S. Penolazzi and A. Jantsch. A high level power model for the nostrum NoC. In *9th EUROMICRO Conference on Digital System Design: Architectures, Methods and Tools, DSD 2006*, pages 673–676, 2006.

[71] Q. Qiu and M. Pedram. Dynamic power management based on continuous-time markov decision processes. In *Proc. of ACM/IEEE Design Automation Conf.*, pages 555–561, New Orleans, LA, June 1999.

[72] T. Sato, Y. Ootaguro, M. Nagamatsu, and H. Tago. Evaluation of architecture-level power estimation for CMOS RISC processors. In *IEEE Symposium on Low Power Electronics*, pages 44–45, October 1995.

[73] Li Shang, Li-Shiuan Peh, and Niraj K. Jha. Power-efficient interconnection networks: Dynamic voltage scaling with links. In *Computer Architecture Letters*, May 2002.

[74] T. Simunic, L. Benini, and G. De Micheli. Dynamic power management for portable systems. In *Proc. of 6th International Conference on Mobile Computing and Networking*, Boston, MA, August 2000.

[75] Amit Sinha and Anantha P. Chandrakasan. Jouletrack – a web based tool for software energy profiling. In *Proc. of ACM/IEEE Design Automation Conf.*, pages 220–225, 2001.

[76] Vassos Soteriou and Li-Shiuan Peh. Design-space exploration for power-aware on/off interconnection networks. In *Proc. of the 22nd Intl. Conf. on Computer Design (ICCD)*, 2004.

[77] M. B. Srivastava, A. P. Chandrakasan, and R. W. Brodersen. Predictive system shutdown and other architectural techniques for energy efficient programmable computation. *IEEE Trans. VLSI Syst.*, 4(1):42–55, March 1996.

[78] A. Stammermann, L. Kruse, W. Nebel, A. Pratsch, E. Schmidt, M. Schulte, and A. Schulz. System level optimization and design space exploration for low power. In *Proc. of Int. Symp. on System Synthesis*, pages 142–146, Quebec, Canada, 2001.

[79] V. Tiwari, S. Malik, A. Wolfe, and M.T.-C. Lee. Instruction level power analysis and optimization of software. *The Journal of VLSI Signal Processing*, 13(2):223–238, January 1996.

[80] C.-Y. Tsui, M. Pedram, and A.M. Despain. Efficient estimation of dynamic power consumption under a real delay model. In *Dig. of IEEE/ACM Int. Conf. on Computer-Aided Design*, pages 224–228, Santa Clara, CA, November 1993.

[81] M. Vasic, O.Garcia, J.A. Oliver, P.Alou, and J.A. Cobos. A DVS system based on the trade-off between energy savings and execution time. In *COMPEL Conference*, pages 1–6, 2008.

[82] Giovanni Vece, Massimo Conti, and Simone Orcioni. PK_tool 2.0: a SystemC environment for high level power estimation. In *Proc. of 12th IEEE Int. Conf. on Electronics, Circuits and Systems. ICECS '05*, Gammarth, Tunisia, December 2005.

[83] Giovanni B. Vece, Simone Orcioni, and Massimo Conti. Bluetooth baseband power analysis with PKtool. In *Proc. of IEEE European Conf. on Circuit Theory and Design. ECCTD '07*, pages 603–606, Seville, Spain, September 2007.

[84] Qing Wu, Qinru Qiu, M. Pedram, and Chih-Shun Ding. Cycle-accurate macro-models for RT-level power analysis. *IEEE Trans. VLSI Syst.*, 6(4):520–528, December 1998.

[85] W. Ye, N. Vijaykrishnan, M. Kandemir, and M. J. Irwin. The design and use of simplepower: A cycle accurate energy estimation tool. In *Proc. of ACM/IEEE Design Automation Conf.*, pages 95–106, 2000.

[86] You-Jian Zhao, Zu-Hui Yue, and Jang-Ping Wu. Research on Next-Generation Scalable Routers Implemented with H-torus Topology. *Journal of Computer Science and Technology*, 2008.

4

Routing Algorithms for Irregular Mesh-Based Network-on-Chip

Shu-Yen Lin and An-Yeu (Andy) Wu

Electrical Engineering Department
National Taiwan University
Taipei, Taiwan
linyan@access.ee.ntu.edu.tw
andywu@cc.ee.ntu.edu.tw

CONTENTS

4.1	Introduction	112
4.2	An Overview of Irregular Mesh Topology	113
	4.2.1 2D Mesh Topology	113
	4.2.2 Irregular Mesh Topology	113
4.3	Fault-Tolerant Routing Algorithms for 2D Meshes	115
	4.3.1 Fault-Tolerant Routing Using Virtual Channels	116
	4.3.2 Fault-Tolerant Routing with Turn Model	117
4.4	Routing Algorithms for Irregular Mesh Topology	126
	4.4.1 Traffic-Balanced OAPR Routing Algorithm	127
	4.4.2 Application-Specific Routing Algorithm	132
4.5	Placement for Irregular Mesh Topology	136
	4.5.1 OIP Placements Based on Chen and Chiu's Algorithm	137
	4.5.2 OIP Placements Based on OAPR	140
4.6	Hardware Efficient Routing Algorithms	143
	4.6.1 Turns-Table Routing (TT)	146
	4.6.2 XY-Deviation Table Routing (XYDT)	147
	4.6.3 Source Routing for Deviation Points (SRDP)	147
	4.6.4 Degree Priority Routing Algorithm	148
4.7	Conclusions	151
	Review Questions	151
	Bibliography	151

4.1   Introduction

In the literature, regular 2D mesh-based network-on-chip (*NoC*) designs have been discussed extensively. In practice, by introducing different sizes of hard IPs (*oversized IPs*, *OIPs*) from various vendors, the original regular mesh-based NoC architecture may be destroyed because the locations of the OIPs invalidate parts of routing paths. The resulting mesh-based NoC becomes irregular and needs new routing algorithms to detour the OIPs. However, some routing algorithms for irregular mesh-based NoC may cause heavy traffic around the OIPs, which also results in nonuniform traffic spots around the OIPs.

In this chapter, the concepts of irregular mesh topology and corresponding traffic-aware routing algorithms are introduced. The irregular mesh topology can support different sizes of OIPs. However, for irregular meshes, existing routing algorithms on 2D meshes may fail. Because the cases of faulty networks and on-chip irregular meshes are similar, direct applications of traditional fault-tolerant routing algorithms can help to deal with the OIP issue. Some previous works apply traditional fault-tolerant routing algorithms to solve routing problems of irregular mesh topology. In Section 4.3, several traditional fault-tolerant routing algorithms [34], [5] in computer networks are reviewed and discussed. However, directly applying these fault-tolerant routing algorithms causes heavy traffic loads around the OIPs and unbalanced traffic in the networks. In Section 4.4, the OIP avoidance pre-routing (*OAPR*) algorithm [21] was proposed to solve the aforementioned problems.

The OAPR can make traffic loads evenly spread on the networks and shorten the average paths of packets. If the NoC design is specialized to a specific application, the routing algorithm can be customized to provide the performance requirements. In [23], the design methodology, called application specific routing algorithm (*APSRA*), was proposed to design the deadlock-free routing algorithm for irregular mesh topology. The APSRA is also introduced in Section 4.4. In irregular mesh topology, the locations of OIP influence the network performance. The best choice of OIP placements heavily depends on the routing algorithms. In [17] and [20], OIP placements based on Chen and Chiu's algorithm [5] and OAPR [21] were analyzed. These analyses are discussed in Section 4.5. Hardware implementation of the routing algorithm is an important issue for NoC designs, too. For irregular meshes, routing tables are often used to accomplish the routing algorithms. However, the number of entries in the routing table equals the number of the nodes in the network. Many efficient implementations use reduced ROM or Boolean logics to achieve an equivalent routing algorithm. In Section 4.6, some hardware-efficient routing algorithms for irregular mesh topologies are reviewed.

4.2 An Overview of Irregular Mesh Topology

System-on-chip (SoC) designs provide the integrated solution to the complex VLSI designs. However, in deep sub-micron (DSM) technology, the existing on-chip interconnections are facing many challenges due to the increasing scale and complexity of the designs. Recently, network on-chips (NoCs, also called on-chip networks) have been proposed as a solution to cope with the problems [19], [30], [15], [29]. Kumar presented a design methodology specifically for a 2D mesh NoC [31]. The nodes in a 2D mesh are connected in a two-dimensional array and each node constitutes an IP connected to a router responsible for message routing. Numerous research works are based on 2D meshes because of their regularity in layout efficiency and good electrical properties. In 2D meshes, each router is connected to a single IP, such as a CPU, a DSP core, or an embedded memory. However, the assumption that each IP has the equal size is not practical since the sizes of outsourced hard IPs from various vendors may be different (such as an ARM processor, a CPU, an MPEG4 decoder etc.). As a result, enhancing a regular 2D NoC to adapt to irregular mesh topology is required to place IPs with different sizes. In this section, an overview of irregular mesh topology is introduced. In Section 4.2.1, we first briefly review the 2D mesh topology. Then, irregular mesh topology is introduced in Section 4.2.2.

4.2.1 2D Mesh Topology

An $n \times n$ 2-dimensional mesh (2D mesh) contains n^2 routers. Each router has an address (x,y), where x and y belong to $\{0, 1, ..., n-1\}$ in an $n \times n$ mesh. We define the coordinate x (y) increasing along east (north) direction. Therefore, the router located in the southwest corner of the 2D mesh has an address $(0,0)$, and the router placed in the northeast corner has an address $(n-1, n-1)$. Each router in 2D mesh contains five ports: four ports connected to neighbor routers (north, east, south, and west) except the routers located in the boundaries of 2D meshes and one port linked to a local IP. Besides, two routers, $u : (u_x, u_y)$ and $v : (v_x, v_y)$, $(u_x, u_y, v_x,$ and v_y belong to $\{0, 1, ..., n-1\}$ in an $n \times n$ mesh) are connected if the addresses differ in only coordinate x or y and the difference is equal to 1. In other words, u and v conform to either $\{|u_x - v_x| = 1, u_y = v_y\}$ or $\{|u_y - v_y| = 1, u_x = v_x\}$ if they are neighbors. Fig. 4.1(a) shows an example of a 6×6 2D mesh. Each router in the 2D mesh is connected to one IP.

4.2.2 Irregular Mesh Topology

In order to place OIPs in 2D mesh topologies (OIPs, which represents that the sizes of the hard IPs are over one tile in a 2D mesh), the concepts of irregular mesh topologies have been proposed in [1] and [16]. By removing some routers

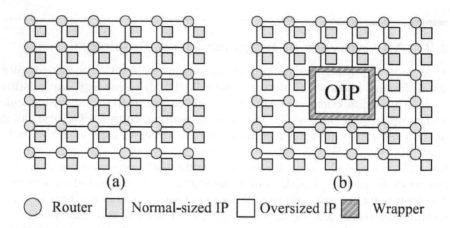

(a)   (b)

⬤ Router   ▣ Normal-sized IP   ▢ Oversized IP   ▨ Wrapper

FIGURE 4.1: (a) A conventional 6 × 6 2D mesh and (b) a 6 × 6 irregular mesh with 1 OIP and 31 normal-sized IPs. (From Lin, S.-Y. et al. *IEEE Trans Computers*, 57(9), 1156–1168. ©IEEE. With permission.)

and links in 2D mesh topologies, some separate regions appear and irregular mesh topologies are formed. OIPs can be placed in these regions by designing suitable wrappers between OIPs and the routers near the separate regions. Fig. 4.1(b) shows a 6 × 6 irregular mesh with 1 OIP and 31 normal-sized IPs. The OIP is connected to eight routers through the wrapper in this example. For irregular mesh topologies, some design issues are pointed out in [17]:

1. Access points: Since each OIP occupies larger area than normal-sized IP in irregular mesh topologies, the communication bandwidth of each OIP may be different. The communication bandwidth between an OIP and the rest of the NoC can be adjusted by the numbers of access points in the wrapper. The wrapper connects the OIP to the rest of the NoC through these access points. Hence, it is useful to use several access points and addresses for the design of the wrapper. The number of access points determines the communication bandwidth between an OIP and the rest of the NoC. For example, consider a NoC system containing one oversized IP, a shared memory. The memory may need huge bandwidth to communicate with other IPs. Therefore the memory perhaps requires many access points around its boundaries. The positions of the access points are also important. If an IP wants to transmit packets to an OIP, the position of access point may influence the latency between the OIP and the rest of the NoC. Hence, the locations and the numbers of access points must be defined according to the performance requirements of the specific application.

2. Routing problems: Because some routers are removed in irregular mesh topologies, routings of packets become more complex. Hence, existing

routing algorithms on 2D meshes may fail because OIPs block some routing paths of the regular 2D mesh. This problem can be solved in two ways. One way is to use physical links in the surroundings of OIPs. These links connect the blocked routing paths. However, this way requires extra resources and these links may cause more serious crosstalk, noise, and delay problems in deep submicron VLSI technologies, as pointed out by [19], [27], [10]. The other way is to find other routing paths around OIPs. Hence, new routing algorithms for irregular mesh topology are needed. The routing algorithms for irregular mesh topologies are discussed in Section 4.4.

3. Placements of OIPs: The methods to place OIP on irregular meshes are also important because different placements can result in different network performances. The best choice of OIP placements is extremely dependent on the routing algorithms. Some analyses based on Chen and Chiu's routing algorithm [5] and OAPR [21] are introduced in Section 4.5.

4.3   Fault-Tolerant Routing Algorithms for 2D Meshes

In computer networks, routing involves selecting a path for a source node to a destination node in a particular topology. The efficiency of routing algorithms can influence the performance of the system heavily. The routing algorithms can be classified into *deterministic routing* and *adaptive routing*. Deterministic routing uses only one fixed path for routes, while adaptive routing makes use of many different paths. The advantage of deterministic routing is ease of implementation in router design. By using simple logics, deterministic routing can provide low latency when the network is not congested. However, the network using deterministic routing may suffer from more degradation of network throughput than the network using adaptive routing. The reason is because deterministic routing cannot avoid congested links, but adaptive routing can use alternative routing paths to reduce the network congestion. Therefore, adaptive routing can result in higher network throughput. However, the implementation of adaptive routing is more complex than deterministic routing. Extra logics are needed to select the routing path with low traffic congestions in adaptive routing. For fault-free networks, some important issues of routing algorithms are high throughput, low latency, avoidance of deadlocks, and the performance requirements under different traffic patterns [12]. For faulty networks, the design of routing algorithms must consider some extra issues: the graceful degradations of network performance, and the complexity of fault-tolerant routing. Many fault-tolerant routing algorithms for 2D meshes have been proposed. These methods can be classified into two categories: 1) meth-

ods using virtual channels and 2) methods using turn models. These methods
are introduced in Sections 4.3.1 and 4.3.2, respectively.

4.3.1   Fault-Tolerant Routing Using Virtual Channels

Virtual channels are often applied to the router design [28], [9], [25]. Virtual channels provide multiple buffers for each physical channel in the network.
Adding virtual channels is similar to adding more lanes to a street network. For
a network without virtual channels, a single blocked packet occupies the whole
channel and all following packets are blocked. By adding virtual channels to
the network, additional buffers allow packets to pass through the blocked
channels. Hence, the network throughput can be increased. In [16], virtual
channels were first discussed to design deadlock-free routing algorithms. Virtual channels provide more freedom of resource allocations to transmit packets. By designing a suitable routing algorithm, virtual channels can be also
applied to tolerate faults in the network. Many related works are discussed
in [22], [7], [3], [4]. In [22], Linder and Harden used virtual channels to
design fault-tolerant routing algorithms for three topologies: 1) unidirectional
k-ary n-cube, 2) torus-connected bidirectional, and 3) mesh-connected bidirectional. Proposed fault-tolerant routing algorithms can tolerate at least one
faulty node with additional numbers of virtual channels. In [7], Chien and Kim
proposed a partial adaptive algorithm, called planar-adaptive routing, which
can avoid deadlocks by using constant numbers of virtual channels. The requirements of virtual channels are fixed and do not grow if the dimension of
the network is increased. The concept of planar-adaptive routing is to restrict
adaptivity of routing in two dimensions. Packets are routed adaptively in a series of two-dimension planes in the k-ary n-cube of more than two dimensions.
By limiting the selections in the routing, the complexity of deadlock prevention
can be reduced. Besides, planar-adaptive routing can be extended to support
the feature of fault tolerance. Planar-adaptive routing handles faulty regions
by misrouting around them. Three virtual channels in each physical channel
are required in their method. In [3], Boppana and Chalasani proposed the
concepts of fault rings (*f-rings*) and fault chains (*f-chains*). An f-ring is a set
of active nodes that enclose a faulty region. An f-chain established around
the faulty block touches the boundaries of the 2D mesh. Besides, a deadlock-
free fault-tolerant e-cube routing algorithm was proposed to misroute faulty
regions by routing packets around f-rings. Four virtual channels per physical
channel are required to tolerate multiple faulty regions. In [4], Boura and
Das proposed an adaptive deadlock-free fault-tolerant routing algorithm for
2D meshes. Messages are routed adaptively in nonfaulty regions. This algorithm can tolerate any number of faults by using three virtual channels in
each physical channel. In the aforementioned methods, routing algorithms
can tolerate faulty nodes with extra virtual channels. However, the extra virtual channels involve adding buffer space and complex switching mechanisms
in the router design. In the analysis of [6], routers with virtual channels re-

quire two to three times as many gates as those without virtual channels. Besides, the setup delays of the routers with virtual channels are 1 to 2 times, and the flow control cycles are 1.5 to 2 times. Moreover, the additional area overheads make the routers with virtual channels more liable to fail. Hence, many researchers focus on designing fault-tolerant routing algorithms without using virtual channels, which are discussed in Section 4.3.2.

4.3.2 Fault-Tolerant Routing with Turn Model

In [13], Glass and Ni presented turn models for designing wormhole routing algorithms without extra virtual channels. The turn model is based on analyzing the directions in which packets can turn in a network. These turns in the network may from the cycles and deadlock happens. By prohibiting some turns in the network, these cycles are broken and the routing algorithms that apply remaining turns can be deadlock-free, livelock-free, minimal or nonminimal, and maximally adaptive for the network. In 2D meshes, eight 90-degree turns can be formed. These turns can form two abstract cycles, as shown in Fig. 4.2. In order to avoid deadlock, at least one turn in a cycle must be prohibited. There are 16 possible cases to break the cycles by prohibiting one turn in each cycle. However, four cases cannot prevent deadlock, as in Fig. 4.3. The three turns allowed in Fig. 4.3(a) are equal to the prohibited turn in Fig. 4.3(b). The allowed turns in Fig. 4.3(b) also form the prohibited turn in Fig. 4.3(a). Hence, deadlock may happen, as shown in Fig. 4.3(c). Three other symmetric cases also result in cycles. Only 12 cases can prevent deadlock. By the analysis of the turn model, Glass and Ni also propose three partially adaptive routing algorithms: west-first, north-last, and negative-first in [13]. These routing algorithms can avoid deadlocks without using virtual channels. Fig. 4.4(a), Fig. 4.4(b), and Fig. 4.4(c) show the prohibited turns for west-first, north-last, and negative-first routing algorithms, respectively.

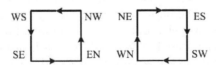

FIGURE 4.2: Possible cycles and turns in 2D mesh. (From Glass, C. J. and Ni, L. M. 1992. *Proceedings.*, The 19th Annual International Symposium on Computer Architecture; Page(s):278–287. With permission.)

These three algorithms are described as follows:

- In west-first routing algorithm, if a packet travels west, it must start from the west direction first. After that, packets are routed adaptively in the south, east, and north directions.

- In north-last routing algorithm, a packet should only travel to the north direction when it is the last direction to travel. First, packets are routed adaptively in west, south, and east directions. If the same column of the destination node reaches north, packets are routed in the north direction if it is necessary.

- In negative-first routing algorithm, a packet must start out in a negative direction to travel in a negative direction. First, packets are routed adaptively in the west and south directions if transmitted in negative directions. Then, these packets are routed adaptively in the east and north directions.

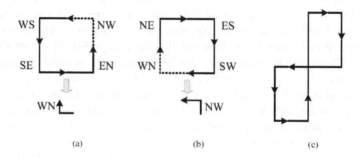

(a)   (b)   (c)

FIGURE 4.3: Six turns form a cycle and allow deadlock. (From Glass, C. J. and Ni, L. M. 1992. *Proceedings.*, The 19th Annual International Symposium on Computer Architecture; Page(s):278–287. With permission.)

In [8], Chiu extended the idea from the Glass and Ni turn model [13] and proposed the odd-even turn model. The odd-even turn model avoids deadlock by prohibiting two turns for odd and even columns and performs fairer routing adaptiveness. Fig. 4.5 shows the restricted turns of the odd-even turn model. In odd columns, south-to-west (SW) turns and north-to-west (NW) turns are avoided. In even columns, east-to-south (ES) turns and east-to-north (EN) turns are restricted. The odd-even turn model prevents the formation of the rightmost column of a cycle. Minimal or nonminimal routing algorithm based on the odd-even turn model is deadlock free as long as 180-degree turns are prohibited. According to the odd-even turn model, Chiu also proposed a partial adaptive routing algorithm in [8], as shown in Fig. 4.6. The algorithm *ROUTE* can route packets in minimal routing paths and avoid deadlock without virtual channels. The *Avail_Dimension_Set* contains the available candidates to forward the packet. If the destination node is located in the west of the source node, packets are prohibited from moving north or south in an odd column unless the destination node is located in the same column. The reason is that the packets may be routed in NW or SW turns to achieve the destination node. The NW and SW turns in odd columns may result in deadlocks. If the destination node is located in the east of the source node, the location of the destination node must be considered in the routing process. For a destination node located in an even column, the packets must complete the

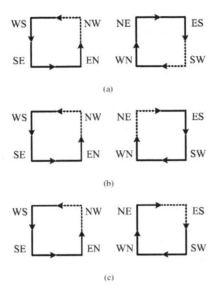

FIGURE 4.4: The turns allowed by (a) west-first algorithm, (b) north-last algorithm, and (c) negative-first algorithm. (From Glass, C. J. and Ni, L. M. 1992. *Proceedings.*, The 19th Annual International Symposium on Computer Architecture; Page(s):278–287. With permission.)

routing in dimension 1 before they reach the column. The packets located one column to the west of the destination node cannot move east unless they are in the same row as the destination node. The restricted EN and ES turns are not allowed in an even column. Besides, if the source node and the destination node are located in the same even column, the packets are allowed to move north or south.

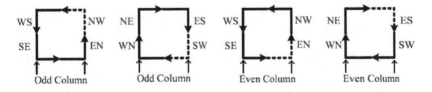

FIGURE 4.5: The six turns allowed in odd-even turn models. (From Chiu, G. M.; *Trans. Parallel and Distributed Systems*; Vol. 11, 729–737, July 2000. ©IEEE. With permission.)

According to the aforementioned turn models, researchers have attempted to design fault-tolerant routing algorithms without virtual channels for 2D meshes. In [14], Glass and Ni proposed a fault-tolerant routing algorithm

```
Algorithm ROUTE
/*Source node: (s₀,s₁); destination mode: (d₀,d₁); current mode (c₀,c₁). */
begin
Avail_Dimension_Set ← Ø;
e₀←d₀ − c₀
e₁←d₁ − c₁
if (e₀ = 0 and e₁ = 0)
    Deliver the packet to the local node and exit;
if (e₀ = 0)   /* currently in the same column as destination */
    if (e₁ > 0)
        Add North to Avail_Dimension_Set;
    else
        Add South to Avail_Dimension_Set;
else
    if (e₀ > 0)       /* eastbound messages */
        if (e₁ = 0)
            Add East to Avail_Dimension_Set;
        else {
                if (c₀ is odd or c₀ = s₀)
                    if (e₁ > 0)
                        Add North to Avail_Dimension_Set;
                    else
                        Add South to Avail_Dimension_Set;
                    if (d₀ is odd or e₀ ≠ 1)   /* odd destination column or ≥ 2 column to destination */
                        Add East to Avail_Dimension_Set;
                }
    else {   /*westbound messages */
            Add West to to Avail_Dimension_Set;
            if (c₀ is even)
                if (e₁ > 0)
                    Add North to Avail_Dimension_Set;
                else
                    Add South to Avail_Dimension_Set;
        }
    Select a dimension from Avail_Dimension_Set to forward the packet;          @[2000]IEEE
end
```

FIGURE 4.6: A minimal routing algorithm *ROUTE* that is based on the odd-even turn model. (From Chiu, G. M.; *Trans. Parallel and Distributed Systems*; Vol. 11, 729–737, July 2000. ©IEEE. With permission.)

based on modifications of negative-first routing algorithm. The modified algorithm has following two phases:

1. Route the packet west or south to the destination node or farther west and south than the destination node, avoiding routing the packet to a negative edge as long as possible. If a faulty node on a negative edge blocks the path along the edge, the packet is routed one hop perpendicular to the edge.

2. Route the packet east or north to the destination, avoiding routing the packet as far east or north as the destination as long as possible. If a faulty node on a negative edge of mesh blocks the path to a destination on the edge, route the packet one hop perpendicular to the edge, two hops toward the destination, and one hop back to the edge.

The proposed algorithm is deadlock-free and fault-tolerant for a single faulty node in 2D mesh topologies. However, this algorithm cannot cope with more faulty nodes in 2D meshes. In [34], Wu proposed an extended X-Y (*E-XY*) routing algorithm based on the dimension-order routing and the odd-even turn model [8]. The E-XY can avoid deadlock without virtual channels. The E-XY can tolerate multiple faulty nodes in 2D meshes if faulty nodes form a set of disjointed rectangular faulty blocks, called an extended faulty block. In the extended faulty block, each fault block must be surrounded by a boundary ring. The boundary ring consists of six lines: two lines at the east side, two lines at the west side, one line at the north side, and one line at the south side. The definition of the extended faulty block and its boundary ring facilitate the fault-tolerant routing based on the odd-even turn model [8]. The localized algorithm to form extended faulty blocks is shown in Fig. 4.7. Each node can exchange and update its status with its neighbors. Four directions are defined as: east $(+ y)$, south $(- x)$, west $(- y)$ and north $(+ x)$. A nonfaulty node is classified as safe and unsafe. First, nonfaulty nodes are marked safe. By executing the steps (1) and (2), each nonfaulty router updates the status based on the status of its neighbors. Eventually, unsafe and faulty nodes form extended faulty blocks. Fig. 4.8 shows three examples of extended faulty blocks. At least two columns or one row between two extended faulty blocks are reserved for E-XY routing.

Safe/unsafe status:
All nonfaulty nodes are initialized to *safe*
repeat
  doall
    (1) Nonfaulty node *u* exchanges its status with its neighbors. In addition, the status of its east (west) neighbor is passed to its west (east) neighbor.
    (2) Change *u*'s status to *unsafe* if
      (a) it has two *unsafe* or faulty neighbors that are not both in the *x* dimension, or
      (b) it has an *unsafe* or faulty neighbor along the *x* dimension and an *unsafe* or
        faulty 2-hop neighbor along the *y* dimension.
  doall
until there is no status change   @[2003]IEEE

FIGURE 4.7: The localized algorithm to form extended faulty blocks. (From Jie Wu; *IEEE Trans. Computers*; Vol. 52, pp. 1154–1169 Sept. 2003. ©IEEE. With permission.)

The E-XY algorithm is shown in Fig. 4.9. The E-XY contains two phases. In phase 1, packets are moved along the x (north or south) dimension until the offset is reduce to zero; in phase 2, packets are moved along the y (east or west) dimension until the offset is also equal to zero. Both phase 1 and phase 2 have two routing modes: 1) normal mode, and 2) abnormal mode. In normal mode, the E-XY is similar to the dimension-order routing and only turns at even columns if no faulty block obstructs the routing paths, as shown in Fig. 4.10. Otherwise, the abnormal mode shown in Fig. 4.11 is selected. In Fig. 4.10, Fig. 4.11, and Fig. 4.12, the symbol E (O) denotes that the E-XY turns at even (odd) columns, and the notation, FBs, stands for faulty

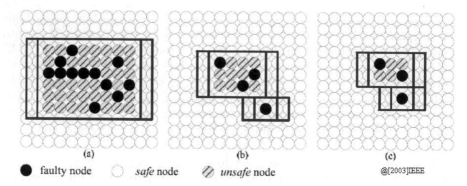

FIGURE 4.8: Three examples to form extended faulty blocks. (From Jie Wu; *IEEE Trans. Computers*; Vol. 52, pp. 1154–1169 Sept. 2003. ©IEEE. With permission.)

blocks Although the E-XY algorithm can tolerate multiple faulty blocks in 2D meshes, some drawbacks are pointed out in [21], which are shown as follows:

- Traffic loads on even columns are more serious than odd columns: the E-XY only turns at even columns in normal mode.

- Traffic loads on the boundaries of faulty blocks are heavy and unbalanced: the routing paths follow the boundaries of faulty blocks in abnormal mode. Moreover, traffic loads on west boundaries of faulty blocks are heavier than east boundaries.

- The E-XY can not solve faulty blocks located at the boundaries of 2D meshes: in the E-XY, each faulty block must be surrounded by a boundary ring. The boundary ring consists of six lines: two lines at the east side, two lines at the west side, one line at the north side, and one line at the south side.

Chen and Chiu proposed their fault-tolerant routing algorithm in [5] (The deadlock problem in [5] was corrected in [18]). The algorithm can tolerate multiple faulty nodes in 2D meshes if faulty nodes form rectangular faulty regions. The procedure to form the faulty regions for Chen and Chiu's algorithm is introduced in [3]. In the procedure, a nonfaulty node can be viewed as an active node, a deactivated node, or an unsafe node. The nonfaulty node X is defined as a deactivated node if X has two or more deactivated nodes. It is a recursive step to define the deactivated nodes. A nonfaulty node which is not deactivated is viewed as an active node. The deactivated nodes and the faulty nodes can form one or many rectangular faulty regions. Besides, a deactivated node can be identified as an unsafe node if it has at least one active neighbor. Fig. 4.12 shows an example to form the faulty regions. In [3], the concepts of the f-ring and the f-chain are proposed according to the positions of faulty

Extended XY routing:
1. /* the packet is sent to an even column first */
    (a) If the source is in an odd column and Δ_x is non zero, then the packet is sent to its west neighbor in an even column.

2. /* phase 1: reduce Δ_x, the offset in the x dimension */
    (a) (Normal mode) reduce Δ_x to zero by sending the packet north (or south) (with no 180° turn).
    (b) (Abnormal mode) when a north-bound (south-bound) packets reaches a boundary node of a faulty block, it is routed around the block, clockwise (counter-clockwise) by following the boundary ring of the faulty block, as shown in Fig. 13(a) (Fig. 13(b)). The packets take the first even column turn whenever possible and step (a) is followed.

3. /* phase 2: reduce Δ_y, the offset in the y dimension */
    (a) Once Δ_x is reduced to zero, a NW or NE turn is preformed for the north-bound packets and a north-to-west or north-to-east turn is performed for a south-bound packet. The selection of a turn depends on the relative location of the destination of the current node.
    (b) (Normal mode) reduce Δ_y to zero by sending the packet east (or west) (with no 180° turn).
    (c) (Abnormal mode) when a east-bound (west-bound) packets reaches a boundary node of a faulty block, it is routed around the block, clockwise or counter-clockwise, alone odd columns of the boundary ring as shown in Fig. 13(c) (even columns of the boundary ring as shown in Fig. 13(d)).Routing around the block is completed when Δ_x is again reduced to zero and step (b) is followed.   @[2003]IEEE

FIGURE 4.9: E-XY routing algorithm. (From Jie Wu; *IEEE Trans. Computers*; Vol. 52, pp. 1154–1169 Sept. 2003. ©IEEE. With permission.)

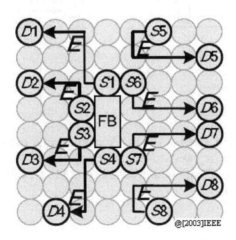

@[2003]IEEE

FIGURE 4.10: Eight possible cases of the E-XY in normal mode. (From Jie Wu; *IEEE Trans. Computers*; Vol. 52, pp. 1154–1169 Sept. 2003. ©IEEE. With permission.)

regions. An f-ring is a faulty region enclosed by a set of nonfaulty routers. An f-chain is a faulty region located at the boundaries of mesh networks. Fig. 4.13(a) shows an example of one f-ring and one f-chain. The disabled routers in faulty blocks can not be used in routing process. Besides, f-chains can be classi-

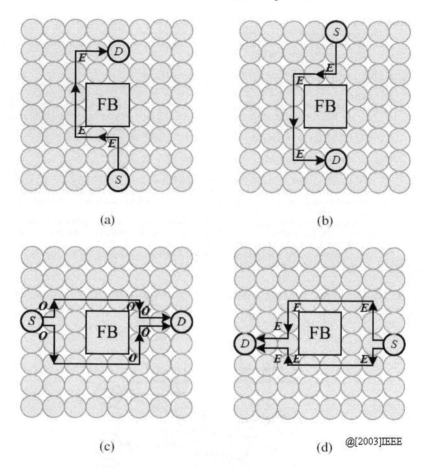

(a)   (b)

(c)   (d)   @[2003]IEEE

FIGURE 4.11: Four cases of the E-XY in abnormal mode: (a) south-to-north, (b) north-to-south, (c) west-to-east, and (d) east-to-west direction. (From Jie Wu; *IEEE Trans. Computers*; Vol. 52, pp. 1154–1169 Sept. 2003. ©IEEE. With permission.)

fied into eight different types according to the boundaries of a mesh network, called NW-chains, NE-chains, SW-chains, SE-chains, N-chains, S-chains, E-chains, and W-chains. Fig. 4.13(b) shows an example of one f-ring and eight different types of f-chains in a 10×10 mesh. According to the classification in Fig. 4.13(b), the Chen and Chiu's algorithm [5] can support both f-rings and f-chains. Chen and Chiu's algorithm prohibits some turns to avoid the formation of the rightmost column segment of a circular waiting path. Hence the algorithm can solve deadlock without using virtual channels. The corrected Chen and Chiu's algorithm is shown as the procedure *Message-Route-Modified* in Fig. 4.14. The procedure *Message-Route-Modified* contains four modes:

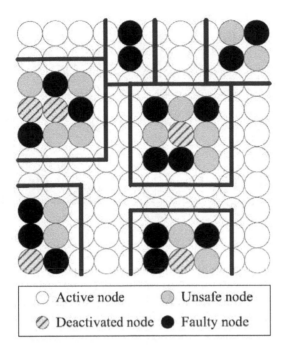

| ○ Active node | ◉ Unsafe node |
| ⦸ Deactivated node | ● Faulty node |

FIGURE 4.12: An example to form faulty blocks for Chen and Chiu's algorithm. (From Chen, K.-H. and Chiu G.-M. *Journal of Information Science and Engineering*, Vol.14, pp.765–783, Dec. 1998. With permission.)

1) *Normal-Route*, 2) *Ring-Route*, 3) *Chain-Route Modified*, and 4) *Overlapped-Ring Chain Route*. If current node is the destination node, the message *mg* is consumed. If the source node *S* is unsafe, *mg* is forwarded to an active neighbor. Each *mg* contains a parameter in the leader flit. The parameter indicates the routing types of the message. If current node is active and is not on any f-ring or f-chain, the routing process is determined by the procedure *Normal-Route*, as shown in Fig. 4.15. In *Normal-Route*, row-first (*RF*), column-first (*CF*), and row-only (*RO*) routing paths are used. Fig. 4.19(a) shows some possible routing paths for RF, CF, and RO. If *mg* encounters a single f-ring or a single f-chain, the routing is determined by the procedure *Ring-Route* or *Chain-Route Modified*. If current node *C* is overlapped by multiple f-rings or f-chains, the procedure *Overlapped-Ring Chain Route* is used. The procedures *Ring-Route*, *Chain-Route Modified*, and *Overlapped-Ring Chain Route* are shown as Figs. 4.16, 4.17, and 4.18. Fig. 4.19(b) shows two examples (*S*1 to *D*1 and *S*2 to *D*2) to misroute the f-rings and f-chains. The Chen and Chiu's algorithm still has a drawback, which is pointed out in [21]: traffic loads around faulty blocks are heavy and unbalanced. The routing paths are asymmetric and along the boundaries of faulty blocks.

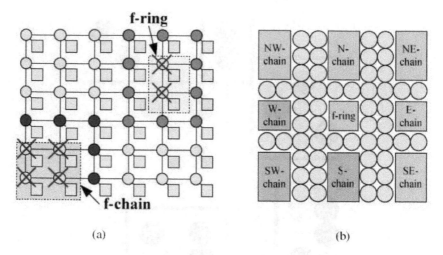

FIGURE 4.13: Two examples of f-rings and f-chains: (a) one f-ring and one f-chain in a 6 × 6 mesh and (b) one f-ring and eight different types of f-chains in a 10 × 10 mesh. (From Chen, K.-H. and Chiu G.-M. *Journal of Information Science and Engineering*, Vol.14, pp.765–783, Dec. 1998. With permission.)

```
Procedure Message-Route-Modified(mg)
/* Message mg is sent from source S to destination D; C is the current node of the header flit */
if (C is the destination D)
    Consume mg;
else
    if (current node C is S, and is unsafe)
        Send mg to an active neighbor;
    else /* active node */
        begin
            Determine the message type (RF, CF, or RO) of mg;
            if (C is not on a fault ring or fault chain)
                Normal-Route(mg);
        else
            if (C is on a single fault ring)
                Ring-Route(mg);
            else
                if (C is on a single fault chain)
                    Chain-Route-Modified(mg);
                else
                    Overlapped-Ring-Chain-Route(mg);            @[2007]IEEE
        end
```

FIGURE 4.14: Pseudo codes of the procedure *Message-Route Modified*. (From Holsmark, R. and Kumar S. *Journal of Information Science and Engineering*, Vol. 23, pp. 1649–1662. May 2007. With permission.)

4.4   Routing Algorithms for Irregular Mesh Topology

In this section, routing algorithms for irregular mesh topology are introduced. In [26] and [17], two fault-tolerant routing algorithms, E-XY [34] and Chen

```
Procedure Normal-Route(mg)
switch   (mg's type)
   case   (RF message)
          Use EW channel to forward mg;
          exit;
   case   (CF message)
          if (mg is NS message)
             Use NS channel to forward mg;
          else
             Use SN channel to forward mg;
          exit;
   case   (RO message)
          Use WE channel to forward mg;                    @[2007]IEEE
          exit;
```

FIGURE 4.15: Pseudo codes of the procedure *Normal-Route*. (From Holsmark, R. and Kumar S. *Journal of Information Science and Engineering*; Vol. 23, pp. 1649–1662. May 2007. With permission.)

and Chiu's algorithm [5] are directly applied on irregular mesh problems. Fault-tolerant routing algorithms are workable because of the similarity between faulty networks and on-chip irregular meshes. However, fault-tolerant routing algorithms are not suitable for irregular meshes. Directly applying fault-tolerant routing algorithms cause heavy traffic loads around the OIP and unbalanced traffic in the networks. In [21], an OIP avoidance pre-routing (*OAPR*) algorithm was proposed. The OAPR is based on the odd-even turn model [13] for routings in irregular meshes without extra virtual channels. The OAPR results in lower and more balanced traffic loads around the OIPs because it can avoid the routing paths around the OIPs and takes all usable turns in the odd-even turn model. Therefore, networks using the OAPR perform better than those using the Chen and Chiu's algorithm [5] and the E-XY [34]. Besides, the design methodology of an application specific routing algorithm for irregular meshes, application specific routing algorithm (*APSRA*), was proposed in [23]. APSRA assumes that the communication among tasks in a specific application is known in advance. The information of the communication can be useful for designing deadlock-free algorithms which are more adaptive in comparison with a general algorithm. APR and APSRA are discussed in Sections 4.4.1 and 4.4.2, respectively.

4.4.1   Traffic-Balanced OAPR Routing Algorithm

The OAPR algorithm is introduced in Section 4.4.1. Fig. 4.20 shows the concept of the OAPR from the experimental results in [21]. If we apply the E-XY [34] and Chen and Chiu's algorithm [5], traffic loads around the OIPs are huge and unbalanced, as shown in Fig. 4.20(a) and (b). However, the OAPR algorithm results in lower and more balanced traffic loads around the OIPs (Fig. 4.20(c)) because it can avoid the routing paths around the OIPs and takes all usable turns in the odd-even turn model [8]. Therefore, the networks using

```
Procedure Ring-Route(mg)
switch    (mg's type)
case      (RF message)
          if (EW channel is available)
              Use EW channel to forward mg;
          else
              Route mg clockwise;
          exit;
case      (CF message)
          if (mg is SN message)
              if (C is on the north boundary of the fault ring)
                  Normal-Route(mg);
              else
                  if (C is on the west boundary of the fault ring and D is in the same column as C)
                      Normal-Route(mg);
                  else
                      if (D is lower than the reference node of the ring)
                          Route mg counter-clockwise;
                      else
                          Route mg clockwise;
          else /* NS message */
              if (C is on the east or south boundary of the ring and EW channel is available)
                  Route mg along EW channel;
              else
                  Route mg counter-clockwise;
          exit;
case      (RO message)
          if (C is in the same row as D and WE channel is available)
              Use WE channel to route mg;
          else
              Route mg counter-clockwise;
exit;
```

@[2007]IEEE

FIGURE 4.16: Pseudo codes of the procedure *Ring-Route*. (From Holsmark, R. and Kumar S. *Journal of Information Science and Engineering*, Vol. 23, pp. 1649–1662. May 2007. With permission.)

the OAPR have better performance than those using the Chen and Chiu's algorithm [5] and the E-XY [34]. The OAPR has two major features, as described as follows:

1. Avoiding routing paths along boundaries of OIPs: in the environment of faulty meshes, we can only know the information of faulty blocks at real time. However, the locations of OIPs are known in advance. Therefore, the OAPR can avoid routing paths along boundaries of OIPs and reduce the traffic loads around OIPs. With these features, the OAPR can achieve more balanced traffic loads in irregular meshes.

2. Supporting f-rings and f-chains for placements of OIPs: the OAPR solves the drawbacks of the E-XY [34] and uses the odd-even turn model [8] to avoid deadlock systematically. However, the E-XY cannot support OIPs placed at boundaries of irregular meshes. In order to solve this problem, the OAPR applies the concepts of f-rings and f-chains [3].

```
Procedure Chain-Route-Modified(mg)
switch   (mg's type)
case     (RF message)
         if (the fault chain is an s-chain)
             if (EW channel is available)
                 Route mg along EW channel;
             else
                 Route mg counter-clockwise;
         else /* not-s-chain */
             if (C is in the same row as D)
                 Route mg along EW channel;
             else
                 if (D is higher than C)
                     Route mg counter-clockwise;
                 else
                     Route mg clockwise;
case     (CF message)
         if (mg is NS message)
             if (the chain is s-chain)
                 if (C (including S) and D is on the west border of the chain)
                     Route mg along NS channel;
                 else
                     Route mg clockwise;
             else /* not s-chain */
                 if (NS channel is available and D is not to the west of C)
                     Route mg along NS channel;
                 else
                     Route mg clockwise;
         else /* SN message */
             if (the chain is an s-chain)
                 if (C is on the north or east boundary of the chain)
                     Normal-Route(mg);
                 else
                     if (C (including S) is on the west boundary of the ring and EW channel is available)
                         Route mg along EW channel;
             else /* not s-chain */
                 if (SN channel is available and D is not to the west of C)
                     Route mg along SN channel;
                 else
                     Route mg counter-clockwise;
                 exit;
case     (RO message)
         if (C is in the same row as D and WE channel is available)
             Use WE channel to route mg;
         else
             Route mg clockwise;                          @[2007]IEEE
             exit;
```

FIGURE 4.17: Pseudo codes of the procedure *Chain-Route Modified*. (From Holsmark, R. and Kumar S. *Journal of Information Science and Engineering*; Vol. 23, pp. 1649–1662. May 2007. With permission.)

With this feature, the OAPR can work correctly if OIPs are placed at the boundaries of the irregular meshes.

The OAPR contains 4 routing modes: 1) default routing, 2) single OIP, 3) multiple OIPs, and 4) f-chain. If no OIP blocks the routing paths in the default routing, packets are routed following default routing. Otherwise, packets are routed following single OIP, multiple OIPs, or f-chain to detour OIPs. The

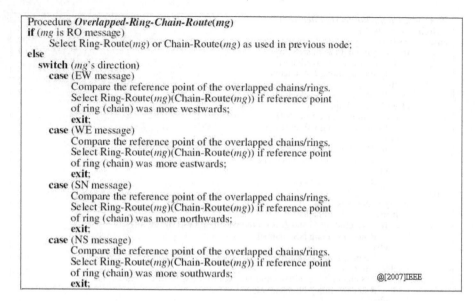

FIGURE 4.18: Pseudo codes of the procedure *Overlapped-Ring Chain Route*. (From Holsmark, R. and Kumar S. *Journal of Information Science and Engineering*, Vol. 23, pp. 1649–1662. May 2007. With permission.)

possible cases in the default routing are shown in Fig. 4.21(a). The symbol E (O) means that packets turn at even (odd) columns. Fig. 4.21(b) shows several routing paths following single OIP ($S1$ to $D1$), multiple OIPs ($S2$ to $D2$), and f-chain ($S3$ to $D3$) to detour OIP. These paths avoid routing paths along boundaries of OIPs and reduce the traffic loads around OIPs. Therefore, these paths alleviate the loads in the boundaries of the OIPs and reduce the network latency. The detailed routings of OAPR are discussed in [21]. Besides, the OAPR contains some restrictions of placements to avoid deadlock, as shown as follows:

1. For an OIP located at $[x_m, x_M, y_m, y_M]$ ($x_m <= x_M$ and $y_m <= y_M$, where $x_m, x_M, y_m,$ and y_M belong to $\{0, 1, ..., n-1\}$ in an $n \times n$ irregular mesh), the routers at range $[x_m - 2, x_M + 2, y_m - 1, y_M + 1]$ can be only linked to normal-sized OIPs. These routers are reserved to satisfy the routings based on the odd-even turn model [8].

2. The routers in the east side of an OIP cannot be connected to normal-sized IPs.

3. All OIPs vertically overlapping must be aligned in the east edge.

4. At most one gap can be greater than 1 at the west boundaries of the irregular meshes.

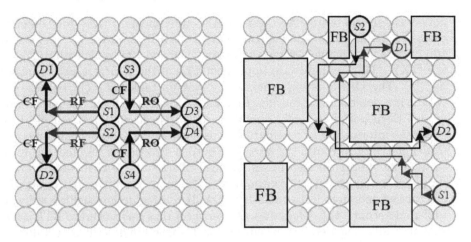

FIGURE 4.19: Examples of Chen and Chiu's routing algorithm: (a) the routing paths (RF, CF, and RO) in *Normal-Route*, and (b) Two examples of *Ring-Route* and *Chain-Route*. (From Chen, K.-H. and Chiu G.-M. *Journal of Information Science and Engineering*, Vol.14, pp.765–783, Dec. 1998. With permission.)

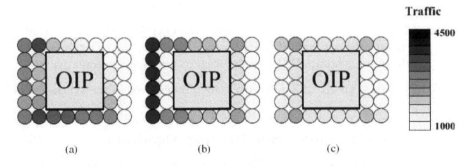

FIGURE 4.20: Traffic loads around the OIPs by using (a) Chen and Chiu's algorithm [5] (unbalanced), (b) the extended X-Y routing algorithm [34] (unbalanced), and (c) the OAPR [21] (balanced). (From Lin, S.-Y. et al. *IEEE Trans Computers*, 57(9), 1156–1168. ©IEEE. With permission.)

Fig. 4.22 shows an example with the rules 1, 2, 3, and 4 described above. Rules 1 and 2 are the same as the E-XY [34] due to the restriction of the routings based on the odd-even turn model. The rule 3 and 4 can prevent deadlock in the networks using the OAPR. Because the designers can control the OIP placements in irregular meshes, it is possible to follow the rules 1 to 4. As long as the rules are followed, the OAPR works correctly and makes the networks perform better. According to the experiments in [21], four different cases are simulated to demonstrate that the OAPR improves 13.3 percent to 100 percent sustainable throughputs than Chen and Chiu's algorithm [5] and the E-XY [34]. The hardware implementation of the OAPR is also discussed in [21]. The OAPR is implemented by look-up tables (*LUTs*) because the OAPR is a deterministic routing. Fig. 4.23(a) shows the basic five-port router model. Each port has a corresponding routing logic and each routing logic keeps the information of destination addresses (*Addr.*) and output directions (*Out*) in LUTs. In routing process, the output direction is selected according to different destination addresses. In Fig. 4.23(b), the OAPR design flow is proposed to implement the routing logic. The input is an irregular mesh with OIP placements from EDA tools. The OAPR routing design tool is a software tool to determine the *Addr.* and *Out* in the LUTs and generate RTL codes automatically. The detailed executions of this flow are described as follows:

1. Software routing function: first, the OAPR routing design tool is executed to determine the *Addr.* and *Out* in the LUTs. Fig. 4.23(c) shows the flowchart to update LUTs. All reachable source-destination pairs are traced in irregular meshes. The path between each source-destination pair is routed once by using the OAPR. In each router, the routing information is recorded by the LUTs in each router if packets pass through. After this phase, all LUTs are obtained. According to the *Addr.* and *Out* in the LUTs, the packets can be routed following the OAPR.

2. LUT coding in Verilog: in this phase, the OAPR routing design tool can generate the synthesizable RTL code. The *Addr.* and *Out* in step 1 are utilized to generate RTL code of each routing logic automatically.

3. Synthesis: finally, the RTL codes are handed over to the synthesis tool.

4.4.2   Application-Specific Routing Algorithm

An NoC system is often specialized for a specific application or for a set of concurrent applications [20] [32]. In [23], the design methodology of an application specific routing algorithm, called application specific routing algorithm (*APSRA*), was proposed. APSRA extends Duato's theory [11] to design deadlock-free adaptive routing algorithms for irregular meshes. APSRA assumes that the communication among tasks in a specific application is known in advance. The information of the communication can be useful for designing deadlock-free algorithms which are more adaptive in comparison with a general algorithm. Fig. 4.24 shows the overview of the APSRA design methodology. APSRA contains three different inputs: 1) the communication graph (*CG*), 2) the topology graph (*TG*), and 3) the mapping function (*M*). In addition, the concurrency information after the task scheduling can also be considered. The output of the APSRA algorithm is the routing table for each node of TG. An application specific channel dependency graph (*ASCDG*)

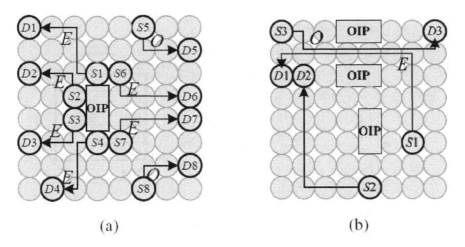

(a)   (b)

FIGURE 4.21: The OAPR: (a) eight default routing cases and (b) some cases to detour OIPs. (From Lin, S.-Y. et al. *IEEE Trans Computers*, 57(9), 1156–1168. ©IEEE. With permission.)

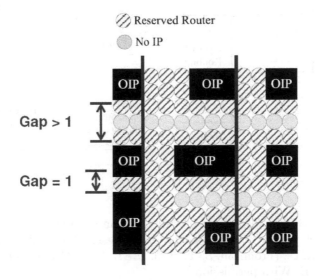

FIGURE 4.22: Restrictions on OIP placements for the OAPR. (From Lin, S.-Y. et al. *IEEE Trans Computers*, 57(9), 1156–1168. ©IEEE. With permission.)

can be built by the actual communication pairs from CG, TG, and M. The ASCDG is a subgraph of the channel dependence graph (*CDG*). In order to

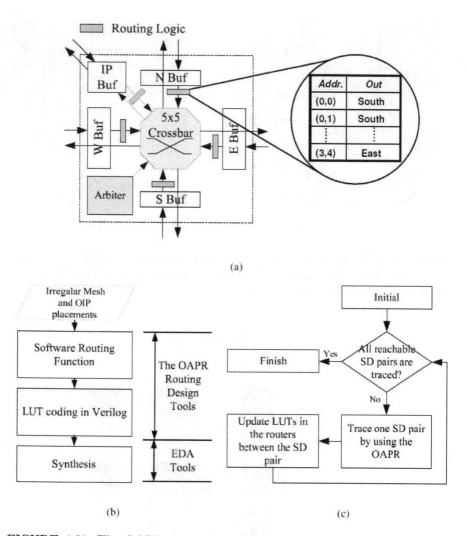

FIGURE 4.23: The OAPR design flow: (a) the routing logic in the five-port router model, (b) the flowchart of the OAPR design flow, and (c) the flowchart to update LUTs. (From Lin, S.-Y. et al. *IEEE Trans Computers*, 57(9), 1156–1168. ©IEEE. With permission.)

guarantee the routing is deadlock-free, the ASCDG must be acyclic. If the ASCDG is not acyclic, ASCDG must follow a heuristic algorithm to break all the cycles. In [23], a heuristic algorithm was proposed. The algorithm can break the cycles and minimize the impact of routing adaptiveness with the constraints to guarantee the reachability of all destination nodes. If ASCDG is acyclic, APSRA extract the routing tables for each node of the TG and

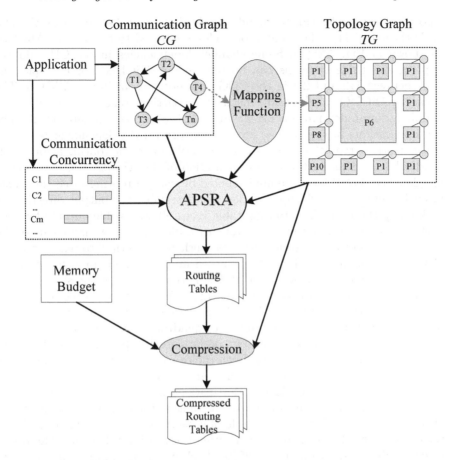

FIGURE 4.24: Overview of APSRA design methodology. (From Palesi, M. et al. *Proceedings International Conference on Hardware-Software Codesign and System Synthesis*; pp. 142–147. Oct. 2007. ©IEEE. With permission.)

stop the algorithm. In addition, a compression technique can be applied to reduce the sizes of routing tables, which are discussed in [24]. Fig. 4.25 shows an example of APSRA. The CG and the TG are depicted in Fig. 4.25(a) and Fig. 4.25(b) respectively. In this example, the TG is assumed to be a 2D mesh. This method can be applied to any network topology without modifications. The mapping function M is assumed as Eq. 4.1:

$$M(Ti) = Pi \qquad (4.1)$$

Ti and Pi denote the node i in the CG, and the node i in the TG, respectively. Fig. 4.25(c) shows the CDG for a minimal fully adaptive routing algorithm. Six cycles can be found and deadlock may be caused by the Duato's theorem [11]. The number of cycles is reduced to two for the ASCDG, as shown in Fig. 4.25(d). Some channel dependencies in the CDG do not appear, and these channels can be removed in ASCDG. For example, the edge between I12 and I23 in the CDG does not present in the ASCDG. The cycles in ASCDG can be also broken by restricting some routing paths and defining communication concurrency. Fig. 4.25(e) shows a possible result. The communications in Fig. 4.25(e) are not concurrent; the dependencies are not concurrently active. Hence, the cycles are broken and the routing algorithm is deadlock-free. The hardware implementation of the APSRA is also discussed in [23]. The APSRA can be implemented by using a routing table embedded in each router and each input packet can determine the output direction by looking up the table. The routing table keeps all admissible outputs for each destination address. However, the routing table occupies a major part of the router area. In order to reduce the area overhead, a method to reduce the size of routing table for APSRA was proposed in [24]. The approach is to store admissible output ports for a set of destinations. Because the shortest path routing is considered, the output port cannot be the same as the input port. For instance, if the router receives the packets from its west input port, the destination will be in the first and forth quadrant. Five possible choices can be selected for the admissible output ports: {north}, {south}, {east}, {north and east}, and {south and east}. Each choice can be represented in one color (e.g. north = red, east = blue, south = green, north and east = purple, and south and east = yellow). Hence, destinations are grouped according to the colors. Fig. 4.26 shows an example of the routing table in the west input port of node X. The original routing table of node X is shown as Fig. 4.26(a). After coloring the destination and clustering the routing table, the compressed routing table is shown as Fig. 4.26(b). Each grouping region is restricted to rectangular shape. In this method, no more information needs to be kept for the set of the regions. Besides, the aforementioned method can further reduce the size of the routing tables by restricting the routing adaptiveness. For instance, A and B can be merged to a new region $R3$ by removing the admissible output north of A, as shown in Fig. 4.27(a). Besides, the region can be also merged. By restricting the admissible outputs of $R1$ from {south, east} to {east}, the regions $R1$ and $R3$ can be merged and the routing table can be further reduced, as shown in Fig. 4.27(b).

4.5   Placement for Irregular Mesh Topology

The problem of OIP placements (the methods to place OIP on irregular meshes) is also important because different placements can result in different network performances. According to the analyses of OIP placements, designers

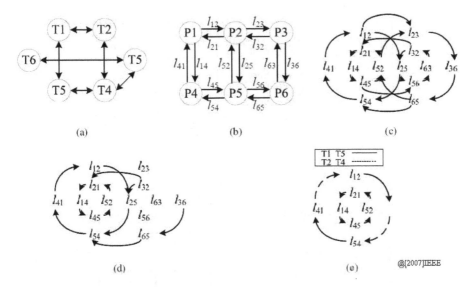

FIGURE 4.25: An example of APSRA methodology: (a) *CG*, (b) *TG*, (c) *CDG*, (d) *ASCDG*, and (e) the concurrency of the two loops. (From Holsmark, R., Palesi, M., and Kumar, S. *Proceedings of the 9th EUROMICRO Conference on Digital System Design*; PP. 696–703. ©IEEE. With permission.)

can determine how to place OIPs and achieve better network performance on irregular mesh-based NoCs. The best choice of OIP placements is extremely dependent on the routing algorithms. In [17], OIP placements based on Chen and Chiu's algorithm [5] were analyzed. In [20], OIP placements based on OAPR [21] were discussed. In this section, the OIP placements based on Chen and Chiu's algorithm [5] and OAPR [21] are introduced in Sections 4.5.1 and 4.5.2, respectively.

4.5.1 OIP Placements Based on Chen and Chiu's Algorithm

This section discusses the OIP placements based on Chen and Chiu's algorithm [5]. In [17], Holsmark and Kumar developed a simulation model using Telelogic's SDL (Specification and Description Language) tool to evaluate the effect of NoC performance for different OIP placement. According to the simulation model, three different experiments of OIP placements are discussed: 1) OIP placements with different sizes, 2) OIP placements with different locations, and. 3) OIP placements with different orientations. These cases are discussed as follows:

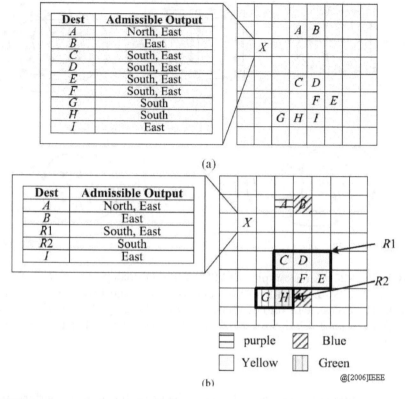

(a)

(b)

purple	Blue
Yellow	Green

@[2006]IEEE

FIGURE 4.26: An example of the routing table in the west input port of node *X*: (a) original routing table and (b) compressed routing table. (From Palesi, M., Kumar, S., Holsmark, R.; *SAMOS VI: Embedded Computer Systems: Architectures, Modeling, and Simulation*; pp. 373–384. July 2006. ©IEEE. With permission.)

1. OIP placements with different sizes: five different cases are considered: 1) region (2,2;6,6), 2) region (3,3;6,6), 3) region (3,3;5,5), 4) non-blocking region (3,3;5,5), and 5) no region. OIP placements with different sizes and locations are shown in Fig. 4.28. Non-blocking region represents that the routers are active but the source and destination nodes are inactive. The "no region" stands for a network without OIP placements. Fig. 4.29 shows the average latency in different cases. The latencies of no region and non-blocking region (3,3;5,5) are almost the same. These cases also perform lower latencies than the cases of regions (2,2;6,6), (3,3;6,6) and (3,3;5,5). The reason is because packets must take longer distances to pass around the region. Considering the cases of regions

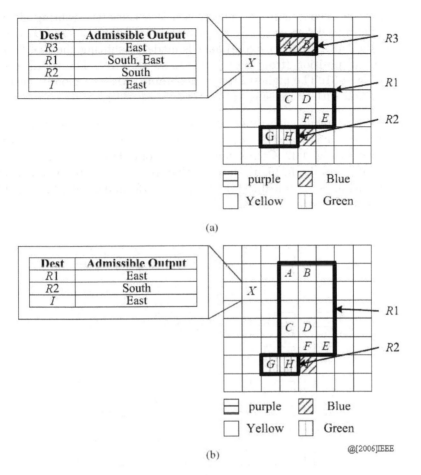

FIGURE 4.27: An example of the compressed routing table in node X with loss of adaptivity: (a) the routing table by merging destinations A and B and (b) the routing table by merging regions $R1$ and $R3$. (From Palesi, M., Kumar, S., Holsmark, R.; *SAMOS VI: Embedded Computer Systems: Architectures, Modeling, and Simulation*; pp. 373–384. July 2006. ©IEEE. With permission.)

(2,2;6,6), (3,3;6,6) and (3,3;5,5), the latency and its sensitivity to load increase with region size.

2. OIP placement with different locations: First, OIP placements with northeast corner at row three and column zero. Then the OIP is moved from west side of the NoC to the east side of the NoC. Fig. 4.30 shows the result with load from 5 percent up to 25 percent. The result shows that the best region is the west-most region. It blocks fewer routers, and the latency is lowest. The worst position is when northeast corner

is in the second column. The reason is because the routing algorithm to detour OIP causes more congestion toward the west and centre parts of the NoC. Another experiment is made by shifting the 2 × 2 OIP in vertical position (from north edge to south edge). The result is shown in Fig. 4.31. The highest latency is obtained with a region in the central position with decreasing values toward the edges.

3. OIP placement with different orientations: two nonquadratic cases, 1) region (2,3;6,5) and 2) region (3,2;5,6), are compared. The results are shown in Fig. 4.32. Comparing the cases of different orientations, region (2,3:6,5) results in higher latency than region (3,2:5,6). The bias of the algorithm is responsible for the poor performance of (3,2;5,6). Other comparisons are discussed in [22].

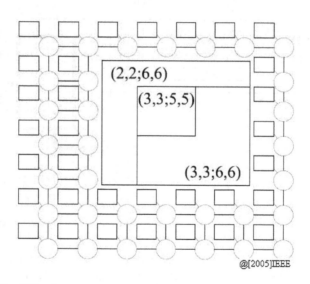

FIGURE 4.28: OIP placement with different sizes and locations. (From Holsmark, R. and Kumar, S.; *Design Issues and Performance Evaluation of Mesh NoC with Regions*; NORCHIP; pp. 40–43, Nov. 2005. ©IEEE. With permission.)

4.5.2　OIP Placements Based on OAPR

In Section 4.5.2, the OIP placement rules based on OAPR [21] will be discussed. In [20], Lin used 2D distribution graphs to show the latencies of an OIP placed at different positions and orientations. Each grid stands for the latency of one OIP placement. Fig. 4.33 shows an example of a 12 × 12 distribution graph. The grids with symbols NE, NW, SE, and SW stand for the

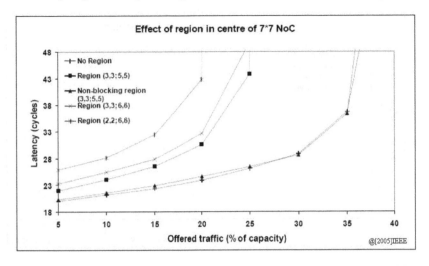

FIGURE 4.29: Effect on latency with central region in NoC. (From Holsmark, R. and Kumar, S.; *Design Issues and Performance Evaluation of Mesh NoC with Regions*; NORCHIP; pp. 40–43, Nov. 2005. ©IEEE. With permission.)

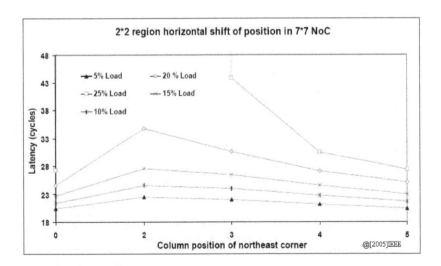

FIGURE 4.30: Latency for horizontal shift of positions. (From Holsmark, R. and Kumar, S.; *Design Issues and Performance Evaluation of Mesh NoC with Regions*; NORCHIP; pp. 40–43, Nov. 2005. ©IEEE. With permission.)

OIP placed at corners of the mesh; the grids with symbols N, E, S, and W represent the OIP placed at the boundaries of the mesh; the grids with oblique lines represent placement restrictions of OAPR [21], which are described in

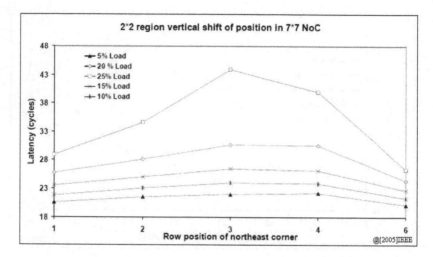

FIGURE 4.31: Latency for vertical shift of positions. (From Holsmark, R. and Kumar, S.; *Design Issues and Performance Evaluation of Mesh NoC with Regions*; NORCHIP; pp. 40–43, Nov. 2005. ©IEEE. With permission.)

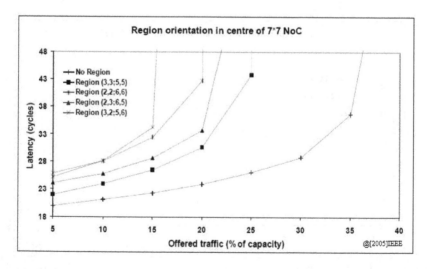

FIGURE 4.32: OIP placements with different orientations. (From Holsmark, R. and Kumar, S.; *Design Issues and Performance Evaluation of Mesh NoC with Regions*; NORCHIP; pp. 40–43, Nov. 2005. ©IEEE. With permission.)

Section 4.5. Each coordinate shows the latency of one OIP placement. For instance, coordinate (5,5) represents the latency of a 3 × 3 OIP placed at

[5,7,5,7]. In [20], two cases of placements are considered: 1) OIP placed at different positions and 2) OIP placed at different orientations, as follows:

1. OIP placed at different positions: one 3×3 OIP is placed on a 12×12 mesh to evaluate how OIP position affects system performance. The experimental result is illustrated in Fig. 4.34. In this experiment, injection rate is fixed to 0.04 flits/IP/cycle and 50,000 packets are collected under a uniform random traffic pattern. The results show that placing the OIP at the corners of the mesh (NW, SW, NE, and SE) results in lowest network latency. Besides, placing the OIP at the boundaries of the mesh (N, S, E, and W) results in lower network latency than placing the OIP in the center of the mesh.

2. OIP placed at different orientations: a four-unit rectangle OIP placed on a 12×12 mesh is evaluated. Two different orientations can the OIP be placed in: 1) vertical placements (1×4 OIP) and 2) horizontal placements (4×1 OIP). In this experiment, injection rate is fixed to 0.04 flits/IP/cycle and 50000 packets are collected under a uniform random traffic pattern. The results are shown in Fig. 4.35(a) and (b), respectively. Comparing OIP placements at different positions, the trends in Fig. 4.35(a) are similar to those in Fig. 4.34. Placing the OIP at the corners can achieve lowest network latency; placing the OIP at the boundaries still results in lower network latency than placing the OIP in the center of the mesh. In Fig. 4.35(b), the situations are similar to Fig. 4.35(a) except placing OIP at the north and south boundaries. These cases result in highest network latency. It means that horizontal placements at the north and south boundaries are not good choices. Comparing OIP placements at different orientations in Fig. 4.35(a) and (b), if the OIP is placed at the corners, the latencies are almost the same. If the OIP is placed in the centers of the mesh, horizontal placements results in lower network latency than vertical placements. If the OIP is placed at boundaries of the mesh, horizontal placements at the north or south boundaries and vertical placements at the east or west boundaries result in lower network latencies. According to the aforementioned results, some placement rules for irregular mesh-based NoCs can be defined. These rules are summarized in Table 4.1. Lin also demonstrated that the placement rules can be extended for the cases of multiple OIPs by some experimental results, which are discussed in [20].

4.6   Hardware Efficient Routing Algorithms

Hardware implementation of the routing algorithm is an important issue for NoC designs. The routing algorithm can be classified into two categories: 1)

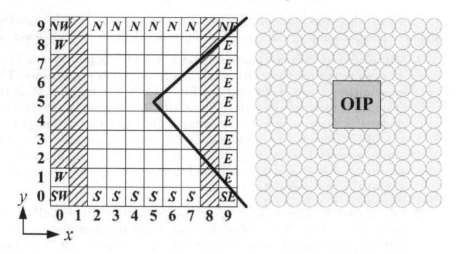

FIGURE 4.33: An example of a 12 × 12 distribution graph. (From Lin, S.-Y.; *Routing Algorithms and Architectures for Mesh-Based On-Chip Networks with Adjustable Topology*; Ph.D. dissertation, Dept. of Electrical Engineering, National Taiwan University. 2009. With permission.)

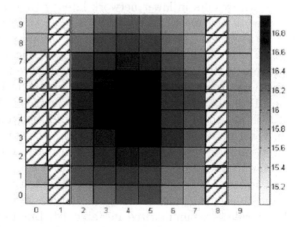

FIGURE 4.34: Latencies of one 3 × 3 OIP placed on a 12 × 12 mesh. (From Lin, S.-Y.; *Routing Algorithms and Architectures for Mesh-Based On-Chip Networks with Adjustable Topology*; Ph.D. dissertation, Dept. of Electrical Engineering, National Taiwan University. 2009. With permission.)

distributed routing and 2) source routing. In distributed routing, each router must embed the routing algorithm whose input is the destination address of

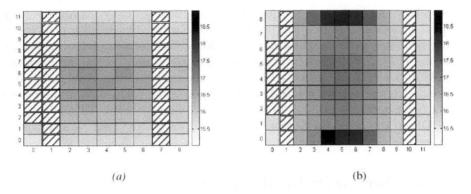

(a)   *(b)*

FIGURE 4.35: Latencies of one four-unit OIP placed on a 12 × 12 mesh: (a) horizontal placements and (b) vertical placements. (From Lin, S.-Y.; *Routing Algorithms and Architectures for Mesh-Based On-Chip Networks with Adjustable Topology*; Ph.D. dissertation, Dept. of Electrical Engineering, National Taiwan University. 2009. With permission.)

TABLE 4.1: Rules for Positions and Orientations of OIPs

Categories	Priority of OIP placements
Position	Corners > boundaries > centers
Orientation (corners)	Horizontal = Vertical
Orientation (boundaries)	Horizontal > Vertical, for north and south boundaries
Orientation (boundaries)	Horizontal < Vertical, for east and west boundaries
Orientation (centers)	Horizontal > Vertical

(From Lin, S.-Y.; *Routing Algorithms and Architectures for Mesh-Based On-Chip Networks with Adjustable Topology*; Ph.D. dissertation, Dept. of Electrical Engineering, National Taiwan University. 2009. With permission.)

the packet and its output is the routing decision. When the packet arrives at the input port of the router, the routing decisions are made either by searching the routing table or by executing the routing function in hardware. In source routing, the predefined routing tables are stored in the network interface of the IP module. When a packet is transmitted from the IP module, it searches the routing information in the routing table and keeps the information in the header of the packet. Hence, the packet follows the routing information to make the routing decision in each hop. Both distributed routing and source routing can be implemented by routing tables. For irregular meshes, routing tables are often used to accomplish the routing algorithms. However, the number of entries in the routing table is equal to the number of nodes in the network. Many efficient implementations use reduced ROM or Boolean logics to achieve an equivalent routing algorithm. Besides, many researchers focus on the hardware efficient routing algorithms for irregular mesh topologies. In [2], two low-cost distributed routing and one low-cost source routing [turns-tables (TT), XY-deviation tables $(XYDT)$, and source routing for deviation points $(SRDP)$] are proposed to reduce the size of routing tables. In addition, a degree priority routing algorithm was also proposed to minimize the hardware overhead in [33]. In Sections 4.6.1 through 4.6.3, TT, XYDT, and SRDP are discussed. In Section 4.6.4, the degree priority routing algorithm is introduced.

4.6.1 Turns-Table Routing (TT)

In TT routing, routing tables keep the information if there is a turn passing through this router toward the destination. Fig. 4.36 shows a simple example. In Fig. 4.36(a), the path from A to D does not make any turns and no routing information needs to be stored in the routing table. In Fig. 4.36(b), the routing tables must keep the information of path B to D and C to D because these paths turn in this router. If a packet arrives at the router, the router searches the routing table according to the destination address. If the entry exists, the routing decision can be made. Otherwise, the packet goes forward without turning. This scheme can reduce the size and the power of routing table in comparison with a full routing table. In [2], a searching algorithm for TT routing was also proposed to minimize the sizes of routing tables. The searching algorithm is described in Fig. 4.37. The algorithm is executed for each destination node. This algorithm uses a greedy approach to select a source node iteratively. The source node is selected if the shortest path from the source node to the destination node or to an already selected path adds minimal number of entries in the routing table.

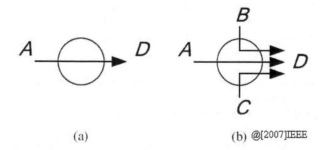

(a) (b) ©[2007]IEEE

FIGURE 4.36: (a) Routing paths without turning to destination D and (b) Routing paths with two turns to D. (From Bolotin, E., Cidon, I., Ginosar, R., and Kolodny, A.; *Routing Table Minimization for Irregular Mesh NoCs*, DATE 2007; pp. 942–947, ©IEEE. With permission.)

```
construct a Turns - Graph TG
∀v∈ V: Dist(v) = ∞, Paved(v) = False, P(v) = nil
   Paved (D) = True; Dist(D) = 0
   while (!(∀s∈ Sources: Paved(s) = true))
      Relax_not_paved (D,TG)
      Pick S_min (∀s', s_min ∈ Sources): /* Heuristic */
   Dist(s_min)<Dist(s') ∩ Paved(s_min) = Paved(s') = false
      Pave_Path(S_min,D)
      foreach node v' on Path:
         Paved(v')= True;
         Distance(v')= hop_num * N;
      end foreach
   end while                              @[2007]IEEE
```

FIGURE 4.37: TT routing algorithm for one destination D. (From Bolotin, E., Cidon, I., Ginosar, R., and Kolodny, A.; *Routing Table Minimization for Irregular Mesh NoCs*, DATE 2007; pp. 942–947, ©IEEE. With permission.)

4.6.2   XY-Deviation Table Routing (XYDT)

In XYDT routing, each entry of routing tables is stored if the routing decision of next hop deviates the XY routing. If a packet arrives at the router, the router searches the routing table according to the destination address. If the entry is found, the packet makes the routing decision following the routing table. Otherwise, the packet is forwarded by XY routing logic. The XY routing logic is a hardware function embedded in the router. Selecting the path of minimal deviations can achieve minimal number of the entries in the routing tables. The searching algorithm for XYDT routing was also described in [2]. Fig. 4.38 shows the searching algorithm. The algorithm is performed for each destination node. For each destination, the routing paths from all source nodes are traced. Among all shortest paths between each source-destination pair, the search algorithm selects a path that makes a minimal number of routing steps that deviate from the XY routing.

4.6.3   Source Routing for Deviation Points (SRDP)

SRDP is a method to reduce the size of the headers in source routing. SRDP combines a fixed routing function and a partial list of SRDP tags. The SRDP

$\forall v \in V$: $Dist(v) = \infty$, $P(v) = nil$; $Dist(D) = 0$

$R_h = \{D\}$, $R_h + 1 = \{\}$, $h = 0$;

while $(!(\forall v \in V: Dist(v) < \infty))$

    foreach node $v_h \in R_h$:

        set_xy_Predecessor(v_h)

        foreach v' *in 1 hop from* v_h :

            if $Dist(v') = \infty$: $R_h \leftarrow \{v'\}$, $Dist(v') = h+1$

        end if

        end foreach

    end foreach

    $R_h = R_{h+1}$, $R_{h+1} = \{\}$, $h=h+1$

end while   @[2007]IEEE

FIGURE 4.38: XYDT routing algorithm for one destination D. (From Bolotin, E., Cidon, I., Ginosar, R., and Kolodny, A.; *Routing Table Minimization for Irregular Mesh NoCs*, DATE 2007; pp. 942–947, ©IEEE. With permission.)

tags keep the routing commands for the traversed nodes between the source node and the destination node. If the routing decision of a traversed node deviates from the fixed routing function (in [2], XY routing is an example), SRDP must keep the SRDP tag in the header of the packet. Otherwise, the routing decision follows the fixed routing function. Hence, the headers of the packets do not keep the SRDP tags for the traversed nodes which do not deviate from the fixed routing function. The selections of the routing paths for SRDP also influence the size of the SRDP tags. The searching algorithm to find minimal deviations from XY routing was also discussed in [2]. All of TT, XYDT, SRDP algorithms can reduce the size of routing tables. In [2], the simulations had demonstrated that these algorithms can achieve 2.9 times to 40 times of the cost reduction of the original source routing and distribution routing. However, deadlock avoidance problems are not considered in these algorithms.

4.6.4 Degree Priority Routing Algorithm

In [33], a degree priority routing algorithm was proposed for irregular mesh topologies. The routing paths are dynamically selected according to the status of the node in the next hop. If the routing decision of the degree priority routing algorithm is different from XY routing, the routing entry must be kept in the routing table. Besides, the entries in routing tables containing the same contents can be combined to further reduce the size of the routing table. Fig. 4.39 shows the degree priority routing algorithm. The optimal path is defined as the path following XY or YX routing path. The output channel is defined as the neighbor node of current node to forward the packet. The degree is defined as the number of output channels of the node. Some examples are shown in Fig. 4.40. The output channels of A, B, C, and D are $\{A_N, A_E, A_S, A_W\}$, $\{B_E, B_S, B_W\}$, $\{C_N, C_E\}$, and $\{D_N\}$. The degrees of A, B, C, and D are 4, 3, 2, and 1. A simple example of the degree priority routing algorithm is shown in Fig. 4.41. The selected routing path from X to A is $\{X \rightarrow 1 \rightarrow 2 \rightarrow 3 \rightarrow 4 \rightarrow 5 \rightarrow 6 \rightarrow A\}$. In the general case, routing tables in node X, 1, 2, 3, 4, 5, 6 must construct the entry for the node A. In order to reduce the routing table, the XYDT routing [2] can be applied. The XYDT is introduced in Section 4.6.2. Besides, the destination nodes with the same next hops can be combined in the routing table. Fig. 4.42 shows the routing tables of a simple case. The source node is X, and destination nodes are A, B, C, D, E, F, G, H, and I in Fig. 4.41. Only the routing tables in the nodes 1, 6, 10, C, and X are kept. However, deadlock problems are not considered in this work. In order to avoid the deadlock problem, virtual channels must be supported.

```
For current node: {
        if   (there are two output channels X and Y on optimal paths)
             if   (the degree of node for X is greater or equal to Y)
                    X is selected as the next node
             else
                    Y is selected as the next node
        else
        if   (one output channel has optimal routing path)
                The channel is selected.
        else
        if   (there is no output channel on optimal path)
                One channel is selected randomly at the next node.
        else
        if   (no output channel exists)
                Go back to the last node.
                                                    @[2008]IEEE
}
```

FIGURE 4.39: Degree priority routing algorithm. (From Bolotin, E., Cidon, I., Ginosar, R., and Kolodny, A.; *Routing Table Minimization for Irregular Mesh NoCs*, DATE 2007; pp. 942–947, ©IEEE. With permission.)

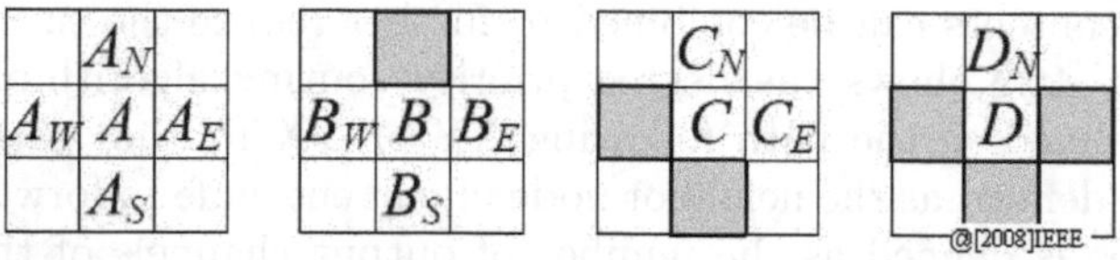

FIGURE 4.40: Examples showing the degrees of the nodes A, B, C, and D. (From Ling Wang, Hui Song, Dongxin Wen, and Yingtao Jiang. *International Conference on Embedded Software and Systems (ICESS '08)*; pp. 293–297, July 2008. ©IEEE. With permission.)

FIGURE 4.41: An example of the degree priority routing algorithm. (From Ling Wang, Hui Song, Dongxin Wen, and Yingtao Jiang. *International Conference on Embedded Software and Systems (ICESS '08)*; pp. 293–297, July 2008. ©IEEE. With permission.)

node 1

Dest	Admissible output
AB	North

node 6

Dest	Admissible output
B	South

node 10

Dest	Admissible output
CDEFHI	South

node C

Dest	Admissible output
EI	South

node X

Dest	Admissible output
AB	North
CDEFGHI	South

FIGURE 4.42: Routing tables of nodes 1, 6, 10, C, and X. (From Ling Wang, Hui Song, Dongxin Wen, and Yingtao Jiang. *International Conference on Embedded Software and Systems (ICESS '08)*; pp. 293–297, July 2008. ©IEEE. With permission.)

4.7 Conclusions

In this chapter, the concept of the irregular mesh topologies was introduced. For irregular mesh topologies, many design issues must be considered, such as the numbers of the access points, the routing problems, and the placements of the OIPs. This chapter introduced many algorithms to solve the routing problems for irregular meshes. Besides, the placements of the OIPs with different sizes, locations, and orientations were also discussed. According to these analyses, the designer can solve the communication problems and achieve better network performance to integrate many hard IPs of different sizes from various vendors in regular 2D mesh-based NoC designs.

Review Questions

[Q_1] What are the differences between the irregular mesh topology and the 2D mesh topology?

[Q_2] List the design concepts for the irregular mesh topologies.

[Q_3] Compare the fault-tolerant routings using virtual channels and the fault-tolerant routings using turn models.

[Q_4] Compare the differences between Extended X-Y routing, Chen and Chiu's routing, and the OAPR.

[Q_5] What are the restrictions of the placements in OAPR?

Bibliography

[1] E. Bolotin, I. Cidon, R. Ginosar, and A. Kolodny. QNoC: QoS architecture and design process for network on chip. *Journal of Systems Architecture*, pages 105–128, Feb 2004.

[2] E. Bolotin, I. Cidon, R. Ginosar, and A. Kolodny. Routing table minimization for irregular mesh NoCs. *Proceedings of the Conference on Design, Automation and Test in Europe*, pages 942–947, Apr 2007.

[3] R. V. Boppana and S. Chalasani. Fault-tolerant wormhole routing algorithms for mesh networks. *IEEE Transactions on Computers*, 44:848–864, Jul 1995.

[4] Y. M. Boura and C. R. Das. Fault-tolerant routing in mesh networks. *Proceedings of 1995 International Conference on Parallel Processing*, pages I.106–I.109, Aug 1995.

[5] K-H. Chen and G-M. Chiu. Fault-tolerant routing algorithm for meshes without using virtual channels. *Journal of Information Science and Engineering*, 14:765–783, Dec 1998.

[6] A. A. Chien. A cost and speed model for k-ary n-cube wormhole router. *Proceedings of Hot Interconnects 93*, Aug 1993.

[7] A. A. Chien and J. H. Kim. Planar-adaptive routing: low-cost adaptive networks for multiprocessors. *Journal of the ACM*, 42:91–123, Jan 1995.

[8] G.M. Chiu. The odd-even turn model for adaptive routing. *IEEE Trans. Parallel and Distributed Systems*, 11:729–737, July 2000.

[9] W. J. Dally and B. Towles. Route packets, not wires: On-chip interconnection networks. *Proceedings of the Design Automation Conference*, pages 684–689, June 2001.

[10] C. Duan, A. Tirumala, and S.P. Khatri. Analysis and avoidance of crosstalk in on-chip buses. *IEEE Symp. High-Performance Interconnects*, pages 133–138, Aug 2001.

[11] J. Duato. A new theory of deadlock-free adaptive routing in wormhole networks. *IEEE Transactions on Parallel and Distributed Systems*, 4:1320–1331, Dec 1993.

[12] S.A. Felperin, L. Gravano, G.D. Pifarre, and J.L. Sanz. Routing techniques for massively parallel communication. *Proceedings of IEEE*, 79:488–503, Apr 1991.

[13] C. J. Glass and L. M. Ni. The turn model for adaptive routing. *Journal of ACM*, 41:874–902, Sept 1994.

[14] C.J. Glass and L.M. Ni. Fault-tolerant wormhole routing in meshes. *23rd Ann. Intl. Symp. Fault-Tolerant Computing*, pages 240–249, Jun 1993.

[15] R. Ho, K.W. Mai, and M.A. Horowitz. The future of wires. *Proc. IEEE*, 89:490–504, Apr 2001.

[16] T. Hollstein, R. Ludewig, C. Mager, P. Zipf, and M. Glesner. A hierarchical generic approach for onchip communication, testing and debugging of SoCs. *Proc. of the VLSI-SoC 2003*, pages 44–49, Dec 2003.

[17] R. Holsmark and S. Kumar. Design issues and performance evaluation of mesh noc with regions. *NORCHIP*, pages 40–43, Nov 2005.

[18] R. Holsmark and S. Kumar. Corrections to Chen and Chiu's fault tolerant routing algorithm for mesh networks. *Journal of Information Science and Engineering*, 23:1649–1662, May 2007.

[19] Intl technology roadmap for semiconductors, 2008. http://public.itrs.net.

[20] Shu-Yen Lin. *Routing Algorithms and Architectures for Mesh-Based On-Chip Networks with Adjustable Topology*. Ph.D. dissertation, Department of Electrical Engineering, National Taiwan University, 2009.

[21] Shu-Yen Lin, Chun-Hsiang Huang, Chih hao Chao, Keng-Hsien Huang, and An-Yeu Wu. Traffic-balanced routing algorithm for irregular mesh-based on-chip networks. *IEEE Trans. Computers*, 57:1156–1168, Sept 2008.

[22] D. H. Linder and J. C. Harden. An adaptive and fault-tolerant wormhole routing strategies for k-ary n-cubes. *IEEE Transactions on Computers*, 40:2–12, Jan 1991.

[23] M. Palesi, R. Holsmark, S. Kumar, and V. Catania. A methodology for design of application-specific deadlock-free routing algorithms for NoC systems. *Proceedings of the 4th International Conference on Hardware/ Software Codesign and System Synthesis*, pages 142–147, Oct 2007.

[24] M. Palesi, S. Kumar, and R. Holsmark. A method for router table compression for application-specific routing in mesh topology NoC architectures. *SAMOS VI: Embedded Computer Systems: Architectures, Modeling, and Simulation*, pages 373–384, July 2006.

[25] L.-S. Peh and W. J. Dally. A delay model for router microarchitectures. *IEEE Micro*, 21:26–34, Jan/Feb 2001.

[26] M.K.F Schafer, T. Hollstein, H. Zimmer, and M. Glesner. Deadlock-free routing and component placement for irregular mesh-based networks-on-chip. *Proc. of ICCAD 2005*, pages 238–245, Nov 2005.

[27] S.R. Sridhara and N.R. Shanbhag. Coding for system-on-chip networks: A unified framework. *IEEE Trans. Very Large Scale Integration (VLSI) Systems*, pages 655–667, June 2005.

[28] Krishnan Srinivasan, Karam S. Chatha, and Goran Konjevod. Application specific network-on-chip design with guaranteed quality approximation algorithms. *Proceedings of the 12th Conference on Asia South Pacific Design Automation*, pages 184–190, Jan 2007.

[29] D. Sylvester and K. Keutzer. A global wiring paradigm for deep submicron design. *IEEE Trans. CAD of Integrated Circuits and Systems*, 19:240–252, Feb 2000.

[30] J.A. Davis *et al.*. Interconnect limits on gigascale integration (gsi) in the 21st century. *Proc. IEEE*, 89:305–324, Mar 2001.

[31] S. Kumar *et al.*. A network on chip architecture and design methodology. *Proc. Intl. Symp. Very Large Scale Integration*, pages 105–112, Apr 2002.

[32] S. Murali *et al.*. Designing application-specific networks on chips with floorplan information. *Proceedings of the 2006 IEEE/ACM International Conference on Computer-Aided Design (ICCAD'06)*, pages 355–362, Nov 2006.

[33] Ling Wang, Hui Song, Dongxin Wen, and Yingtao Jiang. A degree priority routing algorithm for irregular mesh topology NoCs. *International Conference on Embedded Software and Systems (ICESS '08)*, pages 293–297, July 2008.

[34] Jie Wu. A fault-tolerant and deadlock-free routing protocol in 2d meshes based on odd-even turn model. *IEEE Trans. Computers*, 52:1154–1169, Sept 2003.

5

Debugging Multi-Core Systems-on-Chip

Bart Vermeulen

Distributed System Architectures Group
Advanced Applications Lab / Central R&D
NXP Semiconductors
Eindhoven, The Netherlands
bart.vermeulen@nxp.com

Kees Goossens

Electronic Systems Group
Electrical Engineering Faculty
Eindhoven University of Technology
Eindhoven, The Netherlands
k.g.w.goossens@tue.nl

CONTENTS

5.1	Introduction		156
5.2	Why Debugging Is Difficult		158
	5.2.1	Limited Internal Observability	158
	5.2.2	Asynchronicity and Consistent Global States	159
	5.2.3	Non-Determinism and Multiple Traces	161
5.3	Debugging an SoC		163
	5.3.1	Errors	164
	5.3.2	Example Erroneous System	165
	5.3.3	Debug Process	166
5.4	Debug Methods		169
	5.4.1	Properties	169
	5.4.2	Comparing Existing Debug Methods	171
		5.4.2.1 Latch Divergence Analysis	172
		5.4.2.2 Deterministic (Re)play	172
		5.4.2.3 Use of Abstraction for Debug	173
5.5	CSAR Debug Approach		174
	5.5.1	Communication-Centric Debug	175

		5.5.2	Scan-Based Debug	175
5.5.3	Run/Stop-Based Debug	176
5.5.4	Abstraction-Based Debug	176
5.6	On-Chip Debug Infrastructure	178
	5.6.1	Overview	178
	5.6.2	Monitors	178
	5.6.3	Computation-Specific Instrument	180
	5.6.4	Protocol-Specific Instrument	181
	5.6.5	Event Distribution Interconnect	182
	5.6.6	Debug Control Interconnect	183
	5.6.7	Debug Data Interconnect	183
5.7	Off-Chip Debug Infrastructure	184
	5.7.1	Overview	184
	5.7.2	Abstractions Used by Debugger Software	184
		5.7.2.1	Structural Abstraction	184
		5.7.2.2	Data Abstraction	187
		5.7.2.3	Behavioral Abstraction	188
		5.7.2.4	Temporal Abstraction	189
5.8	Debug Example	190
5.9	Conclusions	193
Review Questions	194
Bibliography	194

5.1 Introduction

Over the past decades the number of transistors that can be integrated on a single silicon die has continued to grow according to Moore's law [5]. Higher customer expectations, with respect to the functionality that is offered by a single mobile or home appliance, have led to an exponential increase in system complexity. However, the expected life cycle of these appliances has decreased significantly as well. These trends put pressure on design teams to reduce the time from first concept to market release for these products, the so-called *time-to-market*.

To quickly design a complex system on chip (SoC), design teams have therefore adopted intellectual property block *re-use methods*. Based on customer requirements, pre-designed and pre-verified intellectual property (IP) blocks, or a closely-related set of IP blocks (e.g., a central processing unit (CPU) with its L1 cache), are integrated on a single silicon die according to an application domain-specific *platform template* [15]. Not having to design these IP blocks from scratch and leveraging a platform template significantly reduces the amount of time required to design an system on chip (SoC), and thereby its time-to-market.

Furthermore, during the design of an SoC a structural, temporal, behavioral and data *refinement* process is used to effectively tackle its complexity and efficiently explore its design space within the consumer and technology constraints. During this process, details are iteratively added to a design implementation until it is ready for fabrication. This process is illustrated in Figure 5.1, which is adapted from [38].

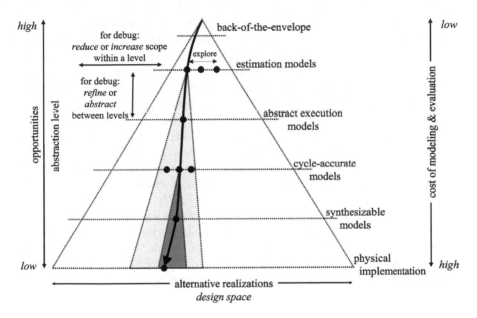

FIGURE 5.1: Design refinement process. (Adapted from A.C.J. Kienhuis. *Design Space Exploration of Stream-based Dataflow Architectures: Methods and Tools.* Ph.D. thesis, Delft University of Technology, 1999.)

The correctness of each refinement step, from one level of design abstraction to a lower level, has to be verified. Techniques such as formal verification, simulation, and emulation provide confidence that no errors were introduced and the resulting design should behave according to its original specification.

The ability to exhaustively verify a design before it is manufactured is severely restricted by the aforementioned increased system complexity. To both timely prepare a design and have sufficient confidence for its release to the market, verification engineers have to make trade-offs between the levels of design abstraction and the number of use cases to verify at each level. Functional problems may go undetected as it is impossible to cover all use cases at the level of the physical implementation before manufacturing. Problems may only manifest themselves after manufacturing test of an SoC, and even worse outside of controlled test and verification environments such as automated test equipment, simulators, and emulators. The root cause of any remaining problem discovered during the initial functional validation of the silicon chip has to be found and removed as quickly as possible to ensure that the product can be sold to the customer on time and for a competitive price. Industry benchmarks [55] show that this validation and debug process consumes over 50 percent of the total project time while the number of designs that are right first time is less than 40 percent.

The focus of this chapter is to describe the *debugging of a silicon implementation of an SoC*, which does not behave as specified in its product environment. During debugging, we need to find the root cause that explains the difference in the implementation's behavior from its specified behavior during a system run. We use the term "run" to mean a single execution of the system. For this we propose to use an iterative refinement and reduction process to zoom in on the location where and the point in time when an error in the system first manifests itself. This debug process requires both observation and control of the system in the environment where it fails.

The remainder of this chapter is organized as follows. Section 5.2 first provides a more in-depth analysis of the fundamental problems that need to be solved to debug an SoC. In particular, it is not easy to observe and control the system to be debugged. Section 5.3 describes how these fundamental problems affect the ideal debug process, and it subsequently defines the debug process used in practice. Section 5.4 presents an overview and comparison of existing debug methods. We introduce our debug method in Section 5.5. Section 5.6 defines the on-chip infrastructure to support our debug method, followed by the off-chip debug infrastructure in Section 5.7. We apply our method on a small example in Section 5.8, and conclude with Section 5.9.

5.2 Why Debugging Is Difficult

In this section, we identify three problems that make debugging intrinsically difficult: (1) limited internal observability, (2) asynchronicity, and (3) non-determinism.

5.2.1 Limited Internal Observability

One of the biggest problems while debugging a system is the volume of data that potentially needs to be examined to find the root cause. Worst case: this volume is equal to the amount of time from start-up of the system to the first manifestation of incorrect behavior on the device pins multiplied by the product of the number of electrical signals inside the chip and their operating frequencies. This data volume is huge for multimillion transistor designs running at hundreds of megahertz. Consider for example a 10 million transistor design running at 100 megahertz. If we sample one signal per transistor per clock, then this design produces 10^{15} bits of data per second.

The exponential increase in the number of transistors on a single chip [5] compared to the (linearly increasing) number of input/output (I/O) pins makes it impossible to observe all electrical signals inside the chip at every moment during its execution. If the same design has 1,000 pins, then even if we could use all these pins to output the data this design produces per second, we would have to operate these pins at speeds of 10^{12} bits per second per pin to output all data, which is clearly beyond current technological capabilities. Typically the number of device pins available for observation is much less as the chip still has to function in its environment and a large number of pins are reserved for power and ground signals.

5.2.2   Asynchronicity and Consistent Global States

In the remainder of this chapter we assume that each IP block in the system operates on a single clock, i.e., is synchronous. However, the clocks of different IP blocks can be *multi-synchronous* or *asynchronous* with respect to each other.

Multi-synchronous clocks are derived from a single base clock by using frequency multipliers and dividers or clock phase shifters. Data transfers between IP blocks take place on common clock edges, where explicit knowledge of the clock frequencies and phase relations of the IP blocks is used to correctly transfer data. Source-synchronous communication that tolerates limited clock jitter also falls in this category.

In contrast, *asynchronous* clocks have no fixed phase or frequency relation. Many embedded systems today use the globally-asynchronous locally-synchronous (GALS) [47] design style. As a consequence, all modern on-chip communication protocols use a so-called valid-accept handshake to safely transfer data between IP blocks, e.g., in the Advanced eXtensible Interface (AXI) [4] protocol, the Open Core protocol (OCP) [50] and the device transaction level (DTL) [54] protocol. As illustrated in Figure 5.2, the initiator prepares the "data" signals and activates its "valid" signal, thereby indicating to the target that the data can be safely sampled. The target samples the data using its own clock and signals the completion of this operation to the initiator by activating its "accept" signal. This handshake sequence ensures that the data are correctly communicated from the initiator to the target, irrespective of their functional clock frequencies and phase. The handshake sequence is part of the communication function of the IP block, and is usually implemented with stall states in an internal finite state machine (FSM). For ease of explanation, we assume the initiator and target stall while transferring data.

Debug requires the sampling of the system state for subsequent analysis. The state of an individual IP block can be safely sampled because it is in

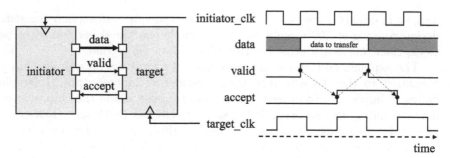

FIGURE 5.2: Safe asynchronous communication using a handshake.

a single clock domain, and an external observer simply has to use the same clock as used by the IP block. Sampling requires synchronicity to the clock of the IP block to prevent capturing a signal while it is making a transition. Proper digital design requires that IP signals are stable around the functional clock edges for an interval defined by the setup and hold times of the flip-flops used. The active edges of the functional clock therefore make good sampling points for external observation.

However, for debugging a system, we may need to inspect the *global state*, i.e., the combined local states of all IP blocks in the system. For multiple IP blocks, their safe sampling points are determined by the greatest common divisor of their frequencies. Only at these points, a *consistent global state* can be sampled, as the state of each IP block can be safely sampled at these points and the combination of all IP states also reflect the global state at these points. At all other points in times, it is not guaranteed safe to sample the state of all IP blocks. One or more local states are therefore unknown at those points preventing debug analysis. With two multi-synchronous clock domains, sampling on the slower clock may lead to missing some possible state transitions in the IP block with the faster clock. Conversely, sampling the state of the IP block running on the slower clock, with the faster clock is unsafe as we may sample in the middle of a state transition.

If two IP blocks are *asynchronous* with respect to each other, then there is no guarantee that their safe sampling points will ever coincide, and no points in time at which the global state can be consistently sampled may exist.

Consider as an example two IP blocks A and B. Block A has a clock period T_A of 2 ns, block B has a clock period T_B of 3 ns. We define the clock phase ϕ_{A-B} between these two clocks as the time between the rising edge of clock A and the rising edge of clock B. If $\phi_{A-B} = 0.5$ ns at a certain point in time $t = t_0$, then there is no point in time where the rising edges of clocks A and B coincide. For this, Equation 5.1 must hold for integer values of m and n. However the left-hand side of Equation 5.3 is always even for integer values of m, while the right-hand side of Equation 5.3 is always odd for integer values of n. Therefore there are no points in time where the rising clock edges for clocks A and B coincide.

$$T_A \times m = \phi_{A-B} + T_B \times n \qquad (5.1)$$
$$2 \times m = 0.5 + 3 \times n \qquad (5.2)$$
$$4 \times m = 1 + 6 \times n \qquad (5.3)$$

This is also illustrated in Figure 5.3.

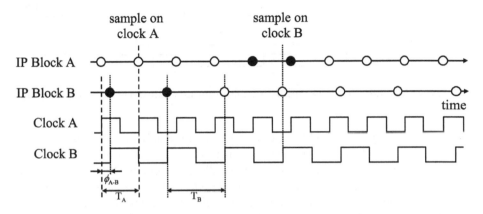

FIGURE 5.3: Lack of consistent global state with multiple, asynchronous clocks.

In general for a GALS system, it may therefore not be possible to correctly sample a globally consistent state at all (or even any) points in time at the clock cycle level. The only points at which the state of multiple IP blocks can potentially be safely captured is during synchronization operations, in which the state of both IP blocks has to be functionally defined and therefore has to be stable. It may therefore be possible to capture a consistent global state at these functional synchronization points. Synchronisation may however take place at different levels of abstraction, and require behavioral knowledge of the design to implement. Examples of using behavioral information to improve the ability to capture a globally consistent state will be introduced in Section 5.5.

5.2.3 Non-Determinism and Multiple Traces

Clock-domain crossings not only complicate the definition of a globally consistent state, but also cause variation in the exact duration of the communication between clock domains. When the initiator and target clocks have different (or even variable) frequencies or phases, then a valid-accept handshake can take a variable number of initiator and/or target clock cycles due to metastability [53, 64] (see A in Figure 5.4).

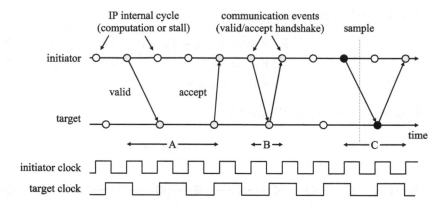

FIGURE 5.4: Non-determinism in communication between clock domains.

Essentially, in a GALS system it is not possible to safely sample a signal from another clock domain using a constant number of local clock cycles, due to metastability [65]. Although statistically it is very likely that the sampled signal is stable quickly, e.g., after one target clock cycle, it is possible that it takes (much) longer. This is illustrated in Figure 5.4 with the two handshakes, labeled B and C, respectively. B takes one initiator clock cycle, and C two cycles, even though in both cases the target responds within a single target clock cycle. This behavior occurs between asynchronous IP blocks in an SoC, but also for communication on the chip pins, for data transfers to and from the chip environment.

Critically, this local (inter-IP) non-determinism in communication behavior propagates to the system level, where it manifests itself in multiple communication traces [31, 60]. With the term "trace" we refer to a unique sequence of observed system states during a run. Figures 5.5a and 5.5b illustrate this phenomenon.

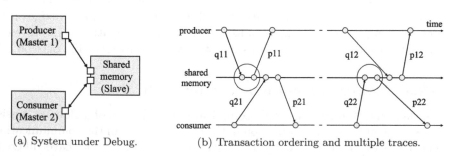

(a) System under Debug.   (b) Transaction ordering and multiple traces.

FIGURE 5.5: Example of system communication via shared memory.

As an example, Figure 5.5a shows two masters, called Producer and Consumer, communicating directly with a shared memory on different ports using transactions, each transaction comprising a request and an optional response message. Examples of transaction requests include read commands with read addresses, and write commands with write addresses and data. Corresponding responses are read data and write acknowledgments, respectively. All modern on-chip communication protocols fit this model [19].

The shared memory in our example only has one execution thread, and therefore can only accept and execute a single request at a time. We will further assume for illustration purposes that a read by the Consumer is only correct if the Producer writes to the shared memory before the Consumer reads from it. Figure 5.5b shows Master 1 initiates a write request "q11," soon followed by a read request from Master 2 "q21." Master 1's request is executed first by the slave, resulting in a response "p11." Afterwards the request of Master 2 is executed by the Slave, resulting in a correct response "p21." Another sequence with a different, incorrect outcome is however also possible and is shown with the subsequent requests ("q12" and "q22"). This time, due to a different non-deterministic delay on the communication path between the masters and the slave, write request "q21" from the Producer is executed after read request "q22" from the Consumer. This response "p22" returned to the Consumer will be incorrect because the Consumer read the response before the Producer could write it.

Executing transactions in different orders can have an impact on the functional behavior of the IP blocks. For example, consider that Master 1 produces data in a first-in first-out (FIFO) data structure for Master 2, and signals that new data is ready by updating a FIFO counter or semaphore in the shared memory [49]. If Master 2 reads the counter from memory using polling, then both sequences are functionally correct. However, in the scenario shown on the right-hand side of Figure 5.5b Master 2 reads the old counter value, and it would require another polling read to observe the new counter value, resulting in a delayed data transfer. Whether this is a problem or not depends on the required data rates. It would definitely be erroneous, however, if the requests of the masters were write operations with different data to the same address. In this case, the functional behavior of the system would be non-deterministic, and possibly incorrect, from this point onward.

5.3   Debugging an SoC

In this section we define errors and explain how the analysis in Section 5.2 of what makes debugging intrinsically difficult affects the ideal debug process. We subsequently describe the debug process that has to be used in practice.

5.3.1   Errors

We assume that the observed global states are consistent in some sense, which is justified in Section 5.5. As shown in Subsection 5.2.3 multiple runs result in the same or different traces due to non-determinism. An error is said to have occurred when a state in a trace is considered incorrect with respect to either the specification or an (executable) reference model. Such a state is called an "erroneous state." Note that we consider errors, i.e., the manifestations of faults, and we consider the objective of debugging to be to find and remove the root cause of these errors (i.e., the faults causing them). Fault classifications and discussions on the relation between faults and errors can be found in [6, 9, 39, 44].

Error observations can be classified in three orthogonal ways: within a trace, between traces, and between systems.

- *Within a trace.* When all states following an erroneous state are erroneous states as well, the error is *permanent*, otherwise the error is *transient*. Transient errors may happen, for example, when erroneous data is overwritten by correct data, before it propagates to other parts of the system.

- *Between traces.* An error is *constant* when it occurs in every run (and hence in every trace). This is always the case when the system is deterministic as deterministic systems have only a single trace. An error is *intermittent* when it occurs in some but not all runs. For a system to exhibit intermittent errors, it has to be non-deterministic, as discussed in Section 5.2.3. It therefore produces different traces over multiple runs.

- *Between systems.* Finally, until now we assumed that the system does not change between runs. This is not necessarily the case. The debug observation or control of the system is often *intrusive*, i.e., it changes the behavior of the system. This phenomenon is also known as the "probe effect." As a result, often the error disappears and/or other errors appear when monitoring or controlling the system. In these cases, we basically generate traces for two different systems, so the resulting traces may be very different and hard to correlate. We call these *uncertain* errors, after the uncertainty principle[1], as opposed to *certain* errors.

For simplicity, we will assume in the remainder of this chapter that all errors are permanent and certain, though they may be intermittent. We use a small example to see how these differences in error types can manifest themselves during debugging of an embedded system.

---

[1]Gray [25] introduced "Bohrbugs" and "Heisenbugs." However, these terms are not used consistently in the literature, and we will therefore not use them.

5.3.2   Example Erroneous System

We illustrate constant, certain and intermittent errors by re-using the simple example system of Figure 5.5a and focus on the states of the individual IP blocks. The possible system traces are illustrated in Figure 5.6.

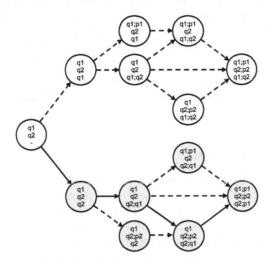

FIGURE 5.6: System traces and permanent intermittent errors.

Each circle corresponds to a consistent global state. The text inside the label indicates from top to bottom the state of Master 1, Master 2, and the Slave respectively. A shaded state indicates that the error has propagated into the global system state. Figure 5.6 also shows the largest scope, i.e., when the consistent global state comprises the local states of both Master 1 and Master 2 and the Slave. "qi" refers to the sending or receiving of a request of Master i, and "pi" for the corresponding response.

We can now illustrate how intermittent errors occur using Figure 5.6. A run proceeds along a certain trace, such as the one that is highlighted by the solid line. In the first state "(q1 q2 -)" both masters generate their request to the slave memory at the same time. As a result of non-deterministic communication between the masters and the shared slave, our example system can have multiple execution traces. Figure 5.5b illustrates this by focusing on the interleaving of transactions. In Figure 5.6 we concentrate on the divergence of the global states and resulting multiple traces instead. As shown in Figure 5.5b, the memory may accept and execute the request of Master 2 first (with global state "(q1 q2 q2)"), and offer an *erroneous* response "p2." Before this response is accepted by Master 2, the memory accepts and executes request "q1," causing the global state "(q1 q2 q2; q1)." Master 2 then accepts "p2" (with global state "(q1 q2; p2 q2; q1)"), followed by Master 1's acceptance of response "p1." The global end state is where both masters have received the response to their request ("(q1; p1 q2; p2 q2; q1)").

In an alternative trace the slave executes request "q1" before request "q2." Master 1 subsequently receives a correct response "p1," followed by a correct response "p2" for master 2. The global end state for this trace is "(q1; p1 q2; p2 q1; q2)," which differs from the end state of the previous trace by the order in which the slave handled the incoming requests ("q1" before or after "q2").

Hence, when executing the system a number of times it can generate different traces. Even with non-intrusive observation (i.e., with certain errors), the error may only be triggered and consequently visible in a subset of the traces and is therefore intermittent. Moreover, the error, i.e. the return of the incorrect response "p2," can become visible at Master 2 at different points in time in the different traces. This makes intermittent errors particularly hard to find [16, 25].

5.3.3 Debug Process

The process of debugging relies on the observation of the system, i.e., its states, for a certain duration of time, and at discrete points in time. This observation results in a state trace. The state can be observed at various levels of *abstraction*, which determines in how much detail we look at the system. We can consider for instance only which applications are running, which transactions are active, which signal transitions occur, or what the voltage levels are on the physical wires.

At a given level of abstraction, the *scope* of the observation determines how much of the system we observe and for how long we observe it. This scope may be varied between runs. For example, Figures 5.7, 5.8, and 5.6 illustrate observations with increasing (spatial) scope.

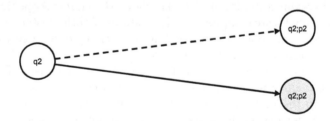

FIGURE 5.7: Scope reduced to include Master 2 only.

Figure 5.7 includes only Master 2 in the scope. We see two distinct end states as the order in which the requests from Master 1 and Master 2 are executed by the slave can still cause the response for Master 2 ("p2") to be different between runs. Figure 5.8 includes both Master 1 and Master 2 where both the request execution ordering by the slave and the order of acceptance of the responses by the individual masters splits the traces in six different traces. Figure 5.6 provides the most detail by including the state of all master and slave IP blocks.

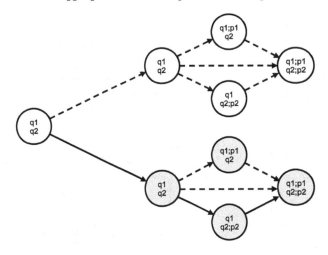

FIGURE 5.8: Scope reduced to include Master 1 and Master 2 only.

The observation and control of the system takes place in the same scope and at the same abstraction level. The debug process essentially involves *iteratively either increasing or decreasing the scope and abstraction level* of observation and control until the root cause of the error is found. In the ideal debug process, we observe only the relevant state to find the root cause for a particular error and for a minimal duration. This process is shown in Figure 5.9a.

First, we reduce the scope, i.e., zoom in on the part of the system where and when the error occurred. Preferably, we "just" walk back in time to when the error first occurred [43, 63], and observe only the state of the relevant IP blocks. Then we refine (lower the level of abstraction) to observe those IP blocks in more detail. For example, we refine the state of an IP block to look at its implementation at register transfer level (RTL) to logic gates or from source code to assembler, or we refine communication events to their individual data handshakes or clock edges. In Figure 5.1 the path from the highest abstraction level down to the physical implementation level can also be interpreted as an instance of the debug process, whereby the reduction of the debug scope takes places within one abstraction level, and the refinement takes place between abstraction levels.

However, in practice, debugging is more challenging due to the lack of internal observability and control, the difficulty involved in reproducing errors, and the problems in deducing their root cause. The effect of these three factors on the debug process is shown in Figure 5.9b.

1. Lack of *observability*. We can inspect given traces, but we need to restart every time we want to observe the trace of a new run. Each trace may take a long time (hours or even days), to trigger the error, resulting in a huge data volume to analyse.

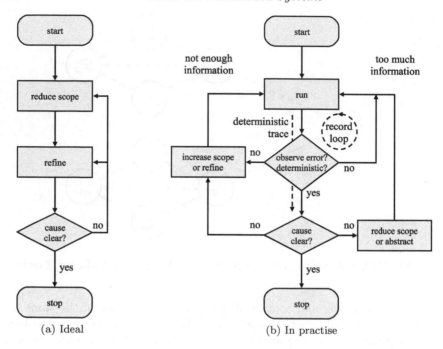

FIGURE 5.9: Debug flow charts.

2. Lack of error *reproducibility*. Non-determinism causes multiple traces and intermittent errors, as discussed in Section 5.3.2. Finding the first state that exhibits the error may take a long time because every run of the system proceeds (non-deterministically) along one of many potential traces, with possibly very different probabilities. For example, the high-lighted trace in Figure 5.8 may only be taken in 0.001 percent of the runs. Consequently the time between two runs that both exhibit the error may be very long.

3. *Deduction of root cause*. At some point during the debug process we arrive at Figure 5.7, where we have a minimal scope that exhibits the error. To deduce why either a good or bad trace is taken, we need to either increase the scope and observing the state of more IP blocks or refine the state of the IP blocks we are already looking at and observe their state in more detail. We need to intelligently guess that adding the state of the slave to the observed state is a good idea. A larger observed state will however usually result in a larger number of possible traces, as illustrated in Figure 5.6. In subsequent runs, the scope will have to be reduced to the relevant parts again. The decision when to increase the scope and when to refine the state is not trivial. Even without non-determinism, the cause of the error is often not evident when a good to bad state transition occurs, as we see an effect but cannot automatically

deduce the cause. We then either increase the information to investigate by increasing the scope or by refining the state. This is illustrated in Figure 5.6, where the state of the slave is added. In a subsequent run it is then possible to observe that executing "q2" before "q1" is the cause (at this abstraction level) of the error.

With this general debug process in mind, we describe in the following section various existing debug methods that have been proposed in literature.

5.4 Debug Methods

To simplify or automate the debug process, several methods have been proposed in the literature. They all assume that it is possible to find a consistent global state. Observing this global state at certain points in time over multiple runs results in a set of traces. Essentially, the existing debug methods differ in how often they observe what state while the system is running, and whether this is intrusive or not. We first define several properties we use to classify common debug methods.

5.4.1 Properties

We compare different, existing debug methods using three important debug properties: their use of abstraction techniques, their scope, and their intrusiveness.

Choosing the right abstraction level helps reduce the volume of data to observe. This reduces the bandwidth requirements for the observation infrastructure as well as the demands on the human debugger. We consider four basic abstractions [45]: (1) structural, (2) temporal, (3) behavioral, and (4) data.

- *Structural abstraction* determines what part of the system we observe within one abstraction level (e.g., all IP blocks, or only the masters) and at what granularity (e.g., subsystem, single IP block, logic gates, or transistors).

- *Temporal abstraction* determines what and how often we observe. For example, traditional trace methods observe the state at every cycle in an interval, or sample the state periodically. Alternatively, only "interesting" relevant state may be observed at or around relevant communication or synchronization events. Examples include the abstraction from clock cycles to handshakes (illustrated by the removal of internal clock cycles in Figure 5.4), moving to transactions, or to software synchronizations using semaphores and barriers.

- *Behavioral abstraction* determines what logical function is executed by a (hardware) module. For example, in a given use case, a processor may be programmed to perform a discrete cosine transform (DCT), and a network on chip (NoC) may be programmed to implement a number of "virtual wires" or connections. In another use case, they may have different logical functions.[2]

- *Data abstraction* determines how we interpret data. At the lowest level we observe voltage levels in a hardware module. We abstract from this voltages first to the bit level and subsequently use knowledge of the module's logical function at that moment in time to interpret the values of these bits. For example, a hardware module that implements a FIFO contains logical read and write pointers defining the valid data. Only with this knowledge can we display the collection of bits as a FIFO. Similarly, a processor's state can be abstracted to its pipeline registers [37], a memory content, for example, to a DCT block, and registers in a NoC to a connection with FIFOs, credit counters, etc.

Existing debug methods also vary in their *scope*, which was introduced in the previous section. Scope uses structural and temporal abstraction, but considers only one abstraction level.

Increased abstraction (and reduced scope) serve to reduce the volume of data that is observed. The system state can either be observed when the system is running, called *real-time trace*, or when it is stopped, called *run/stop debug*, or both.

During real-time trace debugging, the data is either stored on-chip in buffers, streamed off the chip, or both. This is only possible when the volume of data is not too large and hence may require the use of abstraction techniques. This trace process may be intrusive or not.

During run/stop debugging, the system is stopped for observation, which is by definition intrusive. However, in return, it usually allows access to much more system state because ample time and bandwidth are available for inspection, as the system execution has been stopped.

Every debug process relies on the observation of the system, i.e., accessing its state. *Intrusive* observation affects the behavior of the system under observation, and may lead to uncertain errors. *Non-intrusive* observation does not affect the behavior of the system (aside from consuming some additional power), but does require a dedicated and independent debug infrastructure, making it more expensive to implement on-chip than the infrastructure to support intrusive observation.

---

[2]This is a different slant on behavioral abstraction from [45], where it is defined as partial specification. In any case, the distinction of behavioral abstraction and temporal and data abstraction is to some extent arbitrary.

5.4.2   Comparing Existing Debug Methods

Without making changes to the design of a chip, a debug engineer has the classic *physical* and *optical* debug methods at his disposal, such as wafer probing [7], time-resolved photo-emission [48] (also known as picosecond imaging circuit analysis (PICA) [34]), laser voltage probing (LVP) [51], emission microscopy (EMMI) [30], and laser assisted device alteration (LADA) [59]. These physical and optical techniques are *non-intrusive*, provided that removing the package and preparing the sample cause no behavioral side effects. They provide observability at the lowest level of abstraction only, i.e., voltage levels on wires between transistors in real time.

Unfortunately these methods can only access the wires that are close to the surface. Access to other, deeply embedded transistors and wires is often blocked by the many metal layers used today to provide the connectivity inside the chip, and to aid in planarization. Back-side probing techniques help somewhat to reduce the problems of the increasing number of metal layers. In nanometer CMOS processes, these methods still suffer from a number of drawbacks. First, the number of transistors and wires to be probed is too large without upfront guidance. Moreover, the transistors and wires may be hard to access because they are very small. Finally, device preparation for each observation is often slow and expensive.

Hence these methods can only efficiently localize root causes of failures if the error is first narrowed down to the physical domain (such as crosstalk, or supply voltage noise). To reach this point, and walk the debug path in Figure 5.1 all the way down to the level of the physical implementation, we need to reduce the scope and lower the level of system abstraction.

Logical debug methods have been introduced for this purpose. Logical debug methods use built-in support called design for debug (DfD) to increase the internal observability and controllability, and act as a precursor to the physical and optical debug methods by helping to quickly reduce the scope containing the first manifestation of the root cause.

These logical debug methods reduce the data volume by making a trade-off between focusing on the real-time behavior of the system and maximizing the amount of state that can be inspected. Only a small subset of the entire internal state can be chosen for observation when the real-time behavior of the system is to be studied due to the aforementioned I/O bandwidth constraints. Whether this is intrusive or not depends on the infrastructure that is used to transport and/or store the data. ARM's CoreSight Trace [2] and FS2's PDTrace [46] architectures are examples of *non-intrusive*, real-time trace. Sample on the Fly [37] is a real-time trace method used for central processing units (CPUs) that periodically copies part of the CPU state in dedicated scan chains that can then be read out non-intrusively. Memory-mapped I/O can be used to read and write addressable state over the functional/inter-IP interconnect while the system is running, for example, with ARM's debug access port (DAP) [2], or FS2's Multi-Core Embedded Debug (MED) system [41].

This will however be more intrusive than a dedicated observation and control architecture.

By stopping the system at an interesting point in time, a much larger volume of data can be inspected. This run/stop-type approach however is *intrusive*. The infrastructure used to access the state and its implementation cost are then the limiting factors. For example, the manufacturing test scan chains provide a low-cost infrastructure, which can be used to read out the entire digital state when the system is stopped [71].

The majority of published, logical debug methods do not address the problems caused by asynchronicity, inconsistency of global states, non-determinism or multiple traces. However, there are several notable exceptions that we discuss next: latch divergence analysis, deterministic (re)play, and the use of abstraction for debug.

5.4.2.1 Latch Divergence Analysis

Latch divergence analysis [13] aims to automatically pinpoint erroneous states. It does so by running a CPU many times, and recording its state at every clock cycle. The traces that are obtained from runs with a correct end result are then compared with each other. The unstable part of each state, called latch divergence noise, is filtered out. This step yields the stable substate across all good traces. Similarly, the stable substate across traces with an incorrect end result is computed. This substate is then compared with the stable substate of the good traces.

The inference is that the unstable parts are caused by noise, e.g., through interaction with an analog block or uninitialized memory, and can be safely filtered out, as they are not caused by the error. An advantage of this method is that it can be easily automated. However, this method does not distinguish noise in substates due to intermittent errors, i.e., those that only occur in some traces, and correct but only partially specified system behavior. Filtering out the noise caused by the partial specification of the behavior may obscure the root cause of an error.

5.4.2.2 Deterministic (Re)play

Instant replay [42], and deterministic replay [18, 56] aim to reduce the time between runs that exhibit an error. When an error is observed, the system is subsequently placed in "record" mode and restarted. The system is repeatedly run until the error is observed again. This step corresponds to the dashed "record loop" in Figure 5.9b. At this point, the debug process can start by replaying the same run and observing the recorded trace as highlighted in Figure 5.7, provided that the recording contains enough information to deterministically replay the trace containing the error. The key idea is that a previously intermittent error appears in every replayed run ("deterministic trace" in Figure 5.9b). Deterministic replay requires all sources of non-determinism to be recorded at the granularity at which they cause divergence in a trace.

It also requires an additional on-chip infrastructure to force the single trace that triggers the error once it has been recorded.

Deterministic replay has been used successfully for software systems, where the non-determinism is limited to the explicit synchronization of threads or processes. The number of divergence points is relatively small, and the frequency of synchronization is low in these cases [42]. However, for embedded systems with multiple asynchronous clock domains, we have seen in Section 5.2.3 that a clock domain crossing between asynchronous clock domains gives rise to non-determinism. Therefore the delay across this interface needs to be recorded. Since an SoC easily contains more than a hundred IP ports connecting asynchronously to an interconnect [22], running at hundreds of megahertz, the data rate to be recorded quickly reaches gigabits per second. It is expensive in silicon areas to non-intrusively record this data on-chip and expensive in device pins to stream it non-intrusively off-chip. However, an intermediate means of communication, namely source-synchronous embedded systems, has been successfully used for a limited number of processors [60].

Pervasive debugging [29] has been proposed with the same goal as deterministic replay. It proposes to model the entire system in sufficient detail such that non-deterministic effects become deterministic. This may be possible for (source)-synchronous systems. However, it is infeasible for systems that contain asynchronous clock domains, or contain errors relating to physical properties (e.g., crosstalk, or supply voltage noise) and environmental effects (ambient temperature, chip I/O, etc.). Relative debugging [1], where an alternative (usually sequential) version of the system is used as a reference to check observed states against, suffers from the same limitations.

Finally, synchro-tokens [31] may be interpreted as deterministic *play*. All synchronizations of a GALS system are made deterministic in every run (and not only during debug), from the view of the communicating parties. Hence, there is a unique global trace (the "deterministic trace" namely the (software) synchronization points, in Figure 5.9b), and all errors are constant. The main drawback of this method is that it reduces performance by essentially statically scheduling the entire system.

5.4.2.3   Use of Abstraction for Debug

System simulations for debug tend to focus on only one or two abstraction levels at a time. For example, traditional software debug allows observation and control (e.g., single-stepping) per function, per line in the source code, and can show the corresponding assembly code. It is difficult to debug multi-threaded or parallel software programs using conventional software debuggers because the parallel nature of programs is not supported well. However, specialized debuggers make the distinction between inter-process communication and intra-process computation. By abstracting to synchronization events [8] they allow the user to focus on less but more relevant information.

Hardware descriptions define parallel hardware, but traditional hardware simulation does not make a distinction between inter-IP communication (e.g., VHSIC (Very High Speed Integrated Circuit) Hardware Description Language (VHDL) or Verilog signals) and intra-IP computation (e.g., VHDL variables). Traditional hardware simulation is more limited because it simulates either the RTL or the gate-level description, and does not show any relation between them. In recent years, transaction-level modelling and related visualisation techniques have been introduced to abstract away from the signal level IP interfaces and allow a user to focus on the transaction attributes instead [61] or correlate gate level with RTL descriptions [33].

Traditionally, when debugging real hardware that executes software, either functional accesses, real-time trace, or state-dump methods are used to retrieve the system state, as described earlier. Once the state has been collected, it can be interpreted at a higher level, e.g., by re-presenting it at the gate level or RTL level [68]. Recently, DfD hardware has been added to observe and control the system at higher levels of abstraction. Examples include transaction-based debug [24], programmable run-time monitors [11, 73], and observation based on signatures [72].

Overall, we observe that the existing software debug methods are quite mature, especially for sequential software, but less so for parallel software. Existing hardware debug methods are even more limited. Abstraction is currently only applied in a limited fashion, and then almost exclusively for software debug.

5.5 CSAR Debug Approach

In this section we define a debug approach called CSAR and discuss its characteristics. Following this, Sections 5.6 and 5.7 describe how this approach is supported, both on-chip and off-chip. Section 5.8 illustrates how our approach works for a small example.

The CSAR debug method can be characterized as:

- Centered on *C*ommunication

- Using *S*can chains

- Based on *A*bstraction

- Implementing *R*un/stop control

Each characteristic is described in more detail below.

5.5.1   Communication-Centric Debug

Figure 5.10a illustrates traditional computation-centric debug, in which the computation inside IP blocks, especially embedded processors, is observed. When something of interest happens, this is signaled to the debug controller that can take action, such as stopping the computation in some or all IP blocks.

With an increasing number of processors, the communication and synchronization between the IP blocks grow in complexity and become an important source of errors. To complement mature existing computation-centric processor debug methods, we focus on debugging the communication between IP blocks, as shown in Figure 5.10b.

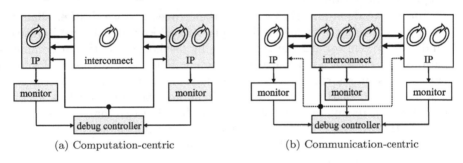

(a) Computation-centric   (b) Communication-centric

FIGURE 5.10: Run/stop debug methods.

Older on-chip interconnects, such as the advanced peripheral bus (APB) and ARM high performance bus (AHB) [3], are single-threaded. This means that only one transaction is processed by the interconnect at any point in time. As a result, the interconnect forces a unique trace for all IP blocks attached to these buses even when using a GALS design style. For scalability and performance reasons, recent interconnects, such as multi-layer AHB and AXI buses [4], and NoCs [14, 36, 52], are multi-threaded. In other words, they allow multiple transactions between a master and a slave (pipelining), and concurrent transactions between different masters and slaves. Moreover, support for GALS operation where the IP-interconnect interface is asynchronous is common. Hence no unique trace exists anymore, as we have seen in Section 5.2.

The aim of communication-centric debug is to observe and control the traces that the interconnect, and hence the IP blocks attached to it, follow. This gives insight in the communication and synchronization between the IP blocks, and allows (partially) deterministic replay.

5.5.2   Scan-Based Debug

As only a limited amount of trace data can be stored on chip or sent off-chip, we only allow the user to observe state when the system has been stopped. We re-use the scan chains that embedded systems use for manufacturing test to create access to all state in the flip-flops and memories of the chip via IEEE Standard 1149.1-2001, Test Access Port (TAP) [71]. This helps minimize the hardware cost.

5.5.3 Run/Stop-Based Debug

As the state can only be observed via the scan chains when the system has been stopped, *non-intrusive monitoring* and *run/stop control* are used to stop the system at interesting points in time. This is implemented by non-intrusively monitoring a subset of the system state, and generating events on programmable conditions.

Ideally we deterministically follow the erroneous trace. Rather than collecting and storing information for replay (recall Figure 5.9b), we iteratively guide the system toward the error trace by disallowing particular communications and thereby forcing execution to continue along a subset of system traces. This allows the user to iteratively refine the set of system traces to a unique trace that exhibits an error. This may be interpreted as *partially* deterministic replay, or "guided replay," although errors may become uncertain, as this process is currently intrusive because the guidance of the system does not occur in real-time, but only after the system has been stopped using off-chip debugger software.

5.5.4 Abstraction-Based Debug

We use temporal abstraction to reduce the frequency and number of observations to those that are of interest. In particular, rather than observing a port between an IP and the interconnect at every clock cycle, we can observe only those clock cycles where information is transferred, i.e., by abstracting to handshakes. In Figure 5.4 this would correspond to observing only the communication behavior at the gray and black clock cycles, and ignoring the internal behavior at the white clock cycles. Conventional computation-centric debug can be used to observe the internal behavior of the IP blocks in isolation.

As an example, a DTL transaction request consists of a command and a number of data words (indicated by the command). Each of these can be individually abstracted to a handshake, called *element*. Similarly, a response consists of a number of data words. A *message* is a request or a response, and a transaction is the request together with the (optional) response. Figure 5.11 shows several temporal abstraction levels: clock cycles, handshakes, messages, transactions, etc. Each time we combine a number of events to a coarser event that is meaningful and consistent by itself.

We also use structural and behavioral abstraction (refer to the left-hand side of Figure 5.11). Our debug observability involves retrieving the functional state (i.e., the bits in registers and memories) from the chip. We re-use the scan chains (the lowest level in Figure 5.11) that are inserted for manufactur-

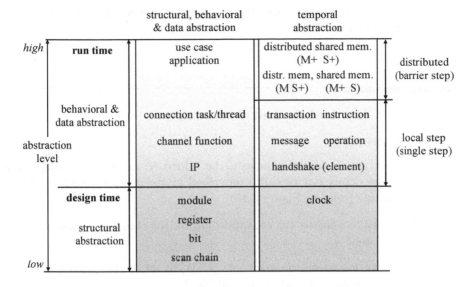

FIGURE 5.11: Debug abstractions.

ing test of the chip, when the system has stopped. This provides an intrusive means to "scan out" all or part of the state from the chip. The resulting state dump is a sequence of bits that still has to be mapped to registers and memories in gate-level and RTL descriptions. One level higher are modules, which correspond to the structural design hierarchy. These abstraction levels only describe structure, i.e., how gates and registers, are (hierarchically) interconnected.

The next level makes a significant step in abstraction by interpreting structural modules as functional IP blocks. In other words, we make use of behavioral information that allows us to interpret a set of registers. For example, a simple IP block, which implements a FIFO contains data registers, and read and write pointers. Without an abstraction from structure to behavior, they are all simply registers. At the functional IP level however, we can interpret the values in the read and write registers and, for example, display only the valid entries in the data registers.

The higher levels of abstraction, from channel to use case, go one step further. They abstract from hardware to software, or from the static design-time view to the dynamic run-time view, in other words, not from what components the system is constructed from, but to how it has been programmed. Because we focus on communication, we move from structural interconnect components such as network interfaces (NIs) and routers to logical communication channels and connections that are used by applications. Processors execute functions, which are part of threads and tasks, which themselves in turn are part of the complete application. The application that runs on the system depends on the use case. The implementation of these abstractions is described in Section 5.7.2.

5.6   On-Chip Debug Infrastructure

5.6.1   Overview

Dedicated debug IP modules have to be added to an SoC at design time to provide the debug functionality described in the previous sections. These modules include (refer to Figure 5.12):

- Monitors to observe the computation and/or communication and generate events

- Computation-specific instruments (CSIs) to act on these events and control the *computation inside* the IP blocks

- Protocol-specific instruments (PSIs) to act on these events and control the *communication between* the IP blocks

- An event distribution interconnect (EDI) to distribute the events from the monitors to the computation-specific instruments (CSIs) and protocol-specific instruments (PSIs)

- A debug control interconnect (DCI) to allow the programming of all debug blocks and querying of their status by off-chip debug equipment (see Section 5.7)

- A debug data interconnect (DDI) to allow access to the manufacturing-test scan chains to read out the complete state of the chip

The following subsections describe the functionality of each of these modules in more detail.

5.6.2   Monitors

Monitors observe the behavior of (part of) a chip while the chip is executing. They can be programmed to generate one or more events when a particular point in the overall execution of the system is reached [58], the system completes an execution step at a certain level of behavioral or temporal abstraction [24], or an internal system property becomes invalid [17]. These events can be distributed to subsequently influence either the system execution or the start or stop of real-time trace.

Monitors can also derive new data from the observed execution data of a system component by, for example, filtering [12] or compressing the information into a signature value using a multiple-input signature register

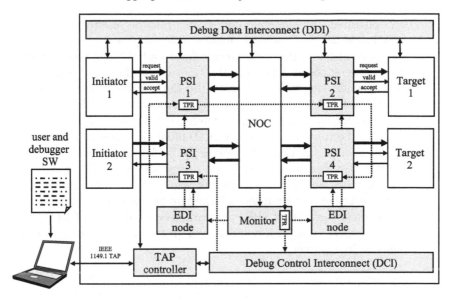

FIGURE 5.12: Debug hardware architecture.

(MISR) [66, 72]. As we focus on run/stop debugging, this type of monitor functionality falls outside the scope of this chapter.

Monitors are specialized to observe either the execution behavior of the *computation* (i.e., intra-IP) or the *communication* (i.e., inter-IP).

- *Computation* monitors can be added to the producers, the consumers, and the communication processing elements inside the communication architecture. CPUs traditionally include on-chip debug support [40], which enables an event to be generated when the program counter (PC) of the CPU reaches a certain memory address. This ability allows the event to be generated on reaching a certain function call, a single source code line, or an assembly instruction. When so required, events can also be generated at the level of clock cycles [28], by counting the number of clock cycles since the last CPU reset. For hardware accelerator IP blocks, custom event logic may be designed [70] that serves the purpose of partitioning the execution interval of an IP block into regular sections at possibly multiple levels of temporal abstraction.

- *Communication* monitors [11, 73] can be added on the interfaces of the producers, the consumers and the communication architecture, or within the communication architecture itself (i.e., in a NoC also on the interfaces between the routers and NIs). They observe the traffic and can generate events when either a transaction with a specific set of attributes is observed, and/or when a certain number of specific transactions have been communicated from a particular producer and/or to a particular consumer. As the communication protocols used in different chips may implement safe communication differently, a communication monitor may utilize a protocol-specific front end (PSFE) to abstract away these differences and provide the transaction data and attributes to a generic back end, which processes this data and determines whether the event condition has occurred. For a bus monitor, the filter criteria typically include an address range, a reference data value, an associated mask value, and optionally a transaction ID identifying the source of the transaction. A network monitor observes the packetized data stream on a link between two routers or between a router and a NI. Filter criteria may include whether the data on the link belongs to a packet header, a packet body, or the end of a packet, information on the quality of service (QoS) of the data (best effort (BE) or guaranteed throughput (GT)), whether a higher-level message has ended, and/or the sequence number of a data element in a packet.

Upon instantiation, the monitor is connected to a specific communication link, at which time the appropriate PSFE can be instantiated, based on the protocol agreed upon between the sender and the receiver [66]. The monitors are programmed and queried via the Debug Control Interconnect (DCI) (see Section 5.6.6 for details).

5.6.3   Computation-Specific Instrument

CSIs are instantiated inside or close to an IP block. Their purpose is to stop the execution of the component at a certain level of behavioral or temporal granularity when an event arrives. CPUs traditionally support interrupt handling, whereby the CPU's program flow is redirected to an interrupt vector look-up table on the arrival of an event. This table contains an entry for each type of interrupt (event) that can occur together with an address from which to continue execution. Debug events can be handled by an IP block as if it is an interrupt. Interrupts on the other hand can also be seen as signals that indicate the IP block's progression and can also be monitored.

Most CPUs support stalling the processor pipeline to halt execution in those cases where data first has to arrive from the communication architecture before its execution can continue. This stalling mechanism can be implemented either in the data path of the pipeline or in the control path (i.e. in the clock signal). In the latter option, special gating logic is added to the clock generation unit (CGU) [28] that prevents the pipeline from being clocked. These functional stalling mechanisms can be re-used for run/stop debugging to halt the execution of the processor at very low additional hardware cost.

Computation-specific Instruments (CSIs) are programmed and queried through the DCI to perform a specific action, such as starting, stopping, or single stepping, at a certain granularity (function entry/exit, source code line, assembly instruction, clock cycle), when an event is received through the Event Distribution Interconnect (EDI).

5.6.4 Protocol-Specific Instrument

Section 5.2.2 described how we cannot always stop multiple IP blocks with asynchronous clocks such that their states are consistent. However, they can communicate safely with each other at different levels of abstraction, e.g., by using a valid-accept handshake as illustrated in Figure 5.2. By using the functional synchronization mechanisms, we can recover a consistent global state for debugging [24]. In Figure 5.2 the initiator raises its valid signal to indicate that the data it wishes to send is valid. The initiator stalls until the target signals that it consumed the data by raising the accept signal. The white circles in Figure 5.4 indicate these stall cycles of an IP block.

Essentially, because the internal state of the IP does not change while it is stalled, it can be safely sampled on any clock. In Figure 5.4 this is illustrated by the two black clock cycles. If the target does not accept the request handshake of the initiator then the dashed synchronization will not occur. The initiator will instead stall, allowing its state to be safely sampled.

We assume that all IP blocks communicate via an interconnect, such as a NoC [21], as shown in Figure 5.13.

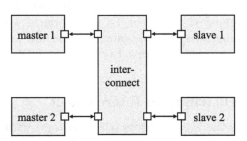

FIGURE 5.13: Example system under debug.

Every IP block will communicate at some point using the interconnect, possibly after some internal computation. If we control the handshakes between the IP blocks and the interconnect, it is possible to stall the IP blocks and the NoC when they offer a request or wait for a response. When all IP blocks are stalled, their states can be safely sampled, and a consistent global state is available.

However, note that the states are consistent in the sense that each IP block is in a stall state, waiting for a request or response. The global state may be inconsistent at a higher level of abstraction. For example, consider inter-IP communication based on synchronized tokens in a FIFO [49], described in Sections 5.2.3 and 5.3.2. Stopping at the level of transactions, many of which constitute the transfer of a single token, does not guarantee that a token is either at the producer or the consumer. It may be partially produced, fully produced but not yet synchronized, etc. This can only be resolved by lifting the abstraction level yet again. In general, the Chandy-Lamport's "snapshot" algorithm [10] or derivatives thereof can be used to ensure that a collection of local states is globally consistent. Sarangi et al. [60] demonstrate this for source-synchronous multiprocessor debugs.

Protocol-specific instruments (PSIs) are instantiated on the communication interfaces of producers and consumers or inside the communication architecture where they control the data communication. A protocol-specific Instrument (PSI) is protocol-specific because it requires knowledge of the communication protocol to determine when a request or response is in progress, and when there are pending responses (for pipelined transactions). Based on this information and its program, a PSI can determine when it should stop the communication on a link after an event arrives from the EDI.

The communication on a bus is stopped by gating the handshake signals, thereby preventing the completion of the communication of the request or response. Communication requests are no longer accepted from the producers and no longer offered to the consumers. Responses are no longer accepted from the consumers nor offered to the producers.

Stopping the communication may take place at various levels of granularity, e.g., individual data elements, data messages, or entire transactions. PSIs are programmed through the DCI to perform a specific action, such as starting, stopping, or single stepping, at a certain behavioral or temporal granularity when an event is received through the EDI.

5.6.5   Event Distribution Interconnect

The EDI connects the event sources (the monitors) with the sinks (the CSIs and PSIs). The EDI acts as a high-speed broadcast mechanism that propagates events to all event sinks. Ideally, when an event is generated anywhere in the SoC, all on-going computation and communication execution steps are stopped as soon as possible, at their specified level of behavioral or temporal abstraction.

There are several possible ways to distribute a debug event:

1. Packet-level event distribution [62] uses the functional interconnect as an EDI. Re-using the functional interconnect does increase the demands on the communication infrastructure as the additional data volume has to be taken into account. This is undesirable because events are only generated during debugging and not during normal operation. Permanent bandwidth reservations can be made if the communication architecture supports this to avoid the "probing" effect the debug data has on the timing of the functional data. However, permanently reserving this bandwidth may be expensive.

2. Cycle-level event distribution [67]. A global, single-cycle event distribution is not scalable and difficult to implement independently from the final chip lay-out. In our solution, a network of EDI nodes is used that follows the NoC topology. The EDI node is parametrized in the number of neighboring nodes. Each node synchronously broadcasts at the NoC functional clock speed any events it receives from neighboring monitors or EDI nodes to the other EDI nodes in its neighborhood. This transport mechanism incurs one clock cycle delay for every hop that needs to be taken to reach the event sinks.

The latter method is the fastest, is scalable and re-uses the communication topology. Therefore it forms the basis of our EDI implementation. Event data travels as fast as or faster than the functional data that caused the event. This is quick enough to distribute an event to all CSIs and PSIs before the data on which the monitor triggered leaves the communication architecture. This is a very important property we can use for debug as it allows us to keep the data that caused the event within the boundaries of the communication architecture for a (potentially) infinite amount of time. The actual processing of this data by the targeted consumer can then be analysed at any required level of detail. This is achieved by subsequently controlling the delivery operation for this data at the required debug granularity by programming the PSI and CSI near the consumer from the debugger software (see Section 5.7).

5.6.6 Debug Control Interconnect

The purpose of the DCI is to allow the functionality of the debug components to be controlled and their status queried.

The DCI allows run-time access to the on-chip debug infrastructure from off-chip debug equipment independently and transparently from the functional operation of the SoC. Examples of debug status information include whether any of the programmed events inside the monitors have already occurred, and/or whether the computation or communication inside the system has been stopped in response.

The state of the monitors, the PSIs and the CSIs becomes observable and controllable via so-called test point registers (TPRs) that connect to a IEEE Standard 1149.1-2001, TAP Controller (TAPC) as user-defined data registers [35]. These TPRs can be accessed and therefore programmed and queried using one or more user-defined instructions in the TAPC.

5.6.7 Debug Data Interconnect

The purpose of the Debug Data Interconnect (DDI) is to allow the system state to be observed and controlled after an event has stopped the relevant computation and communication.

Once the execution of a chip has come to a complete stop, preventing debug accesses from disturbing its execution is no longer a concern. The only concern is storage of the state inside the IP blocks.

We use the manufacturing-test scan chains to implement the DDI, as proposed by [32, 57, 71] and use a standard design flow with commercial, off-the-shelf (COTS) gate-level synthesis and scan-chain insertion. The IEEE 1149.1-compliant scan-based manufacturing test and debug infrastructure are made accessible from the TAP. Using the TAPC, data can be scanned out of the chip for use by the off-chip debug infrastructure described next.

5.7 Off-Chip Debug Infrastructure

5.7.1 Overview

This section presents the off-chip debug infrastructure and describes the techniques it can use to raise the debug abstraction level above the bit- and clock-cycle level, as depicted in Figure 5.11. We also present a generic debug application programmer's interface (API), which allows debug controllability and observability at the behavioral computation and communication level.

Figure 5.14 shows a generic, off-chip debug infrastructure. Our debugger software, called the integrated circuit debug environment (InCiDE) [69], connects to the debug port of the chip in potentially different user environments. Figure 5.14 shows a simulation environment, a field-programmable gate array (FPGA)-based prototyping environment, and a real product environment as three examples. The debugger software gains access the on-chip debug functionality through the debug interface, as described in Section 5.6. The debugger software allows the user to place (parts of) the SoC in functional or debug mode, and to inspect or modify the state of functional IP blocks or debug components.

5.7.2 Abstractions Used by Debugger Software

The InCiDE debugger software is layered and performs *structural, data, behavioral,* and *temporal* abstractions (refer to Figure 5.11) to provide the user with a high-level debug interface to the device-under-debug (DUD). Each abstraction function is described in more detail in the following subsections.

5.7.2.1 Structural Abstraction

Structural abstraction is achieved by applying the following three consecutive steps.

1. *Target Abstraction*

   Target-specific drivers are used to connect the debugger software using

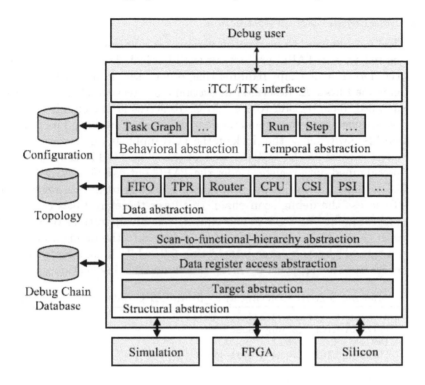

FIGURE 5.14: Off-chip debug infrastructure with software architecture.

the same software API to different implementation types of the DUD. Debug targets include simulation, FPGA prototyping, and product environments. A target driver enables access to the TAPC in its corresponding environment and allows performing capture, shift, and update operations on user data registers connected to the TAPC. An example tool control language (TCL) function call may look like Listing 5.1.

Listing 5.1: Writing and reading a user-defined data register.

```
1  set result [tap_write_read [list 0100 01011]]
```

which will shift the binary string "01011" (right-bit first) into the user-defined data register belonging to the TAPC binary instruction opcode "0100" via the test data input (TDI). The bit-string that is returned contains the values captured on the test data output (TDO) pin of the TAP on successive test clock (TCK) cycles during this shift operation. This layer also provides the `tap_reset` and `tap_nop` n commands to reset the TAPC and have no operation for n TCK cycles, respectively.

2. *Data Register Access Abstraction*

The mechanisms to access the various user-defined data registers connected to the TAPC are not always identical. For example, access to the debug scan chain requires that other user data registers are programmed first. As described in Section 5.6, this scan chain is connected as a user data register to the TAPC. To access it, the circuit first has to be switched from functional mode to debug scan mode and its functional clock(s) switched to the clock on the TCK input. In our architecture [71], a test control block (TCB) is used for this. The TCB is also mapped as a user-defined data register under the TAPC but can be accessed directly, i.e. without having to program another user-defined data register first. To access the debug scan chain, this layer therefore takes care of first programming the TCB to subsequently enable operations on the debug scan chain. For instance, the previous access to the debug scan chain is "wrapped" by this layer into Listing 5.2, while binary instruction opcodes are also replaced by more understandable instruction names.

Listing 5.2: Abstracting away from TAPC data register access details.

```
1  set result [tap_write_read [list \
2    PROGRAM_TCB <debug mode> \
3    DBG_SCAN    01011 \
4    PROGRAM_TCB <functional mode> \
5  ]]
6  set result [lindex $result 1]
```

This layer hides the subtle differences in the exact bit strings that are needed to enable access to the debug scan chain in different SoCs.

3. *Scan-to-Functional Hierarchy Abstraction*

This layer replaces the scan-oriented method of accessing flip-flops in user-defined data registers with a more design(er)-friendly method of accessing flip-flops and registers using their location in the RTL hierarchy. A multi-bit RTL variable or signal may be mapped to multiple flip-flops during synthesis. This layer utilizes this mapping information from the synthesis step to reconstructs the values of RTL variables and signals during debug from the values in their constituent flip-flops. In addition, it groups those signals and variables into hierarchical modules. A designer using this system can refer to signal and variable names using their RTL hierarchical identifiers and retrieve and set their values without needing to know the details about the TAPC, its user-defined instructions and data registers.

For example, the purpose of the previous access, shown in Listing 5.2 may have been to set the value of a five-bit RTL signal "usoc.unoc.u1router.be_queue.wrptr" to 0x0B ("01011"). Using this layer, this can now be accomplished by executing the code in Listing 5.3.

Listing 5.3: Setting and querying a register.

```
1  dcd_set  usoc.unoc.u1router.be_queue.wrptr  0x0B
2  dcd_synchronise
3  puts [dcd_get usoc.unoc.u1router.be_queue.wrptr HEX]
```

This layer takes care of mapping the individual bits of the value 0x0B into the correct bits inside the debug scan chain. The "dcd_synchronise" function is used to send the resulting chain to the chip and retrieve the previous content of the on-chip chain. The "puts" command prints the value of the register just retrieved from the chip.

These three structural abstraction steps are design-independent and are the consequences of our choice to access the state in the design using manufacturing-test scan chains mapped to the TAPC. They can therefore be applied to any digital design that utilizes the same on-chip debug architecture as presented in Section 5.6. They do however require structural information from various stages in the design and design for test (DfT) process, specifically the mapping information of RTL signals and variables to scannable flip-flops in the design, the location of these flip-flops in the resulting user-defined data registers, and specific TAPC instructions to subsequently enable access to these user-defined data registers. In Figure 5.14 all this information is stored in the debug chain database, which is automatically generated by our debugger software InCiDE.

5.7.2.2 Data Abstraction

The second abstraction technique employed by the debugger software is data abstraction. Based on the design's topology information, the debugger software can represent the state of known building blocks at a higher level than individual RTL signals or values.

For example, this layer can represent the state of a FIFO as its set of internal signals, including its memory, its read and its write pointers using the structural abstraction layers. If a design instance called "usoc.unoc.u1router.be_queue" is an 8-entry, 32-bit word FIFO, the user could use the command in Listing 5.4 to display its current state.

Listing 5.4: Querying individual registers of a FIFO.

```
1  dcd_synchronise
2  puts [dcd_get usoc.unoc.u1router.be_queue.mem    HEX ]
3  puts [dcd_get usoc.unoc.u1router.be_queue.wrptr  HEX ]
4  puts [dcd_get usoc.unoc.u1router.be_queue.rdptr  HEX ]
```

resulting in output such as

```
0x00000000
0x00000001
0x00000002
0x00000003
0x00000004
0x00000005
0x00000006
0x00000007
0x3
0x5
```

However, the user can also use the data abstraction layer and use the command in Listing 5.5

Listing 5.5: Printing the state of a FIFO.

```
1  print_fifo usoc.noc.u1router.be_queue VALID_ONLY HEX
```

to get

```
    ---------------------------------
    | usoc.unoc.u1router.be_queue |
    |-----------------------------|
    | Nr |          DATA          |
    |----|------------------------|
    | 03 |            0x00000003 |
    | 04 |            0x00000004 |

    ---------------------------------
```

Note how the software has interpreted the values of the read and write pointer to only print the valid entries in the FIFO ("VALID_ONLY"). Similar data abstraction functions have been implemented for the other standardised design modules, such as the monitors, CSIs, PSIs, routers, NIs and CPUs. In addition, these abstraction functions can be nested, e.g. the data abstraction function for the router may call multiple FIFO data functions to display the state of all its BE queues. The design knowledge required for this is contained in the "topology" file shown in Figure 5.14, which is automatically generated by the NoC design flow [20, 26].

5.7.2.3   Behavioral Abstraction

The previous two abstraction techniques focused on providing an abstracted state view and structural interconnectivity of common IP blocks. Behavioral abstraction targets the abstraction of the programmable functionality of these blocks. For example, two IP blocks communicate via two NIs and several routers. A monitor observes the communication data in Router R3 (refer to Figure 5.15).

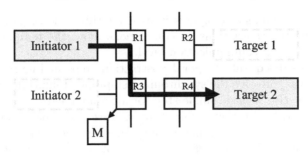

FIGURE 5.15: Physical and logical interconnectivity.

The exact IP modules that are involved depend not only on the physical interconnectivity but also on the programming of these IP blocks. For debugging a problem at the task graph level, we are first interested in the logical connection between these blocks. Only when there appears to be something physically wrong with this logical connection, do we refine the state view and look at their physical interconnectivity. A debug user can for instance issue a command as shown in Listing 5.6.

Listing 5.6: Querying the routers in the NoC.

```
1   set routers [get_router [get_conn {uc3 initiator1 target2}]]
```

This command provides a list of all routers that the logical connection between Initiator 1 and Target 2 uses in Use Case 3. With the data abstraction functions from the previous subsection, the user is able to display the states of these routers at the required level of detail.

Enabling debug at the behavioral level requires knowledge of the active use case, i.e., the programming of the NoC. This information is contained in the "configuration" file shown in Figure 5.14, which is automatically generated by the NoC design flow [20, 27].

5.7.2.4 Temporal Abstraction

A fourth debug abstraction technique is temporal abstraction. Traditionally debugging takes place at the clock cycle level of the CPU that is debugged. A disadvantage of this technique is that in a non-deterministic system the same event is unlikely to occur at the exact same clock cycle in multiple runs. Therefore temporal abstraction couples the debug execution control to events that are more meaningful to measure the progress made in the system's execution. Examples that are enabled using the hardware described in Section 5.6 include "Run until Initiator 1 or 2 initiates a transaction," and "Allow Target 2 to return 5 responses" before stopping the on-chip computation and/or communication [23].

Temporal abstraction first allows multiple clock cycles to be abstracted to one or more data element handshakes (refer to Figure 5.11). Protocol information on the handshake signals is used for this. The steps to messages on channels and to transactions on connections move the temporal abstraction level to the logical communication level.

The two subsequent temporal abstraction steps in Figure 5.11 are more complex as they involve the synchronized stepping of multiple communication channels. For this a basic single step for a communication channel is defined as *all* PSIs involved leaving their stopped state and process one communication request. The TCL command "`step $L -n` S" performs S single steps in succession for all PSIs in List L. For multiple channels, all stopped PSIs of the channels involved will need to process one communication request.

Note that this *single step* method forces a unique transaction order that must be known in advance to accurately represent the original use case. Otherwise there can be unwanted dependencies between the channels that are single-stepped, which potentially can lead to a deadlock. For this reason we also introduce the *barrier stepping* method and a corresponding TCL command extension "`step $L -n` S `-some` N," where *at least* N out of all PSIs in List L must perform a single step [23]. Barrier stepping is equal to single stepping when N is equal to the size of List L.

5.8   Debug Example

In this section we describe the application of the on-chip and off-chip debug infrastructure of Sections 5.6 and 5.7 using the example in Figure 5.12 and the NoC topology shown in Figure 5.15. We run our debugger software InCiDE with its extended API to perform interactive debugging using a simulated target. The following listing and output demonstrate the use of the API to control the communication inside the SoC during debug.

Listing 5.7: Example debug use case.

```
1  tap_reset
2  tap_nop 1000
3  set my_conn [get_conn {uc3 initiator1 target2}]
4  set my_routers [get_router $my_conn]
5  set my_router [lindex $my_conn 1]
6  set my_mon [get_monitor $my_router]
7  set_mon_event $my_mon {-fw 2 -value 0x0E40}
```

Line 1 resets the TAPC and Line 2 provides enough time for the system boot code [27] to functionally program the NoC. Lines 3 and 4 find the connection ("$my_conn") between Initiator 1 and Target 2 for the active use case , and the routers ("$my_routers") involved in the connection between Initiator 1 and Target 2. Note that on Line 5 we select the second router (Router R3) from the list of routers, and retrieve the monitor connected to it (refer to

Figure 5.15). This monitor is programmed on Line 7 to generate an event when the third word in a flit ("-fw 2") is equal to 0x0E40.

```
8   set my_tpr [get_tpr [get_psi $my_conn M req]]
9   set_psi_action $my_tpr −gran e −cond edi
10  dcd_synchronise tpr
11  tap_nop 1000
12  dcd_synchronise tpr
13  print_tpr $my_tpr
```

Lines 8 and 9 find the TPR of the PSI on the master request side of the connection between Initiator 1 and Target 2. This PSI TPR is programmed to stop all communication at the granularity of elements ("-gran e") when an event comes in via the EDI ("-cond edi"). Lines 10 and 11 write the resulting TPR debug program into the chip, and wait 1000 TCK cycles. On Line 12 the chip content is read back and on Line 13 the content of the PSI TPR is printed. This results in the following output.

```
-------------------------------------------------------------
|              {initiator1 pi} -> {core4 pt}                |
|-----------------------------------------------------------|
|Ch. Type | St.En. | St. Gran. | St. Cond. | St.St. | Left |
|---------|--------|-----------|-----------|--------|------|
|   Req   |  Yes   |  Element  |    EDI    |  Yes   | Yes  |
|   Resp  |  No    |  Message  |    EDI    |  No    | No   |
-------------------------------------------------------------
```

This table confirms that between Initiator 1 and its network interface ("core4"), the PSI was programmed to stop the communication on the request channel at the element level when an event comes in from the EDI. The PSI has entered the stop state ("St.St.") on the request channel.

```
14  continue $my_tpr
15  dcd_synchronise tpr
16  print_tpr $my_tpr
```

Line 14 continues the communication on the request channel, while Lines 15 and 16 query the TPR state, resulting in the following output.

```
-------------------------------------------------------------
|              {initiator1 pi} -> {core4 pt}                |
|-----------------------------------------------------------|
|Ch. Type | St.En. | St. Gran. | St. Cond. | St.St. | Left |
|---------|--------|-----------|-----------|--------|------|
|   Req   |  Yes   |  Element  |    EDI    |  No    | Yes  |
|   Resp  |  No    |  Message  |    EDI    |  No    | No   |
-------------------------------------------------------------
```

We observe that the PSI has left the stop state and is currently running, waiting for another event from the EDI. We now retrieve all PSI TPRs on a master request side. We program these to stop at the element level when an event comes in via the EDI. We subsequently generate an event on the EDI via the TAP using the "stop" command.

```
17  set my_tpr_all [get_tpr [get_psi * M req]]
18  set_psi_action $my_tpr_all −gran e −cond edi
19  stop
```

Once all transactions have stopped, we perform barrier stepping. We request that three execution steps are taken (at the granularity of data elements) by at least two PSIs ("-some 2") with verbose output ("-v").

```
20  step $my_tpr_all −n 3 −some 2 −v
```

This results in the following output.

```
- INFO: Checking if all Elements have stopped.....
- INFO: All Elements have stopped.
- INFO: Stepping starts.
- INFO: step 1 finished.
- INFO: step 2 finished.
- INFO: step 3 finished.
- INFO: All Elements are stopped.
```

The printed INFO lines show our barrier stepping algorithm at work. It first checks whether all selected PSIs ("$my_tpr_all") have entered their stopped state. If so, the software continues all PSIs. It subsequently polls whether at least two have since left and returned to their stopped state. When this has happened, the software will issue continue commands for those PSIs only and initiating the second step. This continues until for a third time, at least two PSIs have exited and re-entered their stopped state. Once barrier stepping is completed, we can read the content of the chip and print the content of the router.

```
21  dcd_synchronise
22  print_router $my_router HEX
```

This results for example in the following output.

```
 --------------------
| BE queue of R3_p1  |
|--------------------|
| Q.Nr |    DATA     |
|------|-------------|
|  18  | 0x200000123 |
|  19  | 0x300000124 |
 --------------------
- INFO: No valid data in GT queue of R3_p1.
```

In addition, we can print the state of the network interface.

```
23  print_ni [get_ni conn $my_conn M req ] HEX
```

This results in the following output.

```
----------------------------
| INPUT queue of NI000_p2 |
|--------------------------|
| Q.Nr |        DATA       |
|------|-------------------|
|  21  |      0x08000004   |
|  22  |      0x00000108   |
|  23  |      0x00000109   |
|  24  |      0x0000010A   |
|  25  |      0x0000010B   |
----------------------------
- INFO: No valid data in OUTPUT queue of NI000_p2.
```

5.9   Conclusions

In this chapter, we introduced three fundamental reasons why debugging a multi-processor SoC is intrinsically difficult; (1) limited internal observability, (2) asynchronicity, and (3) non-determinism. The observation of the root cause of an error is limited by the available amount of bandwidth to off-chip analysis equipment. Capturing a globally consistent state in a GALS system may not be possible at the level of individual clock cycles. In addition, an error may manifest itself in some runs of the system but not in others.

We classified existing debug methods by the information (scope), the detail (data abstraction), and the information frequency (temporal abstraction) they provide about the system. Debug methods are either intrusive or not. We subsequently introduced our communication-centric, scan-based, run/stop-based, and abstraction-based debug method, and described in detail the required on-chip and off-chip infrastructure that allows users of our debug system to debug an SoC at several number of levels of abstraction. We also illustrated our debug approach using a simple example system.

The analysis and methods presented in this chapter are only the first steps toward addressing the problem of debugging an SoC using a scientific approach. The use of on-chip DfD components, and debug abstraction techniques implemented in off-chip debugger software are ingredients for an overall SoC debug system. This system should link hardware debug to software debug, for SoCs with distributed computation, and using deterministic or guided replay.

A significant amount of research still needs to be carried out to reach this goal. This includes, for example, understanding and determining what parts of a system need to be monitored, and what parts must be controlled during debug and in what manner. More generally, pre-silicon verification and post-silicon debug methods and tools need to be brought together for seamless verification and debug throughout the SoC design process, and to prevent gaps in the verification coverage, and duplication of debug functionality.

Review Questions

[Q_1] Explain why the internal observability is limited in modern embedded systems.

[Q_2] Using multiple, asynchronous clock domains complicates debugging more than a single clock domain. Explore why designers utilize multiple, asynchronous clock domains when this is the case.

[Q_3] Describe the effect multiple, asynchronous clock domains have on the observation of a consistent global state.

[Q_4] What is the difference between a system run and a system trace?

[Q_5] Which three orthogonal classes of error observation for embedded systems have been explained in this chapter, and what types of errors occur in each class?

[Q_6] Describe how a single, unmodified system can produce multiple traces.

[Q_7] Describe the steps of the ideal debug flow.

[Q_8] List the four abstraction techniques presented in this chapter, and explain their role in the debug process.

[Q_9] Name three optical or physical debug techniques.

[Q_10] Explain the differences between, on the one hand, optical and physical debug techniques, and on the other hand, logical debug techniques.

[Q_11] What is deterministic replay and what are its requirements?

[Q_12] Name the four key characteristics of the CSAR debug approach.

[Q_13] List the required on-chip functionality to support the CSAR debug approach

[Q_14] Describe the functionality of the off-chip debug software in relation to the four abstraction techniques described in this chapter.

Bibliography

[1] D.A. Abramson and R. Sosic. Relative Debugging Using Multiple Program Versions. In *Int'l Symposium on Languages for Intensional Programming*, 1995.

[2] ARM. *CoreSight: V1.0 Architecture Specification*.

[3] ARM. AMBA Specification. Rev. 2.0, 1999.

[4] ARM. *AMBA AXI Protocol Specification*, June 2003.

[5] Semiconductor Industry Association. *The International Technology Roadmap for Semiconductors*. 2008.

[6] Algirdas Avizienis, Jean-Claude Laprie, and Brian Randell. In *Building the Information Society*, ed. René Jacquart. Dependability And Its Threats: A Taxonomy, pages 91–120. Kluwer, 2004.

[7] C. Beddoe-Stephens. Semiconductor Wafer Probing. *Test and Measurement World*, pages 33–35, November 1982.

[8] Michael Bedy, Steve Carr, Xianlong Huang, and Ching-Kuang Shene. A Visualization System for Multithreaded Programming. *SIGCSE Bulletin*, 32(1):1–5, 2000.

[9] British Standards Institute. British Standard BS 5760 on Reliability of Systems, Equipment and Components.

[10] K. Mani Chandy and Leslie Lamport. Distributed Snapshots: Determining Global States of Distributed Systems. *ACM Transactions on Computer Systems*, 3(1):63–75, 1985.

[11] Călin Ciordaş, Kees Goossens, Twan Basten, Andrei Rădulescu, and Andre Boon. Transaction Monitoring in Networks on Chip: The On-Chip Run-Time Perspective. In *Proc. Symposium on Industrial Embedded Systems (IES)*, pages 1–10, Antibes, France, October 2006. IEEE.

[12] Călin Ciordaş, Andreas Hansson, Kees Goossens, and Twan Basten. A Monitoring-aware Network-On-Chip Design Flow. *Journal of Systems Architecture*, 54(3-4):397–410, March 2008.

[13] P. Dahlgren, P. Dickinson, and I. Parulkar. Latch Divergency in Microprocessor Failure Analysis. In *Proc. IEEE Int'l Test Conference (ITC)*, pages 755–763, September/October 2003.

[14] Giovanni De Micheli and Luca Benini, editors. *Networks on Chips: Technology and Tools*. The Morgan Kaufmann Series in Systems on Silicon. Morgan Kaufmann, July 2006.

[15] Santanu Dutta, Rune Jensen, and Alf Rieckmann. Viper: A Multiprocessor SOC for Advanced Set-Top Box and Digital TV Systems. *IEEE Design and Test of Computers*, pages 21–31, September/October 2001.

[16] Marc Eisenstadt. My Hairiest Bug War Stories. *Communications of the ACM*, 40(4):30–37, April 1997.

[17] Jeroen Geuzebroek and Bart Vermeulen. Integration of Hardware Assertions in Systems-on-Chip. In *Proc. IEEE Int'l Test Conference (ITC)*, 2008.

[18] Holger Giese and Stefan Henkler. Architecture-Driven Platform Independent Deterministic Replay for Distributed Hard Real-Time Systems. In *Proc. ISSTA Workshop on the Role Of Software Architecture for Testing and Analysis*, pages 28–39, 2006.

[19] Kees Goossens, Martijn Bennebroek, Jae Young Hur, and Muhammad Aqeel Wahlah. Hardwired Networks on Chip in FPGAs to Unify Data and Configuration Interconnects. In *Proc. Int'l Symposium on Networks on Chip (NOCS)*, pages 45–54. IEEE Computer Society, April 2008.

[20] Kees Goossens, John Dielissen, Om Prakash Gangwal, Santiago González Pestana, Andrei Rădulescu, and Edwin Rijpkema. A Design Flow for Application-Specific Networks on Chip with Guaranteed Performance to Accelerate SOC Design and Verification. In *Proc. Design, Automation and Test in Europe Conference and Exhibition (DATE)*, pages 1182–1187, Washington, DC, USA, March 2005. IEEE Computer Society.

[21] Kees Goossens, John Dielissen, and Andrei Rădulescu. The Æthereal Network on Chip: Concepts, Architectures, and Implementations. *IEEE Design and Test of Computers*, 22(5):414–421, Sept-Oct 2005.

[22] Kees Goossens, Om Prakash Gangwal, Jens Röver, and A. P. Niranjan. Interconnect and Memory Organization in SOCs for Advanced Set-Top Boxes and TV — Evolution, Analysis, and Trends. In Jari Nurmi, Hannu Tenhunen, Jouni Isoaho, and Axel Jantsch, editors, *Interconnect-Centric Design for Advanced SoC and NoC*, chapter 15, pages 399–423. Kluwer, 2004.

[23] Kees Goossens, Bart Vermeulen, and Ashkan Beyranvand Nejad. A High-Level Debug Environment for Communication-Centric Debug. In *Proc. Design, Automation and Test in Europe Conference and Exhibition (DATE)*, 2009.

[24] Kees Goossens, Bart Vermeulen, Remco van Steeden, and Martijn Bennebroek. Transaction-Based Communication-Centric Debug. In *Proc. Int'l Symposium on Networks on Chip (NOCS)*, pages 95–106, Washington, DC, USA, May 2007. IEEE Computer Society.

[25] Jim Gray. Why Do Computers Stop and What Can Be Done about It? In *Proc. Symposium on Reliablity in Distributed Software and Database Systems*, 1986.

[26] Andreas Hansson. A Composable and Predictable On-Chip Interconnect. PhD thesis, Eindhoven University of Technology, June 2009.

[27] Andreas Hansson and Kees Goossens. Trade-offs in the Configuration of a Network on Chip for Multiple Use-Cases. In *Proc. Int'l Symposium on Networks on Chip (NOCS)*, pages 233–242, Washington, DC, USA, May 2007. IEEE Computer Society.

[28] H. Hao and K. Bhabuthmal. Clock Controller Design in SuperSPARC II Microprocessor. In *Proc. Int'l Conference on Computer Design (ICCD)*, pages 124–129, Austin, TX, USA, October 2–4, 1995.

[29] Timothy L. Harris. Dependable Software Needs Pervasive Debugging. In *Proc. Workshop on ACM SIGOPS*, pages 38–43, New York, NY, USA, 2002. ACM.

[30] C.F. Hawkins, J.M. Soden, E.I. Cole Jr., and E.S. Snyder. The Use of Light Emission in Failure Analysis of CMOS ICs. In *Proc. Int'l Symposium for Testing and Failure Analysis (ISTFA)*, 1990.

[31] Matthew W. Heath, Wayne P. Burleson, and Ian G. Harris. Synchro-tokens: A Deterministic GALS Methodology for Chip-level Debug and Test. *IEEE Transactions on Computers*, 54(12):1532–1546, December 2005.

[32] Kalon Holdbrook, Sunil Joshi, Samir Mitra, Joe Petolino, Renu Raman, and Michelle Wong. microSPARC: A Case Study of Scan-Based Debug. In *Proc. IEEE Int'l Test Conference (ITC)*, pages 70–75, 1994.

[33] Yu-Chin Hsu, Furshing Tsai, Wells Jong, and Ying-Tsai Chang. Visibility Enhancement for Silicon Debug. In *Proc. Design Automation Conference (DAC)*, 2006.

[34] William Huott, Moyra McManus, Daniel Knebel, Steven Steen, Dennis Manzer, Pia Sanda, Steven Wilson, Yuen Chan, Antonio Pelella, and Stanislav Polonsky. The Attack of the "Holey Shmoos": A Case Study of Advanced DFD and Picosecond Imaging Circuit Analysis (PICA). In *Proc. IEEE Int'l Test Conference (ITC)*, page 883, Washington, DC, USA, 1999. IEEE Computer Society.

[35] IEEE. *IEEE Standard Test Access Port and Boundary-Scan Architecture.* IEEE Computer Society, 2001.

[36] Axel Jantsch and Hannu Tenhunen, editors. *Networks on Chip.* Kluwer, 2003.

[37] D.D. Josephson, S. Poehhnan, and V. Govan. Debug Methodology for the McKinley Processor. In *Proc. IEEE Int'l Test Conference (ITC)*, pages 665–670, Oct 2004.

[38] A.C.J. Kienhuis. Design Space Exploration of Stream-based Dataflow Architectures: Methods and Tools. PhD thesis, Delft University of Technology, 1999.

[39] Herman Kopetz. The Fault Hypothesis for the Time-Triggered Architecture, In *Building the Information Society,* ed. René Jacquart, pages 221–234. Kluwer, 2004.

[40] Norbert Laengrich. Adapting Hardware-assisted Debug to Embedded Linux and Other Modern OS Environments. *PC/104 Embedded Solutions Journal of Small Embedded Form Factors,* 2006.

[41] Rick Leatherman and Neal Stollon. An Embedded Debugging Architecture for SoCs. *IEEE Potentials,* 24(1):12–16, Feb-Mar 2005.

[42] Thomas J. Leblanc and John M. Mellor-Crummey. Debugging Parallel Programs with Instant Replay. *IEEE Transactions on Computers,* C-36(4):471–482, April 1987.

[43] Bill Lewis. Debugging Backwards in Time. In *International Workshop on Automated Debugging,* October 2003.

[44] Michael R. Lyu, editor. *Handbook of Software Reliability and System Reliability.* McGraw-Hill, Inc., Hightstown, NJ, USA, 1996.

[45] Thomas Frederick Melham. Formalising Abstraction Mechanisms for Hardware Verification in Higher Order Logic. PhD thesis, University of Cambridge, August 1990. Also available as Technical Report UCAM-CL-TR-201.

[46] MIPS Technologies. PDTrace Interface Specification., 2002.

[47] Jens Muttersbach, Thomas Villiger, and Wolfgang Fichtner. Practical Design of Globally-Asynchronous Locally-Synchronous Systems. In *Proc. Int'l Symposium on Asynchronous Circuits and Systems (ASYNC),* April 2000.

[48] N. Nataraj, T. Lundquist, and Ketan Shah. Fault Localization Using Time Resolved Photon Emission and Still Waveforms. In *Proc. IEEE Int'l Test Conference (ITC),* volume 1, pages 254–263, September 30–October 2, 2003.

[49] André Nieuwland, Jeffrey Kang, Om Prakash Gangwal, Ramanathan Sethuraman, Natalino Busá, Kees Goossens, Rafael Peset Llopis, and Paul Lippens. C-HEAP: A Heterogeneous Multi-processor Architecture Template and Scalable and Flexible Protocol for the Design of Embedded Signal Processing Systems. *ACM Tansactions on Design Automation for Embedded Systems,* 7(3):233–270, 2002.

[50] OCP International Partnership. Open Core Protocol Specification, 2001.

[51] M. Paniccia, T. Eiles, V. R. M. Rao, and Wai Mun Yee. Novel Optical Probing Technique for Flip Chip Packaged Microprocessors. In *Proc. IEEE Int'l Test Conference (ITC),* pages 740–747, Washington, DC, USA, October 1998.

[52] Sudeep Pasricha and Nikil Dutt. *On-Chip Communication Architectures.* Morgan Kaufmann, 2008.

[53] Stephen E. Paynter, Neil Henderson, and James M. Armstrong. Metastability in Asynchronous Wait-Free Protocols. *IEEE Trans. Comput.*, 55(3):292–303, 2006.

[54] Philips Semiconductors. *Device Transaction Level (DTL) Protocol Specification. Version 2.2*, July 2002.

[55] Bill Roberts. The Verities of Verification. *Electronic Business*, January 2003.

[56] Michiel Ronsse and Koen de Bosschere. RecPlay: A Fully Integrated Practical Record/Replay System. In *ACM Transactions on Compuer Systems*, volume 17, pages 133–152, May 1999.

[57] G.J. Rootselaar and B. Vermeulen. Silicon Debug: Scan Chains Alone Are Not Enough. In *Proc. IEEE Int'l Test Conference (ITC)*, pages 892–902, Atlantic City, NJ, USA, September 1999.

[58] G.J. van Rootselaar, F. Bouwman, E.J. Marinissen, and M. Verstraelen. Debugging of Systems on a Chip: Embedded Triggers. In *Proc. Workshop on High-Level Design Validation and Test (HLDVT)*, 1997.

[59] J. A. Rowlette and T. M. Eiles. Critical Timing Analysis in Microprocessors Using Near-IR Laser Assisted Device Alteration (LADA). In *Proc. IEEE Int'l Test Conference (ITC)*, volume 1, pages 264–273, September 30–October 2, 2003.

[60] Smruti R. Sarangi, Brian Greskamp, and Josep Torrellas. CADRE: Cycle-Accurate Deterministic Replay for Hardware Debugging. In *Proc. IEEE Int'l Conference on Dependable Systems and Networks*, pages 301–312, Washington, DC, USA, 2006. IEEE Computer Society.

[61] B. Tabbara and K. Hashmi. Transaction-Level Modelling and Debug of SoCs. In *Proc. IP SOC Conference*, 2004.

[62] Shan Tang and Qiang Xu. In-band Cross-trigger Event Transmission for Transaction-based Debug. In *Proc. Design, Automation and Test in Europe Conference and Exhibition (DATE)*, pages 414–419, New York, NY, USA, 2008. ACM.

[63] Radu Teodorescu and Josep Torrellas. Empowering Software Debugging Through Architectural Support for Program Rollback. In *Workshop on the Evaluation of Software Defect Detection Tools*, 2005.

[64] Stephen H. Unger. Hazards, Critical Races, and Metastability. *IEEE Trans. Comput.*, 44(6):754–768, 1995.

[65] H. J. M. Veendrick. The Behaviour of Flip-flops Used as Synchronizers and Prediction of Their Failure Rate. *IEEE Journal of Solid-State Circuits*, 15(2):169–176, April 1980.

[66] Bart Vermeulen and Kees Goossens. A Network-on-Chip Monitoring Infrastructure for Communication-centric Debug of Embedded Multi-Processor SoCs. In *Proc. Int'l Symposium on VLSI Design, Automation and Test (VLSI-DAT)*, 2009.

[67] Bart Vermeulen, Kees Goossens, and Siddharth Umrani. Debugging Distributed-Shared-Memory Communication at Multiple Granularities in Networks on Chip. In *Proc. Int'l Symposium on Networks on Chip (NOCS)*, pages 3–12. IEEE Computer Society, April 2008.

[68] Bart Vermeulen, Yu-Chin Hsu, and Robert Ruiz. Silicon Debug. *Test and Measurement World*, pages 41–45, October 2006.

[69] Bart Vermeulen and Gert Jan van Rootselaar. Silicon Debug of a Co-processor Array for Video Applications. In *Proc. Workshop on High-Level Design Validation and Test (HLDVT)*, pages 47–52, Los Alamitos, CA, USA, 2000. IEEE Computer Society.

[70] Bart Vermeulen, Mohammad Z. Urfianto, and Sandeep K. Goel. Automatic Generation of Breakpoint Hardware for Silicon Debug. In *Proc. Design Automation Conference (DAC)*, pages 514–517, New York, NY, USA, 2004. ACM.

[71] Bart Vermeulen, Tom Waayers, and Sandeep K. Goel. Core-based Scan Architecture for Silicon Debug. In *Proc. IEEE Int'l Test Conference (ITC)*, pages 638–647, Baltimore, MD, USA, October 2002.

[72] Joon-Sung Yang and N.A. Touba. Enhancing Silicon Debug via Periodic Monitoring. In *Proc. Int'l Symposium on Defect and Fault Tolerance of VLSI Systems*, pages 125–133, October 2008.

[73] Pin Zhou, Feng Qin, Wei Liu, Yuanyuan Zhou, and Josep Torrellas. iWatcher: Efficient Architectural Support for Software Debugging. In *Proc. Int'l Symposium on Computer Architecture*, 2004.

6

System-Level Tools for NoC-Based Multi-Core Design

Luciano Bononi

Computer Science Department
University of Bologna
Bologna, Italy
bononi@cs.unibo.it

Nicola Concer

Computer Science Department
Columbia University
New York, New York, USA
concer@cs.columbia.edu

Miltos Grammatikakis

General Sciences Department, CS Group
Technological Educational Institute of Crete
Heraklion, Crete, Greece
mdgramma@cs.teicrete.gr

CONTENTS

6.1	Introduction	202
	6.1.1 Related Work	204
6.2	Synthetic Traffic Models	206
6.3	Graph Theoretical Analysis	207
	6.3.1 Generating Synthetic Graphs Using TGFF	209
6.4	Task Mapping for SoC Applications	210
	6.4.1 Application Task Embedding and Quality Metrics	210
	6.4.2 SCOTCH Partitioning Tool	214
6.5	OMNeT++ Simulation Framework	216
6.6	A Case Study	217
	6.6.1 Application Task Graphs	217
	6.6.2 Prospective NoC Topology Models	218

	6.6.3	Spidergon Network on Chip	219
	6.6.4	Task Graph Embedding and Analysis	221
	6.6.5	Simulation Models for Proposed NoC Topologies . . .	223
	6.6.6	Mpeg4: A Realistic Scenario	227
6.7	Conclusions and Extensions	231
Review Questions	234
Bibliography	235

6.1 Introduction

Networks-on-chips (NoCs) provide a high performance, scalable and power-efficient communication infrastructure to both chip multiprocessor (CMP) and system on chip (SoC) systems [63]. A NoC usually consists of a packet-switched on-chip micro-network, foreseen as the natural evolution of traditional bus-based solutions, such as AMBA *AXI* [2], and IBM's *Core Connect* [35]. Innovative NoC architectures include the LIP6 *SPIN* [1], the M.I.T. *Raw* [79], the VTT (and various Universities) *Eclipse* [24] and *Nostrum* [25], PHILIPS' *Æthereal* NoC [27], and Stanford/Uni-Bologna's *Netchip* [5, 36].

These architectures are mostly based on direct, low-radix, point-to-point topologies, in particular meshes, tori and fat trees, offering simple and efficient routing algorithms based on small area, high frequency routers. In contrast, high-radix, point-to-point networks combine together independent network stages to increase the degree of the routers (making channels wider). At the expense of higher wiring complexity, high-radix NoC topologies reduce network diameter and cost (smaller number of internal channels and buffers) and improve resource sharing, performance, scalability, and energy-efficiency. thus effectively utilizing better available network bandwidth. High radix NoC topologies include the concentrated mesh which connects several cores at each router [4], and flattened butterfly [40] which combines routers in each row of the conventional butterfly topology, while preserving inter-router connections.

A major challenge for predicting performance and scalability of a particular NoC architecture relies on precise specification of real application traffic requirements arising from current and future applications, or scaling of existing applications. For example, it has been estimated that SoC performance varies by up to 250 percent depending on NoC design, and up to 600 percent depending on communication traffic [49], while NoC power dissipation can be reduced by more than 60 percent by using appropriate mapping algorithms [31].

Future MPSoC applications require scalable NoC topologies to interconnect the IP cores. We have developed new system level tools for NoC design space exploration and efficient NoC topology selection by examining theoretical graph properties, as well as application mapping through task graph partitioning. These tools are derived by extending existing tools in parallel processing, graph theory and graphical visualization to NoC domain. Besides enabling efficient NoC topology selection, our methods and tools are important for the design of efficient multi-core SoCs.

Our NoC design space exploration approach explained in Figure 6.1 follows an open-source paradigm, focusing on system-level performance characterization, rather than power dissipation or dynamic power management for low power or power-aware design. The major reason is that although several state-of-the-art, relatively accurate and fast tools can perform behavioral synthesis of cycle-accurate transaction-level SystemC (or C/C++) models to estimate, analyze and optimize total energy (or power evolution with time), they use spreadsheets or back annotation from power-driven high-level synthesis, or corresponding (behavioral and structural) RTL simulation models; these models are rarely available at an early design stage. Moreover, almost all commercial and academic high-level power tools (see list below), are not open source.

- ChipVision's Orinoco is a tool chain estimating system-level performance and power for running algorithms (specified in ANSI-C or SystemC) on different architectures [11] [77]. Components are instrumented with area, dataflow and switching activity using a standard power library for the target technology which consists of functional units, such as adders, subtractors, multipliers, and registers. Algorithms are compiled to hierarchical control data flow graphs (CDFGs) which describe the expected circuit architecture without resorting to complete synthesis.

- Early estimates from RTL simulation can be back annotated through a graphical user interface into system-level virtual platform models created in the Innovator environment, recently announced by Synopsys. These models can help estimate power consumption and develop power management software [78].

- HyPE is a high-level simulation tool that uses analytical power macromodels for fast and accurate power estimation of programmable systems consisting of datapath and memory components [51].

- Web-based JouleTrack estimates power of an instruction-level model specified in C for commercial StrongARM SA 1100 and Hitachi SH-4 processors [72].

- SoftExplorer is similar to JouleTrack, but focuses on commercial DSP processors [74]. Other similar tools are Simunic [73] and Avalanche [30].

- BlueSpec [7], PowerSC [42, 43] and open source Power-Kernel [9] are frameworks built by adding C++ classes on top of SystemC for power-aware characterization, modeling and estimation in multiple levels of abstraction.

In particular, Power-Kernel (see Chapter 3 in this book) is an efficient, open-source, object-oriented SystemC2.0 library, which allows simple introduction of a power macro model in SystemC at RTL level of a complex design. PK achieves much higher simulation speed than lower-level power analysis tools. High-level model instrumentation is based on a SystemC class that uses advanced dynamic monitoring and storage of I/O signal activity of SoC blocks with appropriate signal augmentation, and put_activity and get_activity gathering library functions [9]. Both constant power models and more accurate regression-based models with a linear dependence on clock frequency, gate and flip-flop switching activity are used. As an example, dynamic energy estimation of the AMBA AHB bus is decomposed into arbiter, decoder and multiplexing logic for read and write operations (master to/from slave). The latter operations are estimated to control over 84 percent of the total dynamic power consumption. Similar power instrumentation techniques for synthesizable SystemC code at RTL level are described in [84].

We consider both graph theoretical metrics, e.g., number of nodes and edges, diameter, average distance, bisection width, connectivity, maximum cut and spectra, as well as embedding quality metrics for mapping different synthetic and real applications into NoC resources, such as computing, storage and reconfigurable FPGA elements.

The mapping algorithm of the partitioning tool obtains an assignment of application components into the NoC topology depending on abstract requirements formulated as static or dynamic (run-time) constraints on application behavior components and existing NoC architectural and topological properties. These constraints are expressed using static or dynamic properties of NoC nodes and communications links (e.g., IP type, multi-threading or multiprocessing performance, power, and reliability) or characteristics of computational and storage elements (e.g., amount of memory, number of processors, or task termination deadlines for real-time tasks).

6.1.1 Related Work

Previous research efforts have studied application embedding into conventional symmetric NoC topologies. Hu and Marculescu examined mapping of a heterogeneous 16-core task graph representing a multimedia application into a mesh NoC topology [31, 33], while Murali and De Michelli used a custom tool (called SUNMAP) to map a heterogeneous 12-core task graph representing a video object plane decoder and a 6-core DSP filter application into a mesh or torus NoC topology using different routing algorithms [59, 60].

Other publications focus on application traffic issues, e.g., communication weighted models consider communication aspects (CWM), while communication dependence and computation models (CDCM) simultaneously consider both application aspects. By mapping applications into regular NoCs and computing the NoC execution delay and dynamic energy consumption, (obtained by modeling bit transitions for better accuracy), CDCM is shown to provide

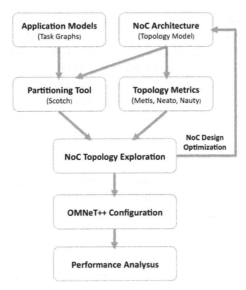

FIGURE 6.1: Our design space exploration approach for system-level NoC selection.

average reductions of 40 percent in NoC execution time, and 20 percent in NoC energy consumption, for current technologies, e.g., refer to [54].

The proprietary SUNMAP tool, proposed by Stanford and Bologna University, performs NoC topology exploration by minimizing area and power consumption requirements and maximizing performance characteristics for different routing algorithms. The XPIPES compiler can eventually extract efficient synthesizable SystemC code for all network components, i.e., routers, links, network interfaces and interconnect, at the cycle- and bit-accurate level.

Other approaches consider generating an ad hoc NoC interconnect starting from the knowledge of the application to support, a given set of constraints (i.e., maximum latency, minimum throughput) and a library of components such as routers, repeaters and network interfaces. Pinto et al. propose a constraint driven communication architecture synthesis of point-to-point links using a *k-way* merging heuristic [71]. In [76] authors propose an application-specific NoC synthesis which optimize the power consumption and area of the design so that the required performance constrains are met.

Quantitative evaluations of mapping through possibly cycle-accurate SystemC-based virtual platforms have also been discussed, and refer to event-driven virtual processing unit mapping networking applications [38]. Finally, notice that topology customization for cost-effective mapping of application-specific designs into families of NoCs is a distinct problem (although it could be solved with similar techniques). Techniques for mapping practical application task graphs into the Spidergon STNoC family have already been examined [13, 68, 69].

Our study generalizes previous studies by considering a plethora of theoretical topological metrics, as well as application patterns for measuring embedding quality. It focuses on conventional NoC topologies, e.g., mesh and torus, as well as practical, low-cost *circulants*: a family of graphs offering small network size granularity and good sustained performance for realistic network sizes (usually below 64 nodes). Moreover, it essentially follows an open approach, as it is based on extending to NoC domain and parameterizing existing open-source (and free) tools coming from a variety of application domains, such as traffic modeling, graph theory, parallel computing, and network simulation.

In Section 6.2 we describe application traffic patterns used in our analysis. In particular, we focus on the *Task Graphs For Free* tool, called TGFF, that we used for generating *synthetic task graphs* in our simulations.

In Section 6.3 we describe the tools that we used to study different NoC architectures in order to understand their topological properties.

In Section 6.4, we describe the problem of *application task graph mapping*. We define the adopted metrics to rate the quality a given mapping and describe the SCOTCH partitioning tool used to map a given task graph into the considered network on chip.

In Section 6.5, we describe the *Objective Modular Network Testbed in C++*, called OMNET++, the simulation framework used to implement our bit- and cycle-accurate network model and perform our system-level design space exploration.

In Section 6.6, we report a case-study consisting of task generation, mapping analysis, and bit- and cycle-accurate system-level NoC simulation for a set of synthetic tree-based task graphs, as well as a more realistic application consisting of an Mpeg4 decoder.

Finally, in Section 6.7, we draw conclusions and consider interesting extensions.

6.2 Synthetic Traffic Models

Parallel computing applications are often represented using *task graphs* which express the necessary computing, communication and synchronization patterns for realizing a particular algorithm.

Task graphs are mapped to basic IP blocks with clear, unambiguous and self-contained functionality interacting together to form a NoC application.

Task graph embedding is also used by the operating system for reconfiguring faulty networks, i.e., providing fault-free virtual sub-graphs in "injured" physical system graphs to maintain network performance (network bandwidth and latency) in the presence of a limited number of faults.

Vertices (or nodes) usually represent computation, while links represent communication. A node numbering scheme in *directed acyclic graphs* (DAGs) takes into account precedence levels. For example, an initial node is labelled node 0, while an interior node is labelled j, if its highest ranking parent is labelled $j-1$.

Undirected and directed acyclic task graphs represent parallelism at both coarse and fine grain. Examples of coarse grain parallelism are inter-process communications, control and data dependencies and pipelining. Fine grain parallelism is common in digital signal processing, e.g., FFTs or power spectra, and multimedia processing, such as common data parallel prefix operations and loop optimizations (moving loop invariants, loop unrolling, loop distribution and tiling, loop fusion, and nested loop permutation).

6.3   Graph Theoretical Analysis

In order to examine inherent symmetry and topological properties in prospective constant degree NoC topologies (especially chordal rings) and compare with existing tables of optimized small degree graphs, we examine available open-source software tools and packages that explore graph theoretical properties, This is particularly important, since the diameter and average distance metrics of general chordal rings are not monotonically increasing and cannot be minimized together. In fact, this methodology helped in evaluating theoretical properties of several families of directed and undirected constant degree circulant graphs. In our analysis, we focus on:

- Small, constant network extendibility

- Small diameter and large, scalable edge bisection for fewer than 100 nodes

- Good fault tolerance (high connectivity)

- Efficient VLSI layout with short, mostly local (small chordal links) wires

- Efficient (wire balanced) point-to-point routing without pre-processing

- Efficient intensive communication algorithms with a high adaptivity factor e.g., for broadcast, scatter, gather, and many-to-few patterns

More specifically, this approach is based on several steps. After METIS and NAUTY analyze automorphisms as explained below, NEATO can display the graph so that certain graph properties and topologically-equivalent vertices are pictured; two vertices are equivalent (identical display attributes), if there is a vertex-to-vertex bijection preserving adjacency. A 4×7 mesh has eight vertex equivalence classes (orbits); all vertices in each orbit have identical colors in NEATO representation; vertices incident to different clusters have different colors. Special colors mark edges that bridge the two clusters forming

FIGURE 6.2: METIS-based NEATO visualization of the Spidergon NoC layout.

the bisection, i.e., from these graphs, we can observe scalability issues, e.g., bisection width. Alternatively, for vertex-symmetric graphs with a single vertex equivalence class (only one orbit), such as ring, torus and hypercubes, NAUTY selects a base vertex (e.g., a red square) and modifies display attributes based on the distance of each vertex from the chosen base vertex. An example of this analysis is Figure 6.2, which shows the NEATO graphical representation for a Spidergon STNoC topology of 32 nodes (without colors); notice that the links resembling to "train tracks" in this figure actually correspond to the cross links of the Spidergon topology.

Next, we describe these open tools, especially METIS and NAUTY in more detail.

- Karypis' and Kumar's METIS provides an extremely fast, multilevel graph partitioning embedding heuristic that can also extract topological metrics, e.g., diameter, average distance, in/out-degree, and bisection width [37]. Concerning edge bisection, for small graphs, ($N < 40$) nodes, a custom-coded version of Lukes' exponential-time dynamic programming approach to partitioning provides an exact bisection if one exists [53]. For larger graphs, METIS partitioning is used to approximate a near-minimum bisection width;

- McKay's NAUTY computes the automorphisms in the set of adjacency-preserving vertex-to-vertex mappings. NAUTY also determines the *orbits* that partition graph vertices into equivalence classes, thus providing symmetry and topological metrics [55];

- AT&T's NEATO is used for visualizing undirected graphs based on spring-relaxation and controlling the layout, while supporting a variety of output formats, such as PostScript and Gif [62].

It is important to mention that our open methodology has led to the development of a Linux-based NoC design space exploration tool suite (*Iput, Imap, Irun,* and *Isee*) at ST Microelectronics.

6.3.1   Generating Synthetic Graphs Using TGFF

In 1998, Dick and Rhodes originally developed *Task Graphs For Free* (TGFF) as a C++ software package that facilitates standardized pseudo-random benchmarks for scheduling, allocation and especially hardware-software co-synthesis [21]. TGFF provides a flexible, general-purpose environment with a highly configurable random graph generator for creating multiple sets of synthetic, pseudo-random directed acyclic graphs (DAGs) and associated resource parameters that model specific application behavior. DAGs may be exported into postscript, VCG graphical visualization or text format for importing them into mapping or simulation frameworks; notice that VCG is a useful graph display tool that provides color and zoom [81].

TGFF users define a source (*.tgffopt) file that determines the number of task graphs, the minimum size of each such graph, and the types of nodes and edges through a set of parameterized commands and database specifications. For example, random trees are constructed recursively using series-parallel chains, i.e., at least one root node is connected to multiple chains of sequentially linked nodes.

Ranges for the number of chains, length of each chain and number of root nodes are set by the user using TGFF commands. Notice that chains may also rejoin with a given probability by connecting an extra (sink) node to the end of each chain. TGFF includes many other support features, such as

- Indirect reference to task data: task attribute information is provided through references to processing element tables for node types or transmission tables for communication edge types.

- User-defined graph attributes: generating statistics for node or edge performance, power consumption, or reliability characteristics.

- Real-time processing through an association of tasks to periods and deadlines.

- Multi-rate task graphs: tasks exchange data at different rates either instantaneously or using queues.

- Multi-level hierarchical task graphs, where each task is actually a task graph; this is possible by interpreting task-graph 1 as the first task in task-graph 0, task-graph 2 as the second task in task-graph 0, etc; there are certain restrictions.

Application graph structures are generated using TGFF in several research and development projects. For example, TGFF is used for application task graph generation in heterogeneous embedded systems, hardware software co-design, parallel and distributed systems and real-time or general-purpose operating systems [21].

Within the NoC domain, TGFF is commonly used in energy-aware application mapping, hw/sw partitioning, synthesis optimization, dynamic voltage scaling and power management. In this respect, all synthetic tree-like benchmarks used in our case study (see Section 6.6) have been generated using our extended version of the TGFF package. Since these task graphs are deterministic, we had to modify TGFF to avoid recursive constructions and impose lower bounds on the number of tasks.

6.4 Task Mapping for SoC Applications

A mapping algorithm selects the most appropriate assignment of tasks into the nodes of a given NoC architecture. In complex, realistic situations, all combinations of task assignments must be considered. In most cases, a near-optimal solution that approximately minimizes a cost function is computed in reasonable time using heuristic algorithms. The heuristic takes into account the type of tasks, the number and type of connected nodes, and related constraints, such as required architecture, operating system, memory latency and bandwidth, or total required memory for all tasks assigned to the same node.

After the mapping algorithm obtains a near optimal allocation pattern for the given task graph, the operating system can initiate automated task allocation into the actual NoC topology nodes.

6.4.1 Application Task Embedding and Quality Metrics

Mapping is a network transformation technique based on graph partitioning. Mapping refers to the assignment of tasks (e.g., specifying computation and communication) to processing elements, thus implicitly specifying the packet routes. Within the NoC domain, mapping can also address the assignment of IP cores to NoC tiles, which together with routing path allocation, i.e., communication mapping, is commonly referred as network assignment. Network assignment is usually performed after task mapping and aims to reduce on-chip inter-communication distance. *Scheduling* refers to time ordering of tasks on their assigned resources, which assures mutual exclusion among different task executions on the same resource. Scheduling can be performed online (during task execution) or offline, in pre-emptive or non-pre-emptive fashion, and it can use static or dynamic task priorities. In non-pre-emptive scheduling tasks are executed without interruption until their completion, while in pre-emptive scheduling, tasks with lower priorities can be suspended by tasks with higher priorities. Pre-emptive scheduling is usually associated with online scheduling, while non-pre-emptive scheduling corresponds to offline scheduling. Static priorities are assigned once at the beginning of scheduling and do not require later updating.

Assuming that tasks are atomic and cannot be broken into smaller tasks, a mapping (or scheduling) scheme is called static if the resource on which each task is executed is decided prior to task execution, i.e., mapping is executed once at compile time (offline), and is never modified during task execution. With dynamic mapping (or scheduling) the placement of a task can be changed during application execution, thus affecting its performance during run-time (online). Quasistatic mapping (or scheduling) is also possible; these algorithms build offline different mappings (or trees of schedules) and choose the best solution during run-time. Dynamic mapping can obviously lead to higher system performance, as well as several other nice properties, such as lower power dissipation and improved reliability, which are particularly important in certain applications, e.g., detection, tracking, and targeting in aeronautics [32]. However, dynamic mapping suffers from overheads, e.g., computational overhead which may increase the run-time delay and energy consumption, and additional complexity for testing. In this work, we deal mainly with static mapping which is usually recommended for embedded systems, especially for NoC where communication overhead can be significant if performed at run-time. However, a more complete and generic system-level view of a multi-core SoC architecture which involves dynamic mapping is provided at the end of this chapter.

Graph *partitioning* decomposes a source (application) or target (architecture) graph into clusters for a broad range of applications, such as VLSI layout or parallel programming. More specifically, given a graph $G(n, m)$ with n weighted vertices and m weighted edges, graph partitioning refers to the problem of dividing the vertices into p cluster sets, so that the sum of the vertex weights in each set is as close as possible (balanced total computation load), and the sum of the weights of all edges crossing between sets is minimized (minimal total communication load).

In the context of multi-core SoC, graph *embedding* optimally assigns data and application tasks (IPs) to NoC resources, e.g., RISC/DSP processors, FPGAs or memory, thus forming a generic binding framework between SoC application and NoC architectural topology. Graph embedding also helps map existing applications into a new NoC topology by porting (with little additional programming overhead) existing strategies from common NoC topologies.

Unfortunately, even in the simple case where edge and vertex weights are uniform and $p = 2$, graph embedding into an arbitrary NoC topology is NP-complete [26]. In general, there is no known, efficient algorithm to solve this problem, and it is unlikely that such an algorithm exists. Hence, we resort to heuristics that partially compromise certain constraints, such as balancing the communication load, or (more typically) using approximate communication load minimization constraints, i.e., maximizing locality and look-ahead time by statically mapping intensive inter-process communication to nearby tasks. These constraints are often specified in an abstract way through a cost function. This function may also consider more complex constraints, such as minimizing the total communication load for all target NoC components, e.g., for optimizing total system-level power consumption during message exchanges. Although this function is clearly application dependent, it is usually expressed as a weighted sum of terms representing load on different NoC topology nodes and communication links, considering also user-defined optimality criteria, e.g., in respect to architecture, such as shortest-path routing and number or speed of processing elements, communication links, and storage elements.

Mapping algorithms for simple application graphs, such as rings or trees have been studied extensively in parallel processing, especially for direct networks, such as hypercubes and meshes [50]. For general graphs, mapping algorithms are usually based on simulated annealing or graph partitioning techniques.

Simulated annealing originates in the Metropolis-Hastings algorithm, a Monte Carlo method to generate sample states of a thermodynamic system, invented in 1953 [56]. Simulated annealing has received significant attention in the past two decades to solve single and multiple objective optimization problems, where a desired global minimum/maximum is hidden among many local minima/maxima. Simulated annealing first defines an initial mapping based on the routing function, e.g., shortest-path, dimension-order or non-minimal path. Then, this algorithm always accepts injection of new disturbances that reduce an appropriately defined cost function that measures the relative cost of the embedding, while it accepts only with a decreasing probability the injection of new disturbances that increase the relative cost function.

Graph partitioning heuristics are usually based on recursive bisection using either global (inertial or spectral) partitioning methods or local refinement techniques, e.g., Kernighan-Lin [39]. Results of global methods can be fed to local techniques, which often leads to significant improvements in performance and robustness. With current state-of-the-art, extremely fast partitioning heuristics are based on bipartitioning, i.e., the graph is partitioned into two halves recursively, until a desired number of sets is reached; notice that quadrisection and octasection algorithms may achieve better results [37].

Popular global partitioning methods are classified into inertial (based on 1-d, 2-d or 3-d geometrical representation) or spectral (using Eigenvectors of the Laplacian of the connectivity graph). For a long time, the Kernighan-Lin algorithm has been the only efficient local heuristic and is still widely used in several applications with some modifications, such as Fiduccia and Mattheyses [23].

Mathematically, an embedding of a source graph G_S into a given target graph G_T is an injective function from the vertex set of G_S to the vertex set of G_T. Quality metrics for embedding includes application-specific mapping criteria and platform-related performance metrics, such as the time to execute the given application using the selected mapping.

Common graph theoretical application-specific embedding quality metrics are listed below.

- **Edge Dilation** of an edge of G_S is defined as the length of the path in G_T into which an edge of G_S is mapped. The dilation of the embedding is defined as the maximum edge dilation of G_T. Similarly, we define average and minimum dilation metrics. These metrics measure latency overhead during point-to-point communication in the target graph G_T. A low dilation is usually beneficial, since most communication devices are located nearby, and hence the probability of higher application throughput increases.

- **Edge Expansion** refers to a weighted-edge graph G_S. It multiplies each edge dilation with its corresponding edge weight. The edge expansion of the embedding is usually defined as the maximum edge expansion of G_T. Similarly, we define average and minimum edge expansion metrics.

- **Edge Congestion** is the maximum number of edges of G_S mapped on a single edge in G_T. This metric measures edge contention in global intensive communication.

- **Node Congestion** is the maximum number of paths containing any node in G_T where every path represents an edge in G_S. This metric is a measure of node contention during global intensive communication. A mapping with high congestion causes many paths to traverse through a single node, thus increasing the probability of a network traffic bottleneck due to poor load balancing.

- **Node Expansion** (also called load factor or compression ratio) is the ratio of the number of nodes in G_T to the number of nodes in G_S. Similarly, maximum node expansion represents the maximum number of nodes of G_S assigned to any node of G_T.

- **Number of Cut Edges**. The cut edges are edges incident to vertices of different partitions. They represent extra (inter-module) communication required by the mapping. This metric is used for comparing target graphs with identical number of edges, the lower its value the better.

In the following sections we will examine edge dilation, edge expansion and edge congestion metrics for a number of traffic patterns particularly interesting in the SoC domain, as well as for communication patterns arising from real applications mapped into Spidergon and other prospective NoC topologies. Through optimized embedding, many algorithms originally developed for common mesh and torus topologies may be emulated on the Spidergon. Furthermore, since embedding of common application graphs, e.g., binary trees on mesh, has already been investigated, we can derive embedding of these graphs into Spidergon by applying graph composition.

```
0
8 24
0 100
6 2 5 4
5 6 6 7 2 3 1 0
7 2 5 4
2 2 5 4
4 6 3 1 0 6 7 2
3 2 4 5
1 2 4 5
0 2 4 5
```

FIGURE 6.3: Source file for SCOTCH partitioning tool.

6.4.2 SCOTCH Partitioning Tool

The SCOTCH project (1994-2001) at *Université Bordeaux I - LaBRI* focuses on libraries for statically mapping any possibly weighted source graph into any possibly weighted target graph, or even into disconnected sub-graphs of a given target graph [70]. SCOTCH maps graphs in linear time to the number of edges in the source graph, and logarithmic time to the number of vertices in the target graph.

SCOTCH has two forms of license: private version licensed for commercial applications, and an open-source version available for academic research. The academic distribution comes with library documentation, sample graphs and free access to source code. SCOTCH builds and validates source and target graphs and then displays obtained mappings in colorful graphs [70]. It easily interfaces to other partitioning or theoretical graph analysis programs, such as METIS or NAUTY, due to standardized vertex/edge labeling formats.

SCOTCH operates by taking as input a *source* file (.src) that represents the application task graph to be mapped. Figure 6.3 shows a snapshot of a sample source file.

The first three lines of the file represent some configuration info such as file version number, number of vertex and edges and other file-related options. From the fourth line onwards, the source file represents the communication task graph, where the first entry column represents the considered node's *id*, the second the number of destinations, and then the list of destination ids. For example the third line in Figure 6.3 says that node 6 communicates with two destinations: nodes 5 and 4. In case of different communication bandwidths, next to each destination id there is the traffic bandwidth between the source node and the specific destination.

Target files are the result of a mapping computation in SCOTCH. Figure 6.4 shows the result of such a mapping. The first element states the number of

```
8
5 5
6 0
7 1
2 3
4 2
3 4
1 7
0 6
S Strat=b{job=t,map=t,poli=S,
strat=m{asc=f{type=b,move=80,
pass=-1,bal=0.005},
low=h{pass=10},type=h,vert=80,rat=0.7}x}

M Processors 8/8 (1)
M Target min=1 max=1 avg=1 dlt=0 maxmoy=1
M Neighbors min=2 max=6 sum=24
M CommDilat=1.666667 (20)
M CommExpan=1.666667 (20)
M CommCutSz=1.000000 (12)
M CommDelta=1.000000
M CommLoad[0]=0.000000
M CommLoad[1]=0.500000
M CommLoad[2]=0.333333
M CommLoad[3]=0.166667
```

FIGURE 6.4: Target file for SCOTCH partitioning tool.

nodes mapped. The following two columns are the pairs:

$$< architecture\ node\ id,\ application\ node\ id > \qquad (6.1)$$

SCOTCH then generates the metrics relative to the mapping that we discussed above.

We have modified the TGFF package for application task generation to adopt SCOTCH format for defining source graphs as follows.

- Source graphs (*.src) are generated either by the user or through the TGFF tool (see Section 6.3.1).

- In addition, geometry files (*.xyz) are generated either by the user, e.g., for Spidergon STNoC, or by the SCOTCH partitioning tool for common graphs, such as mesh or torus. Geometry files have a .xyz extension and hold the coordinates of the vertices of their associated graph. They are used by visualization programs to compute graphical representations of mapping results.

- Finally, target NoC topology graphs (*.tgt) are generated automatically from corresponding source graphs using the SCOTCH partitioning tool. These files contain complex target graph (architecture) partitioning information which is exploited during SCOTCH embedding.

SCOTCH features extremely efficient multi-level partitioning methods based on recursive graph bipartitioning [70]. More specifically, initial and redefined bipartitions use:

- Fiduccia-Mattheyses heuristics that handle weighted graphs

- Randomized and backtracking methods

- Greedy graph-growing heuristics

- A greedy strategy derived from Gibbs, Poole, and Stockmeyer algorithm

- A greedy refinement algorithm designed to balance vertex loads

SCOTCH application developers can select the best partitioning heuristic for each application domain by changing partitioning parameters. However, for symmetric target architectures the default strategy (bipartitioning) performs better than all other schemes.

6.5   OMNeT++ Simulation Framework

OMNET++ is an object-oriented modular discrete-event network simulator [44]. This tool can be used for traffic modeling of queuing networks, communication and protocols, telecommunication networks, and distributed systems, such as multiprocessors or multicomputers. A model is defined by defining and connecting together simple and compound (hierachically nested) modules which implement different model entities. Communication among modules is implemented by exchanging messages. The source code is freely available for the academic community, while it requires a license for commercial use. OMNET++ offers a number of libraries and tools that allow a user to rapidly develop complex simulation projects providing:

- A user-friendly graphical user interface that defines the simulator skeleton: this allows the user to easily define the different agents acting in the environment to be simulated, as well as delineating the relations and hierarchies existing among them; this interface is useful for learning the simulator and debugging.

- A library for automatic handling of inter-process signaling and messaging.

- A library implementing the most important, commonly used statistical probability distribution functions.

- An interesting graphical user interface that allows the user to inspect and interact with the simulation at run-time by allowing the user to modify parameters, inspect objects or plot run-time graphs.

- A number of tools that collect, analyze and plot the simulation results.

- Many freely developed models for wired/wireless network communication protocols like TCP-IP, IEEE 802.11 or ad hoc routing protocols.

In contrast to an already existing SystemC model, the OMNET++ model hides many low-level details relative to NoC implementation in order to concentrate on understanding the effects caused by major issues like core mapping, routing algorithm selection and communication buffer sizing at the router and network interface nodes. Clearly, we do not completely ignore details on these resources (especially for the router) when measuring network performance, but rather treat them as constant parameters in our bit- and cycle-accurate system-level models.

6.6 A Case Study

In this case study, we consider embedding application task graphs into several prospective NoC topologies.

At first, we describe the application traffic patterns and the NoC topologies considered in the analysis. Then, we describe results from embedding the considered applications into the specified NoC topologies. Finally, we present the OMNET++-based simulation results for a selected subset of the considered applications and NoC topologies.

6.6.1 Application Task Graphs

Any application can be modeled using a directed or undirected task graph. In our study, we consider three classes of tree-like benchmarks obtained through the TGFF package. With each task graph a subset of nodes acts as traffic generators (initiators), while the remaining nodes acts as sinks (target nodes).

- **In a single multi-rooted forest** (SRF), the target (bright gray) subset of nodes is addressed by all initiator (dark gray) nodes (see Figure 6.5(a)).

- **In a multiple node-disjoint single-rooted tree** (SRT), initiator nodes are partitioned in subsets each set then communicates to one single target node (see Figure 6.5(b)).

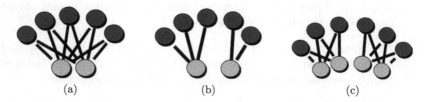

FIGURE 6.5: Application models for (a) 2-rooted forest (SRF), (b) 2-rooted tree (SRT), (c) 2-node 2-rooted forest(MRF) application task graphs.

- **A multiple node-disjoint multi-rooted forest** (MRF) is formed by the combination of the first two traffic patterns: initiator and target nodes are split into disjoint sets. Each set of initiators communicates with a single set of target nodes (see Figure 6.5(c)).

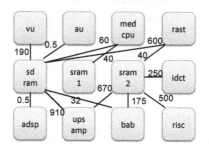

FIGURE 6.6: The Mpeg4 decoder task graph.

We also considered a real 12-node Mpeg4 decoder task graph (shown in Figure 6.6).

All considered task graphs are undirected with unit node weights, and all, with the exception of the Mpeg4 graph, have unit edge weights and scale with the NoC size. Hence, the number of tasks always equals the network size, which ranges from 8 to 64 with step 4.

6.6.2 Prospective NoC Topology Models

The choice of NoC topology has a significant impact on MPSoC price and performance. The bottleneck in sharing resources efficiently is not the number of routers, but wire density which limits system interconnection, affects power dissipation, and increases both wire propagation delay and RC delay for driving the wires. Thus, in this study, we focus on regular, low-dimensional, point-to-point packet-switched topologies with few short, fat and mostly local wires.

As target NoC topology models we have considered low-cost, constant degree NoC topologies, such as one-dimensional array, ring, 2-d mesh and Spidergon STNoC topology.

We also considered the crossbar architecture in order to make a comparison with the classical all-to-all architecture. A large crossbar is prohibitively expensive in terms of its number of links, but it is an optimal solution in terms of embedding quality metrics, with unity edge dilation for all patterns. Modern crossbars connect IP blocks with different data widths, clock rates. and socket or bus standards, such as OCP, and AMBA AHB or AXI. Although system throughput, latency and scalability problems can be resolved by implementing the crossbar as a multistage network based on smaller crossbars and resorting to complex pipelining, segmentation and arbitration, a relatively simple, low-cost alternative is the unbuffered crossbar switch. Thus, we decided to compare the performance of an unbuffered crossbar relative to ring, 2-d mesh (often simply called mesh) and Spidergon topology.

Although multistage networks with multiple layers of routers have good topological properties, e.g., symmetry, small degree and diameter and large bisection, they have small network extendibility, many long wires and large wire area, and thus are not appropriate for NoC realization. High-radix multistage networks, such as flattened butterfly, may be more promising; these networks preserve inter-router connections, but combine routers in different stages of the topology, thereby increasing wire density, while improving network bandwidth, delay, and power consumption [184].

6.6.3 Spidergon Network on Chip

Spidergon is a state-of-the-art low-cost on-chip interconnect developed by ST Microelectronics [15, 8, 16]. It is based on three basic components: a standardized network interface (NI), a wormhole router, and a physical communication link.

Spidergon generalizes the ST Microelectronics' circuit-switched ST Octagon NoC topology used as a network processor architecture. ST Octagon is defined as a Cartesian product of basic octagons with a computing resource connected to each node. Spidergon is based on a simple bidirectional ring, with extra cross links from each node to its diagonally opposite neighbor. It is a *chordal ring* that belongs to the family of *undirected k-circulant graphs*, i.e., it is represented as a graph $G(N; s_1; s_2; ...; s_k)$, where N is the cardinality of the set of nodes, and $0 \leq s_i \leq N$, where s_i is an undirected edge between any node l and node $(l + s_i) mod N$.

Thus, more formally, Spidergon is a vertex-symmetric three-circulant graph with an even number of nodes $N = 2n$, where $n = 1, ..., k = 2$, $s_1 = 1$ and $s_2 = (l + n) mod N$. As shown in Figure 6.7, Spidergon has a practical low-cost, short wire VLSI layout implementation with a single crossing. Notice that VLSI area relates to edge bisection, while the longest wire affects NoC latency.

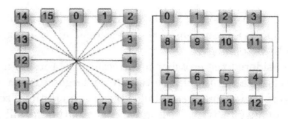

FIGURE 6.7: The Spidergon topology translates to simple, low-cost VLSI implementation.

Chordal rings are circulant graphs with $s_1 = 1$, while double loop networks are chordal rings with $k = \{2, 4, 5, 9, 15, 16, 17\}$. Since the early 1980s with the design of ILLIAC IV, these families have been proposed as simple alternatives to parallel interconnects, in terms of low cost and asymptotic graph optimality, i.e., minimum diameter for a given number of nodes and constant degree; see Moore graphs [82]; in fact the ILLIAC IV parallel interconnect (1980s), often described as similar to 8×8 mesh or torus, was a 64-node chordal ring with skip distance 8. These theoretical studies ignore important design aspects, e.g., temporal and spatial locality, latency hiding and wormhole routing, and NoC-related constraints.

The total number of edges in Spidergon is $\frac{3N}{2}$, while the network diameter is $\lceil \frac{N}{4} \rceil$. For most realistic NoC configurations with up to 60 nodes, Spidergon has a smaller diameter and number of edges than fat-tree or mesh topologies, leading to latency reduction for small packets. For example, the diameter of a 4×5 mesh with 31 bi-directional edges is 7, while that of a 20-node Spidergon with 30 bi-directional edges and less wiring complexity is only 5.

In this chapter, we considered the Across-First (*aFirst*) [15] Spidergon routing algorithm. It is a symmetric algorithm and since the topology is vertex-transitive it can be described at any node. For any arriving packet, the algorithm selects the cross communication port at most once, always at the beginning of each packet route. Thus, only packets arriving from a network resource interface need to be considered for routing. All other packets maintain their sense of direction (clockwise, or counterclockwise) until they reach their destination.

The *aFirst* algorithm can be made deadlock-free by using (at least) two virtual channels that break cycles in the channel dependency graph [20, 17, 22]. Furthermore, optimized, load-balanced virtual channel allocation based on static or dynamic datelines (points swapping of virtual circuits occurs) may provide efficient use of network buffer space, thus improving performance by avoiding head-of-line blocking, reducing network contention, decreasing communication latency and increasing network bandwidth [15]. However, these algorithms are proprietary and are not used in this study.

6.6.4 Task Graph Embedding and Analysis

Through SCOTCH partitioning, we have mapped the application graphs described in Section 6.6.1 into several low-cost NoC topologies (represented with *.tgt target files) using different partitioning heuristics. SCOTCH partitioning was tested and validated with many common well-known examples, such as ring embeddings. We have considered only shortest-path and avoided multi-path routing due to the high cost of implementing packet reordering. Notice that SCOTCH can plot actual mapping results using 2-d color graphical representation; this enhances the automated task allocation phase with a user-friendly GUI.

In Figure 6.8 we compare edge dilation for embedding the previously described master-slave tree-like benchmarks, i.e., single multi-rooted forests, multiple node-disjoint single-rooted trees and multiple node-disjoint multi-rooted forests, into our candidate NoC topologies using the efficient default partitioning strategy (for symmetric graphs) in the SCOTCH partitioning tool; notice however that SCOTCH mapping is not always optimal, even if theoretically possible.

By examining these figures, we make the following remarks and comparisons.

- Ring is the NoC topology with the largest edge dilation.

- For master-slave trees, Spidergon is competitive compared to 2-d mesh for $N \leq 32$. Moreover, for node-disjoint trees or forests, Spidergon is competitive to mesh for larger network sizes (e.g., up to 52 nodes), especially when the number of node-disjoint trees or forests increases, i.e., when the degree of multiprogramming increases. This effect arises from the difficulty to realize several independent one-to-many or many-to-many communication patterns on constant degree topologies.

- Notice that mesh deteriorates for 44 and 52 nodes due to its irregularity. This effect would be much more profound if we had considered network sizes that are multiples of 2 (instead of 4), especially sizes of 14, 22, 26, 30, 34, 38, 46, 50, 54, 58 and 62 nodes.

Figure 6.9 shows our results for edge expansion normalized to the best result, obtained from embedding the Mpeg4 source graph into candidate NoC topologies using the SCOTCH partitioning tool. Notice that the crossbar has the smallest edge expansion so this value is used as reference for this normalization. This is an expected result since in crossbar every node is connected to the others through a single channel. Spidergon and mesh have a very similar edge expansion (where Spidergon has slightly better value), while the ring topology has the highest value of all.

Finally, the NoC mapping considered so far was obtained in seconds on a PENTIUM IV with 2GB of RAM running Linux.

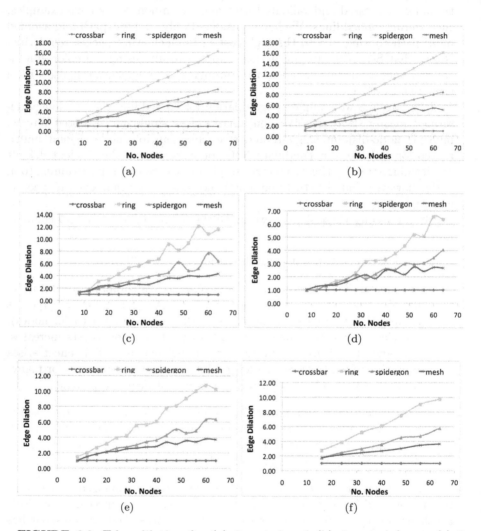

FIGURE 6.8: Edge dilation for (a) 2-rooted and (b) 4-rooted forest, (c) 2 node-disjoint and (d) 4 node-disjoint trees, (e) 2 node-disjoint 2-routed and (f) 4 node-disjoint 4-routed forests in function of the network size.

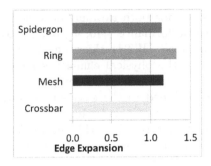

FIGURE 6.9: Relative edge expansion for 12-node Mpeg4 for different target graphs.

6.6.5 Simulation Models for Proposed NoC Topologies

In the NoC domain, IPs are usually connected to the underlying interconnect through a network interface (NI) which provides connection management and data packetization (and de-packetization) facilities.

Each packet is split into data units called *flits* (flow control units) [18, 58]. The size of buffer queues for channels is a multiple of the flit data unit, and packet forwarding is performed using flit-by-flit routing. The switching strategy adopted in our models is wormhole routing. In *wormhole*, the head flit of a packet is actively routed toward the destination by following forward indications of routers, while subsequent flits are passively switched by pre-configured switching functions to the output queue of the channel belonging to the path opened by the head flit. When buffer space is available on the input queue of the channel of the next switch in the path, a flit of the packet is forwarded from the output queue.

In the NoC domain, flit-based wormhole is generally preferred to virtual cut-through or packet-based circuit switching because its pipelined nature facilitates flow control and end-to-end performance, with small packet size overhead and buffer space. However, due to the distributed and finite buffer space and possible circular waiting, complex message-dependent deadlock conditions may arise during routing.

In this respect, the considered mesh architecture uses a simple deadlock avoiding routing algorithm called *dimension order* (or *XY algorithm*) that limits path selection [22]. At first, flits are forwarded toward their destination initially along the X direction (the horizontal link) until the column of the target node is reached. Then, flits are forwarded along the Y direction (vertical link) up to the target node, usually asynchronously.

The bidirectional ring architecture resolves message-dependent deadlock using virtual channels (VC) [20]; this technique maximizes link utilization and allows for improved performance through smart static VC allocation or dynamic VC scheduling. VCs are implemented using multiple output queues and respective buffers for each physical link. A number of VC selection policies have been proposed for both avoiding deadlock and enhancing channel utilization and hence performance [46, 47, 52, 83, 61, 10, 48, 80]. Here we adopt the *winner takes all* (*WTA*) algorithm for VC selection and flit forwarding [19]. Access to the physical channel is handled by a VC arbiter of a single VC through a round robin selection process. Unlike flit interleaving, the channel remains assigned to the selected VC either until the packet is completely transmitted or until it stalls due to lost contention in the following hops. In contrast to flit interleaving, this mechanism performs better than simple round robin, since it allows reducing the average packet transmission time [19].

In this chapter, we also consider an unbuffered crossbar. Each node in this crossbar is directly connected to all others, without any intermediate nodes. Thus, we model an unbuffered (packet-switched) full crossbar switch with round robin allocation of input channels to output ones. A key issue for this interconnect is channel arbitration. In particular, when a first flit is received, the arbiter checks whether the requested output channel is free. If it is, the input channel is associated to the output one until the whole packet is transmitted, otherwise the flit remains in the input register (blocking the relative input channel) until the arbiter assigns the requested channel.

Finally to avoid protocol deadlock [12, 75, 28] caused by the dependency between a target's input and output channels, we configured the network's router with two *virtual networks* (VNs) [6]: one for requests, and a separate one for reply packets. Flits to forward are selected from VNs in a round robin way, and the respective VC is selected with the *WTA* algorithm [19], where flits of a single packet are sent until either the packet stalls or is completely transmitted.

In our experiments, all target input buffers and initiator output buffers are assumed to be infinite: this allows us to focus on network performance by including deadlock avoidance schemes and avoid packet loss due to external devices from playing a bias role in network analysis. However, finite buffers can be treated using the same methodology.

We have modeled the crossbar, ring, 2-d mesh and Spidergon topologies using a number of synchronous (shared clock) network routers, with each router connected to a network interface (NI) through which external IPs with compatible protocols can be connected [34].

In our model, depending on the simulated scenario, each IP acts either as a processing (PE), or as a storage element (SE). Traffic sources (called PEs or initiators) generate packet requests directed to target nodes (SEs) according to their configuration. For all studied topologies, all routers forward incoming flits according to a previously defined shortest-path routing algorithm, provided that the following router has enough room to store them. Otherwise, flits are temporarily stored in the channel output queues. Since crossbar has no intermediate buffers, flits remain in the infinite output buffer of the initiator, until they can be injected into the network.

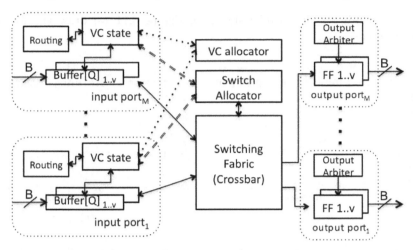

FIGURE 6.10: Model of the router used in the considered NoC architectures.

Figure 6.10 depicts the router model used in all the considered topologies. The number of input and output ports depends on the considered NoC. Also accordingly to the architecture characteristics, input and output ports are equipped with one or more virtual channels handled by a VC allocator. The switch allocator implements the routing function for the considered NoC, while the internal crossbar connects the input to the output channels. All the considered routers implement *look ahead* [57] routing where the routing decision is calculated one hop in advance. In this way routing logic is removed from the critical path and can be computed in parallel with the VC allocation. We also assume that routers have a zero-load latency of one clock cycle, and channels are not pipelined [14].

According to the application type (e.g., Mpeg4), storage elements receive request packets and generate the respective reply packets to be forwarded to the initiator through the same network. PEs/SEs do not implement a computation phase; once a request is completely received, the replay is immediately generated and, in the following cycle, injected into the network. All studied architectures have been modeled using similar routing techniques and PE/SE components are always adapted to specific architecture needs.

Figure 6.11 represents the average throughput of replies for all initiators. For each experiment, the offered load is the initiators' maximum injection rate. In the simulation testbench, requests and replies have the same packet length (5 flits), while each request corresponds to exactly one reply.

Due to traffic uniformity, the reply throughput at each initiator increases when augmenting the real injection rate, until the node saturates. After this point, the router is insensitive to the offered load, but continues to work at the maximum possible rate. Thus, the throughput remains constant (at a maximum point), while the initiators' output queue length (assumed to have an infinite size) actually diverges to infinity very quickly.

By examining the graphs shown in Figure 6.11 we draw the following observations.

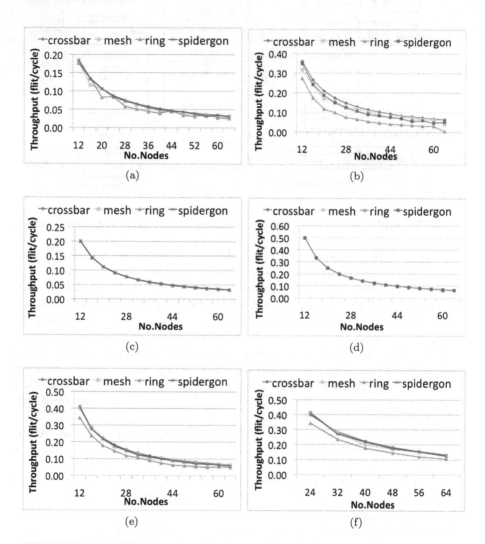

FIGURE 6.11: Maximum throughput as a function of the network size for (a) 2-rooted forest, (b) 4-rooted forest (SRF), (c) 2-rooted tree, (d) 4-rooted tree (SRT), (e) 2-node 2-rooted forest and (f) 4-node 2-rooted forest (MRF) and different NoC topologies.

- As expected, the ring performs generally worse than all studied NoC topologies.

- The Spidergon generally behaves better than mesh for small networks (up to 16 nodes) and remains competitive for larger network sizes in all considered traffic patterns. However, notice that SCOTCH considers equally all minimal paths between any two nodes, while the OMNET++ model uses only the subset of minimal paths defined by the XY routing algorithm. Use of a specific routing algorithm with the SCOTCH mapping tool is an interesting task which requires extra computation and normalization steps prior to computing the actual cost function.

- For the 4-rooted forest, mesh sometimes slightly outperforms Spidergon in larger networks and only for regular 2-d mesh shapes, especially for 36 and 48 nodes.

- Under 2 and 4 node-disjoint single-rooted tree patterns (SRTs), all considered architectures saturate almost at the same point.

- In the 2-rooted and 4-rooted tree cases, considering the total number of injected flits per cycle generated by all initiators, we obtain two constant rates, respectively, 2 and 4 flits/cycle which is exactly what the two and four storage elements can absorb from the network. In this case, the bottleneck arises from the SEs and not by the network architecture which operates under the saturation threshold.

- Crossbar has the best performance in all studied cases, with a smooth and seamless decreasing throughput.

6.6.6   Mpeg4: A Realistic Scenario

In addition to the previous synthetic task graph embedding scenarios, we examined performance of bidirectional ring, 2-d mesh, Spidergon topology and unbuffered crossbar architectures for a real Mpeg4 decoder application modeled by using the task graph illustrated in Figure 6.6. In order to compare these topologies, we set up a transfer speed test where all architectures are mandated to transfer a fixed amount of Mpeg4 packets. Initiators generate requests for SEs (according to the task graph in Figure 6.6), and SEs reply with an instantaneous response message for each received request. Requests and replies have a similar length of four flits. In addition, notice that some PEs in the task graph have a generation rate that heavily differs from others.

In our modeling approach, we have chosen to assign to each intermediate buffer a constant size of three flits. As shown in Figure 6.12, the buffer memory in mesh and Spidergon is comparable (and lower than ring), and Spidergon buffer allocation becomes lower than mesh as the network size increases. The XY routing algorithm used in the mesh NoC, and *aFirst* routing used in Spidergon have the advantage of avoiding deadlock without requiring virtual channels. Ring topology avoids message-dependent deadlock by using two virtual channels for each physical channel in the circular links. Thus, ring requires more buffer space. Notice that the crossbar architecture does not use network buffering; hence, columns in Figure 6.12 are always zero. In fact,

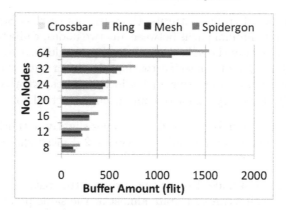

FIGURE 6.12: Amount of memory required by each interconnect.

the crossbar uses buffering only at the network interface. This (like all other architectures) is not considered in the computation.

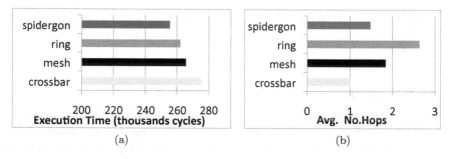

FIGURE 6.13: (a) Task execution time and (b) average path length for Mpeg4 traffic on the considered NoC architectures.

We analyzed the application delay measured as the number of elapsed network cycles from the injection of the first request packet of the load to the delivery of the last reply packet of the same load. In our simulator, the packet size is measured in flits; this essentially relaxes the need to know the actual bit-size of a flit (called phit) in our bit- and cycle-accurate model.

Furthermore, since we focus on topological constraints rather than real system dimensioning, we assume that each channel is able to transmit one flit per clock cycle. As proposed in [41], in order to define a flit injection rate for each different PE of the Mpeg4 task graph, in the transfer speed test we use as reference the highest demanding PE (called UPS-AMP device, see Figure 6.6). All remaining nodes inject flits in a proportional rate with respect to the UPS-AMP device. These rates are reported in tabular form in Table 6.1.

From Figure 6.13 (a), we observe that ring and Spidergon have the best performance while, quite surprisingly, mesh and crossbar perform worse than

TABLE 6.1: Initiator's Average Injection Rate and Relative Ratio with Respect to UPS-AMP Node.

	Offered Load (Mb/sec)	% w.r. UPS	% w.r. Tot
VU	190.0	12.03	5.48%
AU	0.5	0.03	0.01%
MED	100.0	6.33	2.98%
RAST	640.0	40.51	18.47%
IDCT	250.0	15.82	7.21%
ADSP	0.5	0.03	0.01%
UPS	1580	100.0	45.59%
BAB	205.0	12.97	5.91%
RISC	500.0	31.65	14.43%
TOT	3466.0	219.37	100.00%

expected. The explanation can be obtained by considering the allocated buffer size for the 12-node architectures shown in Figure 6.12. mesh and unbuffered crossbar have less buffer memory, i.e., 204 flits for mesh and 0 for crossbar versus 288 flits for ring and 216 for Spidergon. To summarize results, by computing the percentage difference between the data transfer performance reported in the first histogram of Figure 6.13 (a), we conclude that for near optimal Mpeg4 mapping scenarios, Spidergon is faster than ring by 0.6 percent, faster than mesh by 3.3 percent and faster than unbuffered crossbar by 3.2 percent. Next, we obtain more detailed insights on the steady state performance metrics and resource utilization of the proposed architectures, for the considered scenarios.

In the second histogram of Figure 6.13 (b) we illustrate the average path length of flits (and its standard deviation) obtained with the Mpeg4 mapping in the data transfer experiments. Ring forces some packets to follow longer paths than other topologies, but in this way it effectively uses its buffer space more efficiently. Except for unbuffered crossbar (which saturates early), Spidergon provides a good tradeoff among proposed topologies, resulting in shorter and more uniform paths.

In the analysis of node throughput reported in Figure 6.14, we observe that in all topologies the most congested links are those connected to the busiest nodes (SDRAM, UPS-AMP, and RAST of Figure 6.6). Despite the higher number of channels that the mesh disposes, Spidergon and mesh actually forward packets along the same number of links. The mesh XY routing algorithm does not exploit all paths this architecture provides, while Spidergon provides better channel balancing. Because of its shape, the ring exploits its channels much better. The busiest channels in the crossbar are those toward the SDRAM and SRAM2 nodes, the two veritable network hot spots, and the UPS-AMP node which generates more than the 45 percent of the network traffic.

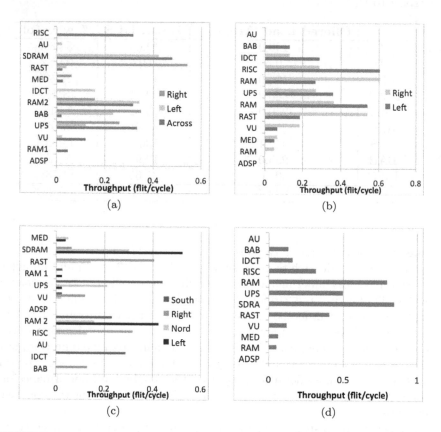

FIGURE 6.14: Average throughput on router's output port for (a) Spidergon, (b) ring, (c) mesh and (d) unbuffered crossbar architecture.

The absence of intermediate buffers makes the crossbar architecture very sensitive to realistic unbalanced traffic. In particular, crossbar may show end-to-end source blocking behavior since a packet addressed to SDRAM may have to wait in the output queue of the initiator, while buffered multi-hop paths could allow initiators to inject more packets into the network (if buffer space is available in the path), thus facilitating an emptying behavior of initiator packets addressed to different targets.

Figure 6.15 shows the average network round trip time (RTT), i.e., the average time required for sending a request packet and obtaining its respective reply packet from the network (only network time is computed, i.e., the time in the infinite size queue of the initiator interface is excluded). Note that in the following figures, the UPS-AMP node injection rate is taken as reference and reported on the X axis, while the injection rate for other nodes can be

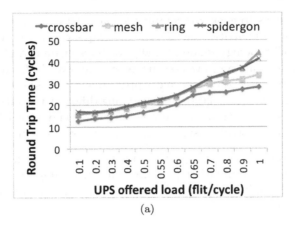

(a)

FIGURE 6.15: Network RTT as a function of the initiators' offered load.

computed proportionally following Table 6.1. The average injection rate of the initiators (total offered load) can be obtained by multiplying this value by a constant factor (percentage of total initiator load) which is 2.1937 for the Mpeg4 scenario. For all NoC topologies, the RTT time slowly increases until congestion starts (rate below 0.6 flits/cycle).

The UPS-AMP network congestion appears for a UPS-AMP injection rate between 0.6 and 0.7 flits/cycle. When the path used by the UPS-AMP IP saturates and becomes insensitive to the offered load (around 0.7 flits/cycle), other initiators using different paths may still augment their input ratio, increasing network congestion and average network RTT. Crossbar has the lowest RTT thanks to the absence of intermediate hops. Spidergon has an average RTT similar to mesh and ring while having shorter paths. This indicates that Spidergon channel buffers are in general better exploited.

6.7   Conclusions and Extensions

In this chapter, we have extended and applied several existing open-source tools from different domains, such as traffic modeling, graph theory, parallel computing, and network simulation, to the analysis and selection of NoC topologies. Our system-level approach is based on abstracting multi=core SoC applications as interacting application components (called tasks), i.e., as intellectual properties (IPs) with clear, unambiguous and self-contained functionality, communicating and synchronizing over a NoC topology model which abstracts the actual NoC architecture into which the application is deployed. More specifically, we have outlined graph theoretical tools for NoC topology exploration, such as METIS, NAUTY, and NEATO. We have also described TGFF:

a tool that allows to generate complex application task graphs and SCOTCH: an embedding and partitioning tool used to map any generated task graph into a selected NoC topology model. Finally, we have examined OMNET++ as a platform for our bit- and cycle-accurate system-level NoC simulation. Using all these open-source tools, we have presented a case study on NoC modeling (embedding and simulation), considering different tree-like synthetic task graphs (representing master-slave combinations) and a real Mpeg4 decoder application. Future work based on this case study, can focus on:

- Improving the SCOTCH partitioning tool to consider more complex cost functions for evaluating non-minimal paths (improving routing adaptivity), or minimizing total edge dilation (optimizing dynamic power dissipation).

- Extending our OMNET++ model to consider dynamic reconfiguration and task migration and optimizing buffer size at the routers or incoming/outgoing network interfaces.

A vast research area can be considered when enhancing the implementation of the network interfaces and cores. In particular, thanks to the described analysis procedure and tools, we intend to explore:

- End-to-end flow controls to be implemented in the network interfaces in order to solve the message-dependent deadlock [45, 75]. In the previous pages this issue has been solved through virtual networks that represent one simple but expensive approach. Literature offers more complex and cost effective solutions that however must be tuned according to the characteristics of the considered application [29, 75].

- Implement more realistic cores in order to model realistic applications [3] and study their synchronization issues.

Although our methodology mainly focuses on system-level modeling, as illustrated in Figure 6.16, it is innovative to extend it to multi-core SoC operating system (or kernel) functions. Since the mapping phase obtains a "near optimal" assignment of the given application task graph into the nodes of the NoC topology, the operating system must initiate processes for online (or offline) application task and data allocation (installation and configuration) into the SoC nodes. Task and data allocation involve scattering and fine/coarse data interleaving algorithms performed by the compiler or the application (e.g., if enough symmetry exists). In general, multi-core SoC applications dynamically request subsystems, thus efficient job scheduling (selecting which jobs will be executed) must also be considered. For predictable performance, all accesses from different multi-core applications to disjoint subsets of independent resources must be through edge-disjoint routes, i.e., not sharing the same physical links.

After the initial allocation of tasks into NoC topology nodes, the application is able to execute. It may also require dynamic task reassignment (called

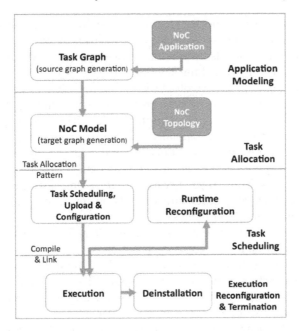

FIGURE 6.16: Future work: dynamic scheduling of tasks.

migration, rescheduling or reconfiguration) to satisfy constraints that develop during execution, thereby requiring a new optimized mapping. Formalizing potential requirements for dynamic reconfiguration is tedious, since reassignment costs are hard to determine in advance without precise monitoring of the application and NoC architecture. This includes monitoring fault tolerance metrics, dynamic metrics for processing and communication load, memory bandwidth and power consumption requirements that largely remain unknown during initial assignment time. A common run-time optimization technique that improves PE utilization and can be used for reconfiguration is load balancing. In its most common implementation, excess elements are locally diffused to neighbor PEs which carry smaller loads. Thus, idle (or under-utilized) PEs can share the load with donor PEs. Donor PEs may be selected in various ways, e.g., through a random variable, a local counter, or a global variable accessed using fetch and add operations. Better load balancing techniques can be based on parallel scan (and scatter) operations. Upon program completion, de-installation takes place, releasing subsystems to the operating system to become available for future requests.

Mapping and scheduling can be followed by voltage scheduling and power management techniques in order to minimize dynamic and static energy consumption of tasks mapped into voltage scalable resources. Voltage scheduling is achieved by assigning a lower supply voltage to certain tasks mapped on voltage scalable resources, effectively slowing them down and exploiting the available slack; however notice that voltage switching overhead is not always negligible. Static voltage scheduling refers to (usually offline) allocation of single or multi-voltage levels to certain voltage scalable resources regardless of its utilization during run-time. Dynamic voltage scheduling (DVS) refers to allocation of single or multivoltage levels for running tasks on voltage scalable resources and can also be performed offline or online. Power management aims to reduce static power consumption by shutting down unutilized or idle resources either totally or partially. Power management can be applied to processing elements and communication links either offline or online, while the manner of applying power management can be static or dynamic. Static scheduling suffers from unpredictable compile-time estimation of execution time, lack of efficient and accurate methods for estimating task execution times and communication delays.

Through our study, we have shown that it is feasible to perform system-level NoC design space exploration using an array of extended and parameterized open-source tools originating from NoC-related application domains. This approach actively promotes interoperability and enhances productivity via coopetition (collaboration among competitors) and quality via increased manpower and broadened expertise. Moreover, it is also valid for system-level multi-core SoC design through adopting (and possibly extending) existing open methods, tools and models in related areas, such as reliability, fault tolerance, performance and power estimation. In fact, several vibrant open SoC standards, as well as system-level methods, tools and IP/core models at different abstraction levels are available [15, 64, 65, 66, 67].

Review Questions

[Q-1] Which application domains are related to multi-core SoC/NoC design?

[Q-2] Which tools from these application domains can be easily transferred to multi-core SoC/NoC design?

[Q-3] Define embedding quality and provide examples from different application domains.

[Q-4] In which way is partitioning related to embedding?

[Q-5] Which type of partitioning algorithms are the fastest?

[Q-6] Study TGFF traffic modeling software and propose an extension toward more general application models invoking synchronization points.

[Q-7] Study the theory of circulant graphs and consider an extension to 3-D circulant topologies as Cartesian products.

[Q.8] Explore SCOTCH software and try to redefine the cost function.

[Q.9] Examine SCOTCH software extensions toward dynamic and possibly real-time reconfiguration.

[Q.10] Study the setup requirements and application performance metrics in a general transfer speed test.

[Q.11] Examine available open source system-level power modeling tools, such as Power-Kernel, discussed in Chapter 3.

Bibliography

[1] A. Adriahantenaina, H. Charlery, A. Greiner, L. Mortiez, et al. SPIN: A scalable, packet switched, on-chip micro-network. In *Proc. Int. ACM/IEEE Conf. Design, Automation and Test in Europe (DATE)*, 2003.

[2] Amba Bus, ARM. Available from http://www.arm.com.

[3] J.M. Arnold. The splash 2 software environment. *Proc. IEEE Workshop on FPGAs for Custom Computing Machines*, 1993.

[4] J Balfour and W. Dally. Design tradeoffs for tiled cmp on-chip networks. *Proc. Int. Conf. Supercomputing*, 2006.

[5] L. Benini and G. De Micheli. Networks on chip: A new SoC paradigm. *IEEE Computer*, 49(2/3):70–71, 2002.

[6] J. C. Bermond, F. Comellas, and D. F. Hsu. Distributed loop computer networks: A survey. *J. Parallel Distrib. Comput.*, 24(1):2–10, 1995.

[7] Bluespec Compiler.

[8] L. Bononi, N. Concer, M. Grammatikakis, M. Coppola, and R. Locatelli. NoC topologies exploration based on mapping and simulation models. In *Proc. IEEE Euromicro Conf. Digital Syst. Design*, 2007.

[9] M. Caldari, M. Conti, M. Coppola, et al. System-level power analysis methodology applied to the AMBA AHB bus. In *Proc. Int. ACM/IEEE Conf. Design, Automation and Test in Europe (DATE)*, 2003.

[10] M. Chaudhuri and M. Heinrich. Exploring virtual network selection algorithms in DSM cache coherence protocols. *IEEE Transactions on Parallel and Distributed Systems*, 15(8):699–712, 2004.

[11] Chipvision, Orinoco.

[12] N. Concer, L. Bononi, M. Soulié, R. Locatelli, and L.P. Carloni. CTC: An end-to-end flow control protocol for SoC architectures. In *Proc. IEEE/ACM Int. Symp. Networks-on-Chips (NOCS)*, 2009.

[13] N. Concer, S. Iamundo, and L. Bononi. aEqualized: a novel routing algorithm for the Spidergon network on chip. In *Proc. Int. ACM/IEEE Conf. Design, Automation and Test in Europe (DATE)*, 2009.

[14] N. Concer, M. Petracca, and L.P. Carloni. Distributed flit-buffer flow control for networks-on-chip. In *Proc. ACM/IEEE/IFIP Int. Workshop on Hardware/Software Codesign and Syst. Synthesis (CODES/CASHE)*, 2008.

[15] M. Coppola, M.D. Grammatikakis, R. Locatelli, G. Maruccia, and L. Pieralisi. *Design of Cost-Efficient Interconnect Processing Units: Spidergon STNoC*. CRC Press, 2008.

[16] M. Coppola, R. Locatelli, G. Maruccia, L. Pieralisi, and A. Scandurra. Networks on chip: A new paradigm for systems on chip design. In *Proc. IEEE Int. Symp. System-on-Chip*, 2004.

[17] W. J. Dally and H. Aoki. Deadlock-free adaptive routing in multicomputer networks using virtual channels. *IEEE Trans. Parallel Distrib. Syst.*, 4(4):466–475, 1993.

[18] W. J. Dally and B. Towles. Route packets, not wires: On-chip interconnection networks. In *Proc. Int. ACM/IEEE Design Automation Conf.*, 2001.

[19] William J. Dally and B. Towles. *Principles and Practices of Interconnection Networks*. Morgan Kaufmann Publishers, 2004.

[20] W.J. Dally. Virtual-channel flow control. In *Proc. ACM/IEEE Int. Symp. Comp. Arch. (ISCA)*, 1990.

[21] R.P. Dick, D.L. Rhodes, and W. Wayne. TGFF: task graphs for free. In *Proc. ACM/IEEE/IFIP Int. Workshop on Hardware/Software Codesign and Syst. Synthesis (CODES/CASHE)*, 1998.

[22] Jose' Duato, Sudhakar Yalamanchili, and Lionel Ni. *Interconnection Networks. An Engineering Approach*. Morgan Kaufmann Publishers, 2003.

[23] C.M. Fiduccia and R.M. Mattheyses. A linear time heuristic for improving network partitions. In *Proc. Int. ACM/IEEE Design Automation Conf.*, 1982.

[24] M. Forsell. A scalable high-performance computing solution for networks on chips. *IEEE Micro*, 22(5):46–55, 2002.

[25] O. P. Gangwal, A. Rădulescu, K.Goossens, S. González Pestana, and E. Rijpkema. Building predictable systems on chip: An analysis of guaranteed communication in the Æthereal network on chip. In *Dynamic and Robust Streaming in and between Connected Consumer-Electronics Devices*, chapter 1, pages 1–36. Springer, 2005.

[26] M. Garey, D. Johnson, and L. Stockmeyer. Some simplified NP-complete graph problems. *Theoretical Computer Science*, 1:237–267, 1976.

[27] K. Goossens, J. Dielissen, J. van Meerbergen, P. Poplavko, A. Rădulescu, E. Rijpkema, E. Waterlander, and P. Wielage. *Guaranteeing the Quality of Services in Networks on Chip.* Kluwer Academic Publishers, 2003.

[28] A. Hansson and K. Goossens. Trade-offs in the configuration of a network on chip for multiple use-cases. In *Proc. IEEE Int. Symp. Networks on Chip (NOCS)*, 2007.

[29] A. Hansson, K. Goossens, and A. Rădulescu. Avoiding message-dependent deadlock in network-based systems on chip. *IEEE J. VLSI*, 2007:1–10, 2007.

[30] J. Henkel and Y. Li. Avalanche: an environment for design space exploration and optimization of low-power embedded systems. *IEEE Trans. VLSI Integr. Syst.*, 10(4):454–468, 2002.

[31] J. Hu and R. Marculescu. Exploiting the routing flexibility for energy/performance aware mapping of regular NoC architectures. In *Proc. Int. ACM/IEEE Conf. Design, Automation and Test in Europe (DATE)*, 2003.

[32] J. Hu and R. Marculescu. Energy-aware communication and task scheduling for network-on-chip architectures under real-time constraints. In *Proc. Int. ACM/IEEE Conf. Design, Automation and Test in Europe (DATE)*, 2004.

[33] J. Hu and R. Marculescu. Energy- and performance-aware mapping for regular NoC architectures. *IEEE Trans. Computer-Aided Design of Integr. Circ. and Syst.*, 24(4):551–562, 2005.

[34] F.K. Hwang. A complementary survey on double-loop networks. *Theoretical Computer Science*, 263(1-2):211–229, 2001.

[35] IBM On-chip CoreConnect Bus, IBM Research Report. Available from http://www.chips.ibm.com/products/coreconnect.

[36] A. Jalabert, S. Murali, L. Benini, and G. De Micheli. xpipesCompiler: A tool for instantiating application specific networks on chip. In *Proc. Int. ACM/IEEE Conf. Design, Automation and Test in Europe*, 2004.

[37] G. Karypis and V. Kumar. METIS: a software package for partitioning unstructured graphs, meshes, and computing fill-reducing orderings of sparse matrices (version 3.0.3). Technical Report, University of Minnesota, Department of Computer Science and Army HPC Research Center, 1997.

[38] T. Kempf, M. Doerper, R. Leupers, G. Ascheid, et al. A modular simulation framework for spatial and temporal task mapping onto multiprocessor SoC platforms. In *Proc. Int. ACM/IEEE Conf. Design, Automation and Test in Europe (DATE)*, 2005.

[39] B.W. Kernighan and S. Lin. An efficient heuristic procedure for partitioning graphs. *Bell System Tech. Journal* 49, AT&T Bell Laboratories, Murray Hill, NJ, USA, 1970.

[40] J. Kim, J. Balfour, and W. J Dally. Flattened butterfly topology for on-chip networks. In *Proc. IEEE/ACM Int. Symp. Microarchitecture*, 2007.

[41] M. Kim, D. Kim, and G.E. Sobelman. MPEG-4 performance analysis for CDMA network on chip. In *Proc. IEEE Int. Conf. Comm. Circ. and Syst.*, pages 493–496, 2005.

[42] F. Klein, R. Leao, G. Araujo, et al. An efficient framework for high-level power exploration. In *Proc. Int. IEEE Midwest Symp. Circ. and Syst. (MWSCAS)*, 2007.

[43] F. Klein, R. Leao, G. Araujo, et al. PowerSC: A systemC-based framework for power estimation. Technical report, University of Campinas, 2007.

[44] D. E. Knuth. *The Art of Computer Systems Performance Analysis*. Wiley Computer Publishing, 1991.

[45] H. D. Kubiatowicz. Integrated Shared Memory and Message Passing Communications in the Alewife Multiprocessor. PhD thesis, Massachusetts Institute of Technology. Boston, 1997.

[46] A. Kumar and L. Bhuyan. Evaluating virtual channels for cache-coherent shared-memory multiprocessors. *Proc. Int. Conf. on Supercomputing*, 1996.

[47] A. Kumar, L.-S. Peh, P. Kundu, and N. Jha. Express virtual channels: towards the ideal interconnection fabric. In *Proc. ACM/IEEE Int. Symp. Comp. Arch. (ISCA)*, 2007.

[48] A. Kumar, L.-S. Peh, P. Kundu, and N. K Jha. Toward ideal on-chip communication using express virtual channels. *IEEE Micro*, 28(1):80–90, 2008.

[49] K. Lahiri, A. Raghunathan, and S. Dey. Evaluation of the traffic performance characteristics of system-on-chip communication architectures. In *Proc. Int. Conf. VLSI Design*, 2001.

[50] T. F. Leighton. *Introduction to Parallel Algorithms and Architectures: Algorithms and VLSI*. Morgan Kaufmann Publishers, 2006.

[51] X. Liu and M.C. Papaefthymiou. HyPE: hybrid power estimation for IP-based programmable systems. In *Proc. Asia and South Pacific Design Automation Conf. (ASPDAC)*, 2003.

[52] J. Lu, B. Kallol, and A.M.Peterson. A comparison of different wormhole routing schemes. *Proc. Int. Workshop on Modeling, Analysis, and Simulation on Computer and Telecom. Syst.*, 1994.

[53] J.A. Lukes. Combinatorial solution to the partitioning of general graphs. *IBM Journal of Research and Development*, 19:170–180, 1975.

[54] C. Marcon, N. Calazans, F. Moraes, A. Susin, et al. Exploring NoC mapping strategies: an energy and timing aware technique. In *Proc. Int. ACM/IEEE Conf. Design, Automation and Test in Europe (DATE)*, 2005.

[55] B. McKay. NAUTY User's Guide (version 1.5). Technical Report, Australian National University, Department of Computer Science, 2003.

[56] N. Metropolis, A.W. Rosenbluth, M.N. Rosenbluth, A.H. Teller, and E. Teller. Equations of state calculations by fast computing machines. *Journal of Chemical Physics*, 21(6):1087–1092, 1953.

[57] M.Galles. Scalable pipelined interconnect for distributed endpoint routing. In *Proc. IEEE Symp. Hot Interconnects*, 1996.

[58] G. De Micheli and L. Benini. *Networks on Chips: Technology and Tools (Systems on Silicon)*. Morgan Kaufmann Publishers, 2006.

[59] S. Murali and G. De Micheli. Bandwidth-constrained mapping of cores onto NoC architectures. In *Proc. Int. ACM/IEEE Conf. Design, Automation and Test in Europe*, 2004.

[60] S. Murali and G. De Micheli. SUNMAP: a tool for automatic topology selection and generation for NoCs. In *Proc. Int. ACM/IEEE Design Automation Conf.*, 2004.

[61] C. Nicopoulos, D. Park, J. Kim, and N. Vijaykrishnan. Vichar: A dynamic virtual channel regulator for network-on-chip routers. In *Proc. IEEE/ACM Int. Symp. Microarchitecture*, 2006.

[62] S.C. North. NEATO User's Guide. Technical Report 59113-921014-14TM, AT&T Bell Laboratories, Murray Hill, NJ, USA, 1992.

[63] J. Nurmi, H. Tenhunen, J. Isoaho, and A. Jantsch. *Interconnect-Centric Design for Advanced SOC and NOC.* Springer, 2004.

[64] Open Source, GEDA. Available from http://geda.seul.org.

[65] Open Source, Linux Softpedia. Available from http://linux.softpedia.com.

[66] Open Source, Open Cores. Available from http://opencores.org.

[67] Open Source, Sourceforge. Available from http://sourceforge.net.

[68] G. Palermo, G. Mariani, C. Silvano, R. Locatelli, et al. Mapping and topology customization approaches to application-specific STNoC designs. In *Proc. Int. Conf. Application-Specific Syst. Arch. and Processors (ASAP)*, pages 61–68, 2007.

[69] G. Palermo, C. Silvano, G. Mariani, R. Locatelli, and M. Coppola. Application-specific topology design customization for STNoC. In *Proc. IEEE Euromicro Conf. Digital Syst. Design Arch.*, pages 547–550, 2007.

[70] F. Pellegrini and J. Roman. SCOTCH: A software package for static mapping by dual recursive bi-partitioning of process and architecture graphs. In *Proc. Int. Conf. on High Perf. Computing and Networking.* Springer, 1996.

[71] A. Pinto. *A platform-based approach to communication synthesis for embedded systems.* PhD thesis, University of California at Berkeley, 2008.

[72] T. Simunic, L. Benini, and G. De Micheli. Cycle-accurate simulation of energy consumption in embedded systems. In *Proc. IEEE Design Automation Conf. (DAC)*, 1999.

[73] A. Sinha and A. Chandrakasan. JouleTrack: a web-based tool for software energy profiling. In *Proc. IEEE Design Automation Conf. (DAC)*, 2001.

[74] SoftExplorer.

[75] Y. H. Song and T.M. Pinkston. A progressive approach to handling message-dependent deadlock in parallel computer systems. *IEEE Trans. Parallel Distrib. Syst.*, 14(3):259–275, 2003.

[76] K. Srinivasan, K. S. Chatha, and G. Konjevod. Application specific network-on-chip design with guaranteed quality approximation algorithms. In *Proc. Int. ACM/IEEE Design Automation Conf.*, 2007.

[77] A. Stammermann, L. Kruse, W. Nebel, et al. System level optimization and design space exploration for low power. In *Proc. Int. Symp. System Synthesis*, 2001.

[78] Synopsys Innovator.

[79] M. Taylor, J. Kim, J. Miller, et al. The raw microprocessor: A computational fabric for software circuits and general purpose programs, 2002.

[80] A. Vaidya, A. Sivasubramaniam, and C. Das. Performance benefits of virtual channels and adaptive routing: an application-driven study. *Proc. Int. Conf. on Supercomputing*, 1997.

[81] VCG. Available from http://rw4.cs.uni-sb.de/users/sander/html/gsvcg1.html.

[82] E. Weisstein. Moore graphs. Available http://mathworld.wolfram.com/MooreGraph.html.

[83] D. Wu, B.M. Al-Hashimi, and M.T. Schmitz. Improving routing efficiency for network-on-chip through contention-aware input selection. *Proc. IEEE Conf. Asia South Pacific on Design Automation*, 2006.

[84] S. Xanthos, A. Chatzigeorgiou, and G. Stephanides. Energy estimation with SystemC: a programmer's perspective. In *Proc. WSEAS Int. Conf. Circ.*, 2003.

7

Compiler Techniques for Application Level Memory Optimization for MPSoC

Bruno Girodias

École Polytechnique de Montréal
Canada
Bruno.Girodias@PolyMtl.ca

Youcef Bouchebaba, Pierre Paulin, Bruno Lavigueur

ST Microelectronics
{Youcef.Bouchebaba, Pierre.Paulin, Bruno.Lavigueur}@st.com

Gabriela Nicolescu

École Polytechnique de Montréal
Canada
Gabriela.Nicolescu@PolyMtl.ca

El Mostapha Aboulhamid

Université de Montréal
Canada
EM.Aboulhamid@UMontreal.ca

CONTENTS

7.1   Introduction   244
7.2   Loop Transformation for Single and Multiprocessors   245
7.3   Program Transformation Concepts   246
7.4   Memory Optimization Techniques   248
    7.4.1   Loop Fusion   249
    7.4.2   Tiling   249
    7.4.3   Buffer Allocation   249
7.5   MPSoC Memory Optimization Techniques   250
    7.5.1   Loop Fusion   251
    7.5.2   Comparison of Lexicographically Positive and Positive
        Dependency   252

        7.5.3   Tiling 253
        7.5.4   Buffer Allocation 254
7.6   Technique Impacts 255
        7.6.1   Computation Time 255
        7.6.2   Code Size Increase 256
7.7   Improvement in Optimization Techniques 256
        7.7.1   Parallel Processing Area and Partitioning 256
        7.7.2   Modulo Operator Elimination 259
        7.7.3   Unimodular Transformation 260
7.8   Case Study 261
        7.8.1   Cache Ratio and Memory Space 262
        7.8.2   Processing Time and Code Size 263
7.9   Discussion 263
7.10   Conclusions 264
Review Questions 265
Bibliography 266

7.1   Introduction

The International Technology Roadmap for Semiconductors (ITRS) defines
multiprocessor systems-on-chips (MPSoCs) as one of the main drivers of the
semiconductor industry revolution by enabling the integration of complex
functionality on a single chip. MPSoCs are gaining popularity in today's high
performance embedded systems. Given their combination of parallel data processing in a multiprocessor system with the high level of integration of system-
on-chip (SoC) devices, they are great candidates for systems such as network
processors and complex multimedia platforms [15]. The important amount of
data manipulated by these applications requires a large memory size and a
significant number of accesses to the external memory for each processor node
in the MPSoC architecture [34]. Therefore, it is important to optimize, at the
application-level, the access to the memory in order to improve processing time
and power consumption. Embedded applications are commonly described as
streaming applications which is certainly the case of multimedia applications
involving multi-dimensional streams of signals such as images and videos. In
these applications, the majority of the area and power cost arise from global
communication and memory interactions [4, 5]. Indeed, a key area of concen-
tration to handle both real-time and energy/power problems is the memory
system [34]. The development of new strategies and techniques is necessary
in order to decrease memory space, code size and to shrink the number of
accesses to the memory.

Multimedia applications often consist of multiple loop nests. Unfortunately, today's compilation techniques for parallel architectures (not necessarily MPSoCs) consider each loop nest separately. Hence, the key problem associated with these techniques is that they fail to capture the interaction among different loop nests.

This chapter focuses on applying loop transformation techniques for MPSoC environments by exploiting techniques and some adaptation for MPSoC characteristics. Section 7.2 overviews the literature domain in loop transformation for single and multiprocessor environments. Section 7.3 initiates the lecture to some basic concepts in program transformation. Section 7.4 introduces some memory transformation techniques. Section 7.5 goes into detail about memory transformation techniques in MPSoCs. Section 7.6 discusses the impact of these techniques in multiprocessor environments. Section 7.7 brings forward some improvements and adaptations to memory transformation techniques for MPSoC environments. Section 7.8 shows some results. Finally, a discussion and concluding remarks are found in Sections 7.9 and 7.10 respectively.

7.2   Loop Transformation for Single and Multiprocessors

In single processor environments (e.g., SoCs), there has been extensive research in which several compiler techniques and strategies have been proposed to optimize memory. Among them, one can point out scalar replacement [3], intra array storage order optimization [12], pre-fetching [24], locality optimizations for array-based codes [32, 6], array privatization [30] and array contraction [8]. The IMEC group in [5] pioneered code transformation to reduce the energy consumption in data dominated embedded applications. Loop transformation techniques like loop fusion and buffer allocation have been studied extensively [7, 19]. Fraboulet et al. [10] and Marchal et al. [23] minimize the memory space in loop nests by using loop fusion. Kandemir et al. [17] studied inter-nest optimizations by modifying the access patterns of loop nests to improve temporal data locality. Song et al. [29]proposed an aggressive array-contraction and studied its impact on memory system performances. Song et al. [28] used integer programming for modeling the problem of combining loop shifting, loop fusion and array contraction. [21, 26] proposed a memory reduction based on a multi-projection. [8] developed a mathematical framework based on critical lattices that subsumes the approaches given by [21] and [26].

Tiling was introduced by Irigoin et al. [14], who studied a sufficient condition to apply tiling to a single loop nest. Xue [35] generalized the application of this technique and gave a necessary and sufficient condition to apply it. Anderson et al. [1] and Wolf et al. [32] addressed this technique in more detail by proposing a mathematical model for evaluating data reuse in affine data access functions (single loop nest). Kandemir et al. introduced DST [16] for data space-oriented tilling, which also aims at optimizing inter-nest locally by dividing the data space into data tiles.

This chapter presents compiler techniques targeted to MPSoCs. Techniques discussed in this chapter can be used on a large scale system; however they have more impact on an MPSoC environment. MPSoC is a more sensitive environment, because it has limited resources and is more constrained in area and energy consumption.

Most efforts in the MPSoC domain focus on architecture design and circuit related issues [7, 20]. Compilation techniques in this domain target only single loop nests (i.e., each loop nest independently). Recently, Li et al. [22] proposed a method with a global approach to the problem. However, it is limited when the partitioned data block sizes are larger than the cache [2]. To circumvent this problem, a new approach is proposed. It consists of applying loop fusion to all loop nests and partitioning the data space across the processors. [31, 13] present a methodology for data reuse exploration. It gives great detail with formalism and presents some cost functions and trade-offs. [13] presents an exploration and analysis with scratchpad memories (SPM) instead of caches. Using SPM requires the use of special instructions that are architecture dependent. Some works might also use additional architectural enhancements for performance purposes.

This chapter completes the existing works by presenting an adapted version of the loop fusion and buffer allocation techniques in an MPSoC environment. A modulo optimization and unimodular transformation technique are presented as well to optimize the processing time in a buffer allocation technique. All transformations presented in our current work require no changes to the architecture and can still obtain significant performance enhancement.

7.3   Program Transformation Concepts

Program transformations like loop fusion, tiling and buffer allocation have been studied extensively for data locality optimizations in a mono-processor architecture [12, 33, 9]. The loop fusion technique generates a code where several loop nests are merged together (merged code). This enables array elements already in the cache to be reused immediately since the loop fusion brings the computation operations using the same data closer together. Tiling divides the array lines into subsets if they do not fit inside the cache. Buffer allocation keeps only the useful data in the cache [12]. To ease the understanding of concepts presented in this chapter, the compiler techniques are adapted for multimedia applications with a code structure similar to the one presented in Figure 7.1. This code has the following characteristics:

- There is no dependency within a loop nest (each loop nest is parallel).

- The dependencies are between two consecutive loop nests via one array. We considered this constraint to simplify the chapter presentation. Thus, four techniques can be applied to a code where a loop nest L_k can use the elements produced in all preceding loop nests $(L_1, .., L_{k-1})$.

- All the loop nests have the same depth (n).

- The loop bounds are constants.

- All the arrays have the same dimensions (n).

- The access functions to the arrays are uniform (the same access function, except for constants).

$$L_1 : \text{do } \vec{i} = (i_1, ..., i_n) \in D_1$$
$$S_1 : A_1(\vec{i}) = F_1(A_0)$$
$$\text{end}$$

$$\vdots$$

$$L_k : \text{do } \vec{i} = (i_1, ..., i_n) \in D_k$$
$$S_k : A_k(\vec{i}) = F_k(A_{k-1})$$
$$\text{end}$$

FIGURE 7.1: Input code: the depth of each loop nest L_k is n (n loops), A_k is n dimensional.

Our techniques can be applied to more general code forms, but this will complicate the automation. This will also introduce more overhead in the optimized code without any particular interest, because our target applications (imaging, video) and most multimedia applications found in the industry respect the above conditions. In some cases, if an application does not meet one of the previous conditions, we can transform it in order to meet this condition.

Throughout this chapter, the polyhedral model [18, 27] is used to represent the loop nest computations. For example, the loop nest of depth 2 in the code of Figure 7.2 (a) can be represented by a two-dimensional domain with axis i and j (Figure 7.2 (b)). The axis i corresponds to loop i and the axis j corresponds to loop j. At each iteration (i, j), three statements S_1, S_2 and S_3 are computed. The computation order of the iterations is given by a lexicographic order (i.e., the iteration (i, j) will be computed before the iteration (i', j'), if and only if $(i, j) \prec (i', j')$). The vector operators often used throughout this chapter are in lexicographic order $(\prec, \preceq, \succ, \succeq)$ and the usual component-wise operators $(<, \leq, >, \geq)$. Note that the lexicographic order is a complete order, (i.e., any two vectors are comparable), while the component-wise comparator defines only a partial order. A complete order is very important, since it can be used to schedule computations; given two elements indexed by vectors, one knows which element computation precedes the other. The transposition of vector $(i_1, \ldots, i_n)$ will be noted by $(i_1, \ldots, i_n)^t$.

```
do i = 0, N
   do j = 0, M
      S₁ : A(i + 2, j + 2) = F(INPUT)
      S₂ : B(i + 1, j + 1) =
         = A(i + 1, j + 1) + A(i + 2, j + 2)
      S₃ : C(i, j) = B(i + 1, j) + B(i, j + 1)
   end
end
```

(a) Code

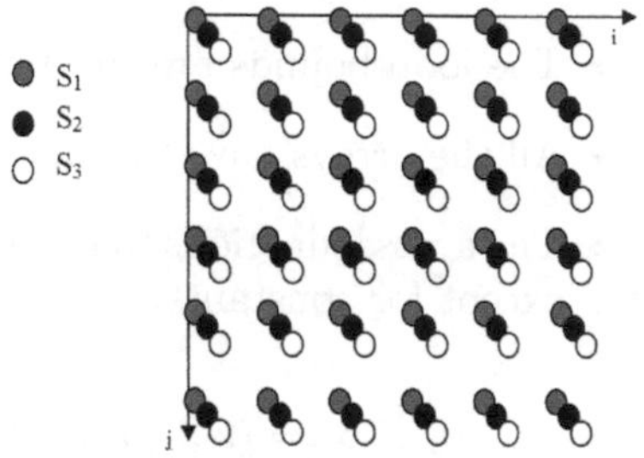

(b) Iteration

FIGURE 7.2: Code example and its iteration domain.

For the code given in Figure 7.1, each loop nest body execution can be represented by an iteration vector $\vec{i}$, with each vector entry corresponding to a loop. An iteration $\vec{j}$ of loop nest L_k, is said to depend on an iteration $\vec{i}$ of loop nest L_{k-1}, if $\vec{i}$ produces an element of array A_{k-1} which is used by $\vec{j}$. The difference between these two vectors $(\vec{j}-\vec{i})$ is called the data dependency vector. This work is mostly of interest in the case where all the entries of a dependency vector are constants, in which case it is also referred to as the distance vector. Since there are no dependencies inside the loop nests of code in Figure 7.1, one way to parallelize this code is to start by computing the loop nest L_1 in parallel, and then L_2 in parallel, etc. This solution brings more parallelism to the code, but considerably decreases the data locality. To avoid this problem, this chapter proposes to start by applying loop fusion with or without tiling. However, this introduces new dependencies inside the resulting code which complicates the parallelization step. Later in this chapter, we will present a new approach to solve these dependency problems.

7.4 Memory Optimization Techniques

This section will review different techniques used in the compilation field, particularly in program transformations. As described earlier, loop transformation techniques such as loop fusion, tiling and buffer allocation have been studied extensively and will be reviewed later in this chapter. While more techniques exist, this chapter will emphasize the optimization of three selected techniques.

7.4.1   Loop Fusion

L_1 do: $i = 0, 7$
  do $j = 0, 7$
    $S_1 : A(i,j) = F(INPUT)$
  end
end

L_2 do: $i = 0, 7$
  do $j = 0, 7$
    $S_2 : B(i,j) = A(i,j) + A(i-1,j-1)$
  end
end

$L_{1,2}$ do: $i = 0, 7$
  do $j = 0, 7$
    $S_1 : A(i,j) = F(INPUT)$
    $S_2 : B(i,j) = A(i,j) + A(i-1,j-1)$
  end
end

(a) Initial Code   (b) Loop Fusion

FIGURE 7.3: An example of loop fusion.

Loop fusion is often used in applications with numerous loop nests. This technique replaces multiple loop nests by a single one. It is widely used in compilation optimization since it increases data locality in a program. It enables data already present in the cache to be used immediately. Figure 7.3 illustrates the loop fusion technique. Details on the requirements needed to accomplish a loop fusion will be presented subsequently.

7.4.2   Tiling

Tiling is used for applications using arrays of significant size. It partitions a loop's iteration space into smaller blocks. It makes loop execution more efficient and like the loop fusion technique, it ensures the reusing of data. Figure 7.4 presents an example of tiling.

7.4.3   Buffer Allocation

The third and last technique is buffer allocation. It is often utilized for applications with temporary arrays which store intermediate computations. Buffer allocation is a technique which reduces the size of temporary arrays. It decreases memory space and reduces the cache miss ratio. The buffer size is defined by the dependencies among statements. The buffer will contain only useful elements (also called live elements). An element of an array is considered live at time t, if it is assigned (written) at t_1 and last used (read) at t_2

do: $i = 0,7$	do: $l_1 = 0,1$
do: $j = 0,7$	do: $l_2 = 0,1$
$S_1 : A(i,j) = F(INPUT)$	do: $l_3 = 0,3$
$S_2 : B(i,j) = A(i,j) + A(i-1,j-1)$	do: $l_4 = 0,3$
end	$i = 4*l_1 + l_3$
end	$j = 4*l_2 + l_4$
	$S_1 : A(i,j) = F(INPUT)$
	$S_2 : B(i,j) = A(i,j) + A(i-1,j-1)$
	end
	end
	end
	end
(a) Initial Code	(b) Tiling

FIGURE 7.4: An example of tiling.

whereas $t_1 \leq t \leq t_2$. Figure 7.5 illustrates an example of the buffer allocation technique where array A is replaced by the buffer BUF of size 10.

do: $i = 0,7$	do: $i = 0,7$
do: $j = 0,7$	do: $j = 0,7$
$S_1 : A(i,j) = F(INPUT)$	$S_1 : BUF[(8*i+j)\%10] = F(INPUT)$
$S_2 : B(i,j) = A(i,j) + A(i-1,j-1)$	$S_2 : B(i,j) = BUF[(8*i+j)\%10]$
end	$+BUF[(8*(i-1)+(j-1))\%10]$
end	end
	end
(a) Initial Code	(b) Buffer Allocation

FIGURE 7.5: An example of buffer allocation.

7.5 MPSoC Memory Optimization Techniques

The following section will demonstrate how to combine these loop transformation techniques in an MPSoC environment. These techniques apply to any sequence of loop nests where loop nest k depends on all previous loop nests.

7.5.1 Loop Fusion

Loop fusion cannot be applied directly to a code. All dependencies among loop nests must be positive or null. Therefore, a loop shifting technique must be applied to the code.

```
//Loop 1
L₁ do: i = 2, N + 2
    do: j = 2, M + 2
        S₁ : A(i, j) = F(INPUT)
    end
end
//Loop 2
L₂ do: i = 1, N + 1
    do: j = 1, M + 1
        S₂ : B(i, j) = A(i, j) + A(i + 1, j + 1)
    end
end
//Loop 3
L₃ do: i = 0, N
    do: j = 0, M
        S₃ : C(i, j) = B(i + 1, j) + A(i, j + 1)
    end
end
```

L_1 do: $i = 2, N + 2$
  do: $j = 2, M + 2$
    $S_1 : A(i, j) = F(INPUT)$
  end
end

L_2 do: $i = 1, N + 1$
  do: $j = 1, M + 1$
    $S_2 : B(i, j) = A(i, j) + A(i + 1, j + 1)$
  end
end

L_3 do: $i = 0, N$
  do: $j = 0, M$
    $S_3 : C(i, j) = B(i + 1, j) + A(i, j + 1)$
  end
end

(a) Initial Code

//Loop 1
L_1 do: $i = 0, N$
  do: $j = 0, M$
    $S_1 : A(i + 2, j + 2) = F(INPUT)$
  end
end

//Loop 2
L_2 do: $i = 0, N$
  do: $j = 0, M$
    $S_2 : B(i + 1, j + 1) = A(i + 1, j + 1) + A(i + 2, j + 2)$
  end
end

//Loop 3
L_3 do: $i = 0, N$
  do: $j = 0, M$
    $S_3 : C(i, j) = B(i + 1, j) + A(i, j + 1)$
  end
end

(b) Loop shifting

//Loop 1,2 and 3 are merged after the loop shifting
$L_{1,2,3}$ do: $i = 0, N$
  do: $j = 0, M$
    $S_1 : A(i + 2, j + 2) = F(INPUT)$
    $S_2 : B(i + 1, j + 1) = A(i + 1, j + 1) + A(i + 2, j + 2)$
    $S_3 : C(i, j) = B(i + 1, j) + A(i, j + 1)$
  end
end

(c) Loop Fusion

FIGURE 7.6: An example of three loop nests.

Figure 7.6 illustrates the sequence of loop transformations going from the initial code (a) to a loop shifting (b) and finally a loop fusion (c). As shown in this example, one must shift the iteration domain of the loop nest L_1 by a vector $(-2, -2)$ and the iteration domain of loop nest L_2 by a vector $(-1, -1)$ in order to make all dependencies positive or null (≥ 0). A lexicographically positive ($\succeq 0$) dependency is a satisfactory condition to apply loop fusion.

However, in a parallel application it is advantageous to have positive or null dependencies. This will be discussed later in this section.

The code generated after a loop fusion cannot be automatically parallelized. Border dependencies appear when partitioning the application. To avoid these dependencies, elements at the border of each processor block must be pre-calculated before each processor computes its assigned block (initialization phase).

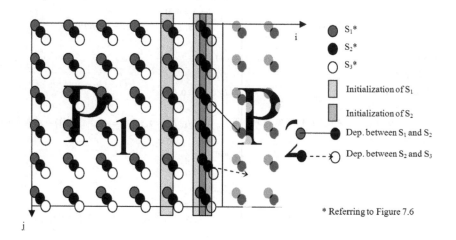

FIGURE 7.7: Partitioning after loop fusion.

Figure 7.7 illustrates the partitioning of the code in Figure 7.6 (c) across two processors where the left-hand side of array calculations is assigned to P_1 and the right-hand side to P_2. As shown in Figure 7.7, dependencies between S_1 and S_2 and dependencies between S_2 and S_3 do not allow one to parallelize the code directly. In the border, processor P_2 cannot compute statements S_3 and S_2 before P_1 computes statements S_1 and S_2. Therefore, this chapter proposes an initialization phase where statement S_1 and S_2 on the border of processor P_1 block must be pre-calculated before processing concurrently each processor assigned block. This solution is possible since S_1 does not depend on any statement which is the general scenario in the types of applications studied in this research.

7.5.2 Comparison of Lexicographically Positive and Positive Dependency

In order to apply a fusion, it can be seen later in this chapter that one must force all dependencies to be lexicographically positive or null ($\succeq 0$). However, to simplify code generation for parallel execution, one can force them to be positive or null (≥ 0). This is illustrated in Figure 7.8 which demonstrates the data parallelization of a merged code involving two statements S_1 and

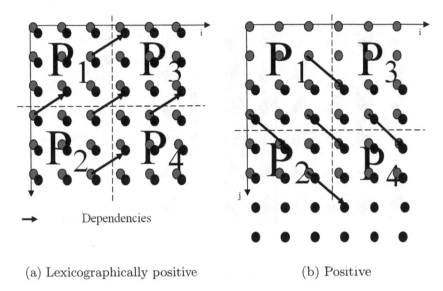

→ Dependencies

(a) Lexicographically positive   (b) Positive

FIGURE 7.8: Difference between positive and lexicographically positive dependence.

S_2 across four processors. In Figure 7.8 (a), the dependency $(1, -1)$ from S_1 to S_2 is lexicographically positive. This dependency implies that on the vertical border, initial data is located at the ends of the blocks assigned to processors P_1 and P_2 while at the horizontal border, initial data is located at the beginning of the blocks assigned to processors P_2 and P_4. However, in Figure 7.8 (b), the dependency $(1, 1)$ is positive and the initializations on both axes are located at the ends of the blocks. Theoretically, both solutions are equivalent. The difference between these two figures lies in the ease of code generation when the code transformation is automated. In Figure 7.8 (b), at the horizontal border, initial data is needed by the blocks P_2 and P_4. When generating code, it is easy to regroup in an initial phase, the processing of initial data with the beginning of the normal processing of blocks P_2 and P_4. However, in Figure 7.8 (a), at the horizontal border, initial data is needed by the blocks assigned to P_1 and P_3. Since the required values will normally only be calculated at the ends of the blocks P_1 and P_3, it is not easy to include an initial phase to process these initial data in the generated code.

7.5.3   Tiling

Tiling is applied after fusion in a multiprocessor architecture. Parallelized tiled code needs an initialization phase for each processor block border (as does loop fusion). The main difference is the additional phase which consists of dividing each processor block into several sub-blocks.

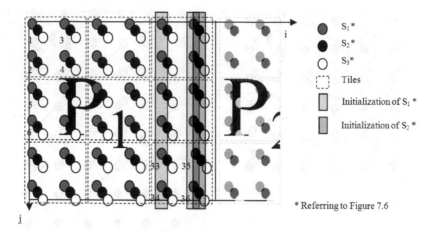

FIGURE 7.9: Tiling technique.

Figure 7.9 illustrates the tiling technique on a two processors architecture. The numbers on this figure refer to the execution order of the iterations.

7.5.4   Buffer Allocation

Multimedia applications often use temporary arrays to store intermediate computations. To reduce memory space in this type of application, several techniques are proposed in the literature like scalar replacement, buffer allocation and intra array storage order optimization. Nevertheless, these techniques are used in monoprocessor architectures.

In a monoprocessor architecture, each array is replaced by one buffer containing all live elements. However in a multiprocessor architecture, the number of buffers replacing each temporary array depends on: (1) the number of processors, (2) the depth of the loop nest, (3) the division of the iteration domain and (4) the dependencies among loop nests. Two types of buffers are needed: (a) buffers for inner computation elements of blocks assigned to each processor and (b) buffers for the computation of elements located at the borders of these blocks. The last buffer type is needed for the initialization phase as seen in the previous sub-section.

Figure 7.10 illustrates the buffer allocation for one temporary array (array B) of the code in Figure 7.6 (c) partitioned across four processors. The initialization phase is needed to compute these blocks in parallel (the vertical initialization for processors P_2 and P_4 and horizontal initialization for processors P_3 and P_4). In this example, a total of six buffers are needed, one for each of the four processors and two buffers for the initialization phase.

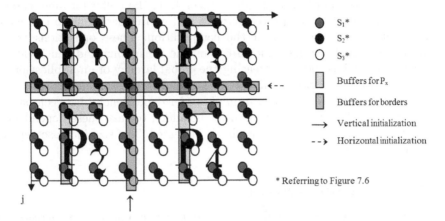

FIGURE 7.10: Buffer allocation for array B.

Since type (a) buffers can be seen as circular structures, a modulo operator (%) is used to manage them. Using buffer allocation reduces memory space but increases processing time. This issue will be revisited later.

7.6 Technique Impacts

Optimizing a specific aspect of an application does not come without cost. The techniques described in the last section have increased the hit cache ratio tremendously in a multiprocessor architecture, but computation time and code size have increased as well. It is certainly a major concern since MPSoCs are often chosen for their high data processing capacity while having limited memory space for applications.

7.6.1 Computation Time

Buffer allocation uses modulo operators extensively, which are very time consuming operations for any processor. Every time one must read or write into a buffer, it uses one or more modulo operators.

As seen in the previous section, two types of buffers are needed in a multiprocessor architecture. This means that when a processor is processing a statement, it must be aware of the location of the elements that will be used for computation. If the computation is located close to the border of the processor's block, some data needed to compute will be located in one of the buffers used to pre-calculate the borders' elements. If the computation is not located close to a border, data will be taken from the buffer for inner computation of the block. Each processor must add extra operations to test which buffer will receive the data for the computation. These operations are "if statements" which are also known to be time consuming and can break the processor's pipeline, hence increasing the global latency of an application.

Using fusion and buffer allocation in a multiprocessor increases data locality, but requires the consideration of border dependencies between parallel processor data blocks. Some code must be added to take these dependencies into account. Therefore, the code size is increased. The size increase depends on the size of the application.

7.6.2 Code Size Increase

In the buffer allocation technique, the size of the application is increased by adding extra code to the tests described in the last sub-section. However, the size is mostly increased by all the code needed to manage and partition the parallel application across processors and extra code to pre-calculate the elements located in each border of the processor's block (initialization phase). Using buffer allocation decreases memory space, but requires modulo operators for buffer management. Using modulo operators increases processing time, especially on platforms like MPSoC, where the embedded processors are more limited and where co-processors may be used for special instruction like modulo.

7.7 Improvement in Optimization Techniques

As discussed in the previous section, the optimization techniques described earlier significantly improve the hit cache ratio, but this is done at the expense of processing time. This section depicts improvements to save significant processing time by (1) changing the partitioning, (2) eliminating modulo operators and (3) changing the order of the iterations with a unimodular transformation.

7.7.1 Parallel Processing Area and Partitioning

This section presents a novel manner to partition the code across the processors to eliminate supplementary tests needed to manage and locate data in multiple buffers. Despite the fact that this new block assignment confers a great advantage at the level of the processing time, it also makes the code easier to parallelize. This is important if these techniques are automated in a parallel compiler.

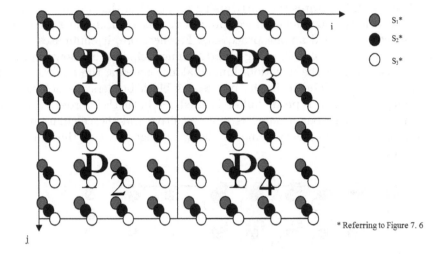

FIGURE 7.11: Classic partitioning.

Figure 7.11 illustrates what one can expect to see in the literature. This division affects the processing time since each processor may interact with several others. This fact is even more significant in a buffer allocation scenario, where broad computation is needed to manage buffers.

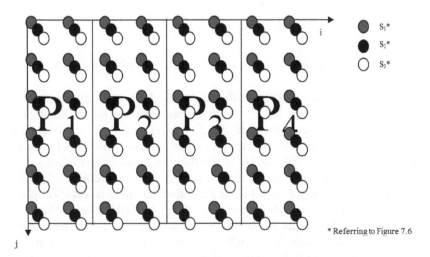

FIGURE 7.12: Different partitioning.

Figure 7.12 illustrates the partitioning proposed in this section (along one axis). This block assignment reduces processing time by decreasing the number of interactions needed among processors. Furthermore, it eliminates vertical dependencies introduced by using a partitioning technique as shown in Figure 7.12. By separating the iteration domain along one axis, the buffers used to store the elements located in the processor's block borders are not required any longer. A single type of buffer is used for both the border and block computations. This partitioning is more intuitive and the calculation of the processor's block boundary remains the same regardless of the number of processors.

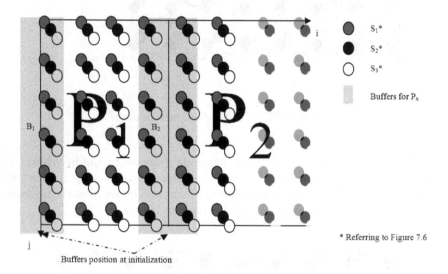

FIGURE 7.13: Buffer allocation for array B with new partitioning.

Figure 7.13 illustrates the buffer allocation for one temporary array (array B) of the code in Figure 7.6 (c) divided across two processors. Only one buffer is required for each processor (B_1 for P_1 and B_2 for P_2). The total number of buffers is strictly equal to the number of processors. The size of each buffer is equal to $M_j * (d + 1)$ where M_j is the number of iterations of axis j and d is the highest dependency along the i axis. As seen in Figure 7.13, buffers are located at the border of each processor's block at the initialization phase, and then shift along the axis during computation. No further supplementary tests are needed to determine from which buffers data should be recovered. Each processor recovers data from only one buffer only because there is presently only one type of buffer. Using different data partitioning may reduce the code size and facilitate data parallelization. However, if one restricts oneself to one type of data partitioning, data locality may decrease depending on the application, image size and cache size. Using a two axis data partitioning ensures a better data fit in the memory cache; however, it necessitates more work for border dependencies. Using a one axis data partitioning facilitates and eliminates the border dependencies. However, the partitioned block has a better chance of not fitting in the cache.

7.7.2   Modulo Operator Elimination

To eliminate modulo operators, each block assigned to a processor is divided into sub-blocks of the same width as the buffer (also equivalent to the largest dependency). The buffer shifts from sub-block to sub-block.

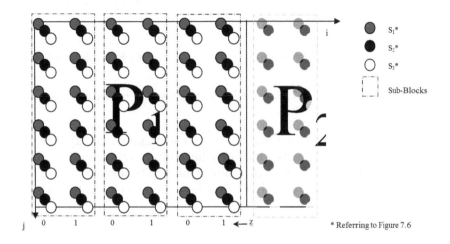

FIGURE 7.14: Sub-division of processor P_1's block.

Figure 7.14 demonstrates the division of processor P_1's block into sub-blocks of equal size as the buffer. The loop z which goes across the sub-block is completely unrolled. Here, the unrolling technique is used to optimize the time spent computing the modulo operator.

Figure 7.15 illustrates the elimination of the modulo operators. The loop i in Figure 7.15 (b) executes an equivalent of two loops at each iteration (computing sub-blocks). By unrolling the loop scanning the sub-blocks, the access to the buffer is done with constants which are defined by dependencies (elimination of modulo operators).

Traditionally, loop unrolling is used to exploit data reuse and explore instruction parallelism; however, this chapter uses this technique to eliminate the modulo operator.

Using the modulo operator elimination technique reduces the processing time, however the technique proposed uses loop unrolling which, depending on the unrolling factor (in our case the modulo factor), increases the code size. For this reason, we have proposed to combine the next optimization to correct this potential problem.

```
for (i = 6; i < 12; i + +)
    for (j = 0; j < 6; j + +)
        S₁ : A(i%2, j) = ...
        S₂ : B(i, j − 1) = A((i − 1)%2, j − 1) + A(i%2, j)
    end
end
```

NOTE: Modulo of number which is a power of 2, can be done by shifting, but this technique works with any numbers.

(a) Example with modulo

```
for (i = 6; i < 12; i + +)
    // Loop z=0
    for (j = 0; j < 6; j + +)
        S₁ : A(0, j) = ...
        S₂ : B(i, j − 1) = A(1, j − 1) + A(0, j)
    end
    // Loop z=1
    for (j = 0; j < 6; j + +)
        S₁ : A(1, j) = ...
        S₂ : B(i + 1, j − 1) = A(0, j − 1) + A(1, j)
    end
end
```

(b) Example without modulo

FIGURE 7.15: Elimination of modulo operators.

A technique to remove modulo operator is proposed in [11]. The solution uses conditional statements which may introduce overhead cost on most SoC architecture. Our work proposes a similar solution to remove modulo operator in conjunction with a unimodular transformation to eliminate any overhead cost.

7.7.3 Unimodular Transformation

A final optimization can be done on the code (Figure 7.15 (b)) obtained after the fusion and the buffer allocation without modulo. This optimization is the fusion of the two innermost j loops which will decrease processing time and increase the cache hit ratio. However, this transformation cannot be applied directly.

Merging j loops in Figure 7.15 (b) changes the execution order of the statements inside a sub-block which corresponds to the application of a unimodular transformation $T = \begin{bmatrix} 0 & 1 \\ 1 & 0 \end{bmatrix}$ on each sub-block. Figure 7.16 (b) illustrates the issue by applying fusion directly. Elements generated by iteration 2 in sub-block 1 are needed by iteration 3 of the second sub-block. However, this element will have been erased by iteration 2 in the second sub-block since all sub-blocks share the same buffer. Figure 7.16 (c) illustrates the execution order to avoid this issue. One must apply a unimodular transformation $T = \begin{bmatrix} 1 & 1 \\ 1 & 0 \end{bmatrix}$ to each sub-block. This matrix is a function of dependencies. Through this transformation, the processing time has decreased and the cache hit ratio has increased.

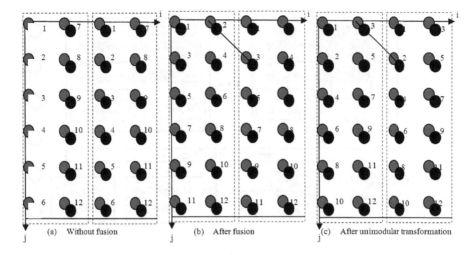

FIGURE 7.16: Execution order (a) without fusion (b) after fusion and (c) after unimodular transformation.

Using the unimodular transformation reduces the code size led by the previous modulo operator elimination technique. However, it may require more effort to find the appropriate unimodular transformation to respect all dependencies for some applications.

7.8   Case Study

Experiments were carried out on the MultiFlex multiprocessor SoC programming environment.

The MultiFlex application development environment was developed specifically for multiprocessor SoC systems. Two parallel programming models are supported in the MultiFlex system, the first is the distributed system object component (DSOC) model. This model supports heterogeneous distributed computing, reminiscent of CORBA and Microsoft DCOM distributed component object models. It is a message-passing model and it supports a very simple CORBA-like interface definition language (dubbed SIDL in our system). The other model is symmetric multi-processing (SMP), supporting concurrent threads accessing shared memory. The SMP programming concepts used here are similar to those embodied in Java and Microsoft C#. The implementation performs scheduling, and includes support for threads, monitors, conditions and semaphores.

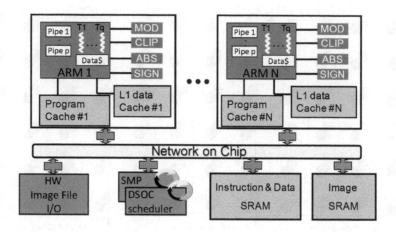

FIGURE 7.17: StepNP platform.

The MultiFlex tools map these models onto the StepNP multiprocessor SoC platform [25]. The architecture consists of processors, with local cache, connected to a shared memory by a local bus (see Figure 7.17).

The multimedia application simulated consists of an imaging application used in medical applications (e.g., cavity detection). It is composed of five computations where each of them corresponds to a loop nest. The first computation is done on an input image. Then, the computation results are passed to the following computation.

The experiments consisted of four different simulations: (1) the initial code without any transformations, (2) the initial code with fusion and buffer allocation using modulo operators, (3) the initial code with fusion and buffer allocation without using modulo operators and (4) the initial code with fusion and buffer without using modulo, but with a unimodular transformation.

7.8.1   Cache Ratio and Memory Space

Figure 7.18 shows the data cache hit ratio of the multimedia application on a multiprocessor architecture (four CPUs) with a four-way set-associative cache with a block size of four bytes. As one can see, most of the techniques presented in this chapter considerably increase the cache hit ratio compared to the initial application. The best results are obtained by the fusion with buffer allocation using modulo operators. One can observe an average increase of 20% of the data cache hit ratio.

The combination of the loop fusion, buffer allocation and mainly the partitioning reduce the memory space by approximately 80 percent.

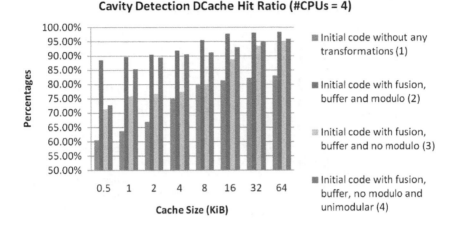

FIGURE 7.18: DCache hit ratio results for four CPUs.

7.8.2   Processing Time and Code Size

Figure 7.19 shows the processing time of the multimedia application on a multiprocessor architecture (four CPUs) with a four-way set-associative cache with a block size of four bytes. As discussed in the previous section, the fusion with buffer allocation using modulo operators is great for cache hit but at the expense of prolonging processing time (see Figure 7.19). However, the two other techniques show great improvements in processing times while still increasing the data cache hit ratio. The best results are seen with the fusion with buffer allocation using no modulo operators with a unimodular transformation. One can observe an average decrease of 50 percent of the processing time. The partitioning proposed here reduces the code size by approximately 50 percent in the case of the fusion.

7.9   Discussion

The targeted applications are composed of several loop nests and each loop nest produces an array which will be used by the following ones (each array is read and written in different loop nests). This implies that the application of tiling alone to each loop nest would have no impact on the data locality, since tiling is primarily used when a loop nest reads and writes in the same array. The application of loop fusion has an impact because it brings the computation operations using the same data (reads and writes) closer together. This enables array elements already in the cache to be reused. However, when the

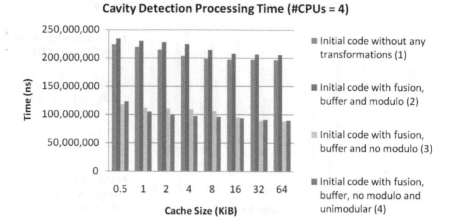

FIGURE 7.19: Processing time results for four CPUs.

array lines are bigger than the cache line, loop fusion may take advantage of the tiling technique which will divide these array lines into smaller lines which will fit in the cache line. Therefore, the combination of loop fusion and tiling is appropriate for these types of applications. These techniques cannot be applied to any MPSoC architecture. They cannot be applied to MPSoC with scratchpad memory, because the scratchpad memory needs explicit instructions to load the data. This was not taken into account in this chapter. Finally, these techniques are applied on SMP architecture, and for other architectures, the code generation approach will need be to be adapted adequately.

7.10 Conclusions

This chapter presented an approach and techniques to significantly reduce the processing time while increasing the data cache hit ratio for a multimedia application running on an MPSoC. All experiments were carried out on the MultiFlex platform with SMP architecture.

From the results presented, one can see that the best results are obtained with the fusion with buffer allocation using no modulo operators with a unimodular transformation. This technique displays excellent balance among data cache hit, processing time, memory space and code size. Data cache hit ratio is increased by using fusion and buffer allocation and processing time is decreased mainly by avoiding modulo operators. In addition, these techniques reduce memory space. This technique demonstrates a global approach to the problem of data locality in MPSoCs, somewhat different from the techniques found in the literature which concentrate on single loop nests separately.

Review Questions

[Q.1] Why and what makes MPSoCs great candidates for systems such as network processors and complex multimedia?

[Q.2] Why is it important to optimize, at the application level, the access to the memory?

[Q.3] Name several compiler techniques and strategies proposed to optimize memory.

[Q.4] Name the issues on which most efforts in the MPSoC domain focus initially.

[Q.5] What are the characteristics shared by multimedia codes targeted in this chapter?

[Q.6] What is a polyhedral model?

[Q.7] What is the difference between lexicographic order and component-wise operator order?

[Q.8] What is loop fusion?

[Q.9] What is tiling?

[Q.10] What is buffer allocation?

[Q.11] Describe why loop fusion cannot be applied directly on a code in every situation.

[Q.12] What is the difference between lexicographically positive and positive dependency?

[Q.13] Of what should a designer be aware regarding the data when applying memory optimization in MPSoC?

[Q.14] What is the impact that memory optimization can have on performance?

[Q.15] Explain the improvements on existing memory optimization techniques presented in this chapter.

Bibliography

[1] Jennifer M. Anderson and Monica S. Lam. Global optimizations for parallelism and locality on scalable parallel machines. In *Proceedings of SIGPLAN '93 Conference on Programming Language Design and Implementation (PLDI '93)*. ACM, 1993.

[2] Y. Bouchebaba, B. Girodias, G. Nicolescu, E.M. Aboulhamid, P. Paulin, and B Lavigueur. MPSoC memory optimization using program transformation. *ACM Trans. Des. Autom. Electron. Syst.*, 12(4):43, 2007.

[3] S. Carr and K. Kennedy. Scalar replacement in the presence of conditional control flow. *Software - Practice and Experience*, 24(1):51–77, 1994.

[4] F. Catthoor, K. Danckaert, K.K. Kulkarni, E. Brockmeyer, P.G. Kjeldsberg, T. van Achteren, and T. Omnes. *Data Access and Storage Management for Embedded Programmable Processors*. Springer, 2002.

[5] F. Catthoor, F. Franssen, S. Wuytack, L. Nachtergaele, and H. De Man. Global communication and memory optimizing transformations for low. In *Workshop on VLSI Signal Processing, VII, 1994*, pages 178–187, 1994.

[6] Michal Cierniak and Wei Li. Unifying data and control transformations for distributed shared-memory machines. In *Proceedings of the ACM SIGPLAN 1995 Conference on Programming Language Design and Implementation*, pages 205–217, 1995.

[7] Alain Darte. On the complexity of loop fusion. In *Proceedings of the International Conference on Parallel Architectures and Compilation Techniques (PACT '99)*, pages 149–157, 1999.

[8] Alain Darte and Guillaume Huard. New results on array contraction. In *13th IEEE International Conference on Application-Specific Systems, Architectures and Processors (ASAP'02)*. IEEE Computer Society, 2002.

[9] Alain Darte, Yves Robert, and Frederic Vivien. *Scheduling and Automatic Parallelization*. Birkhauser, Boston, 2000.

[10] A. Fraboulet, K. Kodary, and A. Mignotte. Loop fusion for memory space optimization. In *Proceedings of the 14th International Symposium on System Synthesis, 2001*, pages 95–100, 2001.

[11] C. Ghez, M. Miranda, A. Vandecappelle, F. A. Catthoor F. Catthoor, and D. A. Verkest D. Verkest. Systematic high-level address code transformations for piece-wise linear indexing: illustration on a medical imaging algorithm. In *IEEE Workshop on Signal Processing Systems, 2000*, pages 603–612, 2000.

[12] Eddy De Greef. *Storage Size Reduction for Multimedia Application*. PhD thesis, Katholieke Universiteit, 1998.

[13] Issenin Ilya, Brockmeyer Erik, Miranda Miguel, and Dutt Nikil. Drdu: A data reuse analysis technique for efficient scratch-pad memory management. *ACM Trans. Des. Autom. Electron. Syst.*, 12(2):15, 2007.

[14] F. Irigoin and R. Triolet. Supernode partitioning. In *Proceedings of the 15th ACM SIGPLAN-SIGACT Symposium on Principles of Programming Languages*. ACM, 1988.

[15] A. Jerraya, H. Tenhunen, and W. Wolf. Guest editors' introduction: Multiprocessor systems-on-chips. *Computer*, 38(7):36–40, 2005.

[16] M. Kandemir. Data space oriented tiling. In *Programming Languages and Systems. 11th European Symposium on Programming, ESOP 2002*, Lecture Notes in Computer Science, Vol. 2305. Springer, 2002.

[17] M. Kandemir, I. Kadayif, A. Choudhary, and J. A. Zambreno. Optimizing inter-nest data locality. In *Proceedings of the 2002 International Conference on Compilers, Architecture, and Synthesis for Embedded Systems*, pages 127–135, 2002.

[18] Richard M. Karp, Raymond E. Miller, and Shmuel Winograd. The organization of computations for uniform recurrence equations. *J. ACM*, 14(3):563–590, 1967.

[19] K. Kennedy. Fast greedy weighted fusion. *International Journal of Parallel Programming*, 29(5):463–91, 2001.

[20] V. Krishnan and J. Torrellas. A chip-multiprocessor architecture with speculative multithreading. *Computers, IEEE Transactions on*, 48(9):866–880, 1999.

[21] Vincent Lefebvre and Paul Feautrier. Automatic storage management for parallel programs. *Parallel Comput.*, 24(3-4):649–671, 1998.

[22] F. Li and M. Kandemir. Locality-conscious workload assignment for array-based computations in MPSoC architectures. In *Proceedings of the 42nd Design Automation Conference*, pages 95–100, 2005.

[23] P. Marchal, F. Catthoor, and J.I. Gomez. Optimizing the memory bandwidth with loop fusion. In *International Conference on Hardware/Software Codesign and System Synthesis*, pages 188–193, 2004.

[24] Kunle Olukotun, Basem A. Nayfeh, Lance Hammond, Ken Wilson, and Kunyung Chang. The case for a single chip multiprocessor. In *Proceedings of the 7th International Conference on Architectural Support for Programming Languages and Operating Systems*, pages 2–11, 1996.

[25] P.G. Paulin, C. Pilkington, M. Langevin, E. Bensoudane, and G. Nico-lescu. Parallel programming models for a multi-processor SoC platform applied to high-speed traffic management. In *International Conference on Hardware/Software Codesign and System Synthesis*, pages 48–53, 2004.

[26] Fabien Quiller and Sanjay Rajopadhye. Optimizing memory usage in the polyhedral model. *ACM Trans. Program. Lang. Syst.*, 22(5):773–815, 2000.

[27] Patrice Quinton. The systematic design of systolic arrays. Princeton University Press, 1987.

[28] Yonghong Song, Cheng Wang, and Zhiyuan Li. A polynomial-time algorithm for memory space reduction. *Int. J. Parallel Program.*, 33(1):1–33, 2005.

[29] Yonghong Song, Rong Xu, and Cheng Wang. Improving data locality by array contraction. *IEEE Trans. Comput.*, 53(9):1073–1084, 2004.

[30] Peng Tu and David A. Padua. *Automatic Array Privatization*. Springer, 1994.

[31] T. Van Achteren, G. Deconinck, F. Catthoor, and R. Lauwereins. Data reuse exploration techniques for loop-dominated applications. In *Proceedings of the Design, Automation and Test in Europe Conference and Exhibition, 2002*, pages 428–435, 2002.

[32] M. E. Wolf and M. S. Lam. A data locality optimizing algorithm. In *Proceedings of the ACM SIGPLAN 1991 conference on Programming Language Design and Implementation*, pages 30–44, 1991.

[33] M. E. Wolf and M. S. Lam. A loop transformation theory and an algorithm to maximize parallelism. *IEEE Trans. Parallel Distributed System*, 2(4):452–471, 1991.

[34] W. Wolf. The future of multiprocessor systems-on-chips. In *Proceedings of the Design Automation Conference 2004*, pages 681–685, 2004.

[35] J. Xue. *Loop tiling for parallelism*. Kluwer Academic, 2000.

8

Programming Models for Multi-Core Embedded Software

Bijoy A. Jose, Bin Xue, Sandeep K. Shukla

Fermat Laboratory
Bradley Department of Electrical and Computer Engineering
Virginia Polytechnic Institute and State University
Blacksburg, Virginia, USA
{bijoy,xbin114,shukla}@vt.edu

Jean-Pierre Talpin

Project ESPRESSO
INRIA
Rennes, France
Jean-Pierre.Talpin@irisa.fr

CONTENTS

8.1   Introduction   270
8.2   Thread Libraries for Multi-Threaded Programming   272
8.3   Protections for Data Integrity in a Multi-Threaded Environment   276
    8.3.1   Mutual Exclusion Primitives for Deterministic Output   276
    8.3.2   Transactional Memory   278
8.4   Programming Models for Shared Memory and Distributed Memory   279
    8.4.1   OpenMP   279
    8.4.2   Thread Building Blocks   280
    8.4.3   Message Passing Interface   281
8.5   Parallel Programming on Multiprocessors   282
8.6   Parallel Programming Using Graphic Processors   283
8.7   Model-Driven Code Generation for Multi-Core Systems   284
    8.7.1   StreamIt   285
8.8   Synchronous Programming Languages   286
8.9   Imperative Synchronous Language: Esterel   288
    8.9.1   Basic Concepts   288

8.9.2   Multi-Core Implementations and Their Compilation
Schemes 289
8.10   Declarative Synchronous Language: LUSTRE 290
8.10.1   Basic Concepts 291
8.10.2   Multi-Core Implementations from LUSTRE
Specifications 291
8.11   Multi-Rate Synchronous Language: SIGNAL 292
8.11.1   Basic Concepts 292
8.11.2   Characterization and Compilation of SIGNAL 293
8.11.3   SIGNAL Implementations on Distributed Systems . . 294
8.11.4   Multi-Threaded Programming Models for SIGNAL . . 296
8.12   Programming Models for Real-Time Software 299
8.12.1   Real-Time Extensions to Synchronous Languages . . . 300
8.13   Future Directions for Multi-Core Programming 301
Review Questions 302
Bibliography 305

8.1   Introduction

Introduction of multi-core processors is one of the most significant changes in the semiconductor industry in recent years. The shift to multi-core technology was preceded by a brief stint with the use of multiple virtual processors on top of a uniprocessor machine. Virtual processor techniques like the Intel Hyper-Threading technology [5] depended heavily on the distribution of computation between virtual processes. Improvement in performance due to the new techniques notwithstanding, the software which runs on these machines has remained the same. However it was soon realized that driving processor speed or simultaneous multi-threading did not solve power problems. The responsibility for driving up efficiency of multi-core systems now rests on the utilization of processing power by the software. Efficient distribution of work using multi-threaded programming or other parallel programming models must be adopted for this purpose. Embedded systems are following the lead of multi-core processors and will very soon transform themselves into parallel systems which need new programming models for design and execution. ARM Cortex-A9 [1] and Renesas SH-2A DUAL [15] are examples of such embedded processors. The buzz about parallel programming has resurfaced due to these developments and a rethinking about programming models for real-time software targeting embedded systems is underway.

Programming for multi-core systems is not a natural transition for software developers. Multi-threaded programming or system-level concurrent programming requires a sea change in the mental execution model to obtain good performance results. The steps in achieving this goal involve identifying concurrency in the specification, coalescing the parallelizable codes, converting them into different flows of control, adding synchronization points between computations and finally optimizing the code for target platforms. Moving into multi-threaded software programming requires tackling issues which are new to programmers. Deadlocks, corruption of data, and race conditions are some of the issues which can result in the failure of a safety-critical system. Handling real-time response to stimuli on a multi-core system can generate priority issues between processes that are new to designers. Dealing with these issues and verifying the correctness of a program requires a deep understanding of the programming model for multi-core systems [44].

A programming model usually refers to the underlying execution model of a computational device on which a program is run. For example, a programming model that an assembly language programmer assumes is based on the concept of instructions fetched from memory, executed in ALU, data transferred between architectural registers, and memory. However, a C programmer assumes a slightly higher level programming model where memory, registers and other parameters are not distinguished as such, and control of the program goes back and forth between procedures/functions and the main body of the program. Programming models for multiprocessors have been studied for several decades now. In a shared memory model, the software techniques adopted for multiprocessor and multi-core technology are very similar and will be used interchangeably in this chapter. The parallel programming models can be classified according to the layer of abstraction where the parallelism is expressed. Figure 8.1 shows a general view of the abstraction levels of the helping libraries, parallel architectures, software languages and tools available in the industry. Threading application programming interfaces (APIs) like POSIX [4] and Windows threads [11] form the lowest level of multi-threading models. The directives for parallelizing code like OpenMP [14], thread building blocks (TBBs) [6] and architectures like CUDA [13] and CellBE [2] programmed using extended C languages form a higher abstraction level. Model-driven software tools such as LabVIEW [12], SIMULINK® [9] etc. have their own programming languages or formalisms which can be transformed to a lower level code. The lower the abstraction level, the more control the user can have on the handling of tasks. The disadvantage about this approach is that identifying parallelism at a lower level is harder and optimization opportunities inherent in the specification stage will be missed.

Multi-core architecture is a new concept which was designed to break the power and frequency barriers reached by single core processors. But the programming model for multi-core was not conceived or developed with the same interest. Multiprocessor programming models which are suitable for multi-core processors are being proposed as candidates to extract parallel execution. In this chapter, we look at some of the programming concepts which are suitable for writing software targeting multi-core platforms. We scale abstraction levels and for each level of abstraction, we examine the capabilities and vulnerabili-

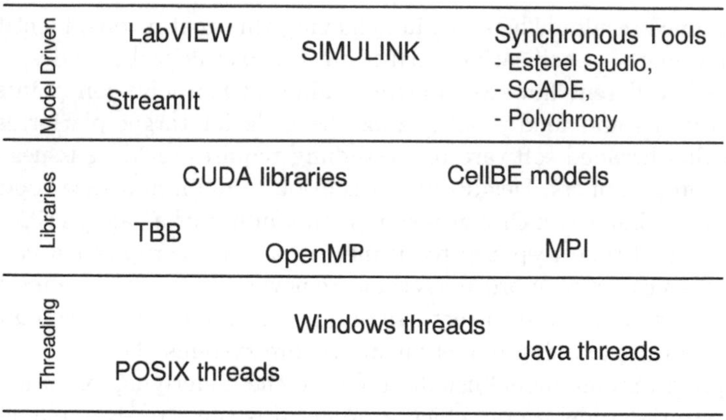

Model Driven	LabVIEW	SIMULINK	Synchronous Tools
	StreamIt		- Esterel Studio, - SCADE, - Polychrony
Libraries	CUDA libraries		CellBE models
	TBB	OpenMP	MPI
Threading	POSIX threads	Windows threads	Java threads

FIGURE 8.1: Abstraction levels of multi-core software directives, utilities and tools.

ties of the programming paradigm and discuss the multi-core implementations available in the literature. We choose the family of synchronous programming languages for detailed discussion due to its appealing features like determinism, concurrency, reactive response etc. which are crucial in ensuring safe operation. Multi-threaded implementations of synchronous languages such as Esterel [22], LUSTRE [31] and SIGNAL [30] are discussed in detail along with proposed real-time extensions.

8.2   Thread Libraries for Multi-Threaded Programming

Thread libraries are one of earliest APIs available to perform multi-tasking at operating system level. POSIX threads [4], and Windows threads [11] are the APIs for multi-threading used in Unix-like systems, and Microsoft Windows respectively. With the help of the specialized functions defined in these libraries, threads (or flows of control) can be generated which can execute concurrently. The level of abstraction is low for thread libraries which makes the programmer's task harder, but gives him more control over the parallel execution.

The implementations of POSIX threads and Windows threads are different, but their overall programming model is the same. A single flow of control or main thread is forked out into separate flows of execution. The **fork and join** threading structure for libraries such as POSIX threads or Windows threads is shown in Figure 8.2. The main thread shown in Figure 8.2 has been separated into five flows, each thread with a unique identification. These threads have associated function calls which specify the operations they will perform and attributes which is the data passed onto the function. The main thread (Thread A) can be used for computation as well, but in common practice the intention is to control the fork and join process. The join is used to wait for the completion of execution in different flows. The functions executed will return the data and a single flow of control is resumed.

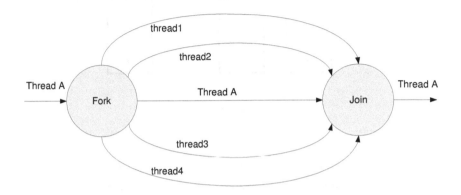

FIGURE 8.2: Threading structure of fork-join model.

Tasks in an operating system (OS) are modeled as processes which follow a threading model. They are scheduled by the kernel based on criteria like priority, data integrity, etc. OS executes these kernels, leaving the user with less control over the execution. The hardware thread runs on each core or a virtual core is called a *kernel thread* and the code provided by the user is called a *user thread*. The parallel execution of the threads can be one-to-one, many-to-one, or many-to-many ratio between user and kernel threads. In the absence of multi-core processors, the threading has to be performed by time sharing of a hardware kernel thread between the software user threads. This can still outperform the single threaded execution model, because a thread that is not running can still be performing a memory operation in parallel using peripherals of the processor. A **work distribution** model from user threads to kernel threads for multi-core or multiprocessor systems is shown in Figure 8.3. In the figure, n threads are distributed among m execution cores by the scheduler. Here the focus is on maximizing the utilization of the processing cores by an efficient scheduling algorithm. The cores (homogeneous or heterogeneous) are not allowed to remain idle by the scheduler. Another programming model using threading libraries is the **pipeline** model. The work done by each stage in the pipeline is modeled as a thread and the data acted upon changes over time. Every pipeline stage needs to be executed simultaneously for optimal performance results. Figure 8.4 shows a three-stage pipeline, each stage having its own execution thread. Since this model involves transfer of data from one stage to the other, additional synchronization constraints need to be considered. The difference between the two programming models is in the flow of data. A pipeline model has separate data and instructions, with the data moving across stages performing repeated operations. The work distribution model has data tied to the complete set of operations assigned to one or many cores.

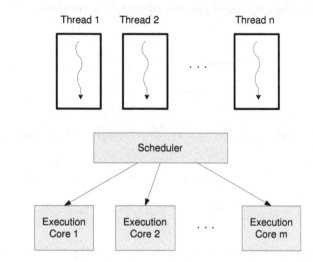

FIGURE 8.3: Work distribution model.

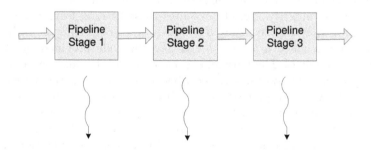

FIGURE 8.4: Pipeline threading model.

POSIX threads (portable operating system interface for Unix) or Pthreads are APIs for operating systems like Linux, MacOS etc. It consists of header files and libraries which have Pthread functions to create, join and wait for threads. Each thread will have its own *threadID*, which is useful for the user to allocate functions and data for their tasks. The listing of a POSIX-based threaded code for upcount and downcount of a protected variable in no fixed order is shown in Listing 8.1. The master thread is the `main` function used to fork and join threads and is devoid of any functional computation. *pthread_create* is used to create threads which call functions `countUp` and `countDown`. There are no attributes to be sent to the functions, hence only the function names are associated with the threadIDs, i.e., *thread1*, *thread2*. The `countUp` function increments the variable a and `countDown` function decrements it. The protection for the shared variable a is provided by POSIX primitives which will be discussed in a later section. Please note that this example is a simplified form to show the threading functions. The threads create and join operations are usually accompanied with check error statements to abort operation in the case of an error. Windows threads are sets of APIs provided by Microsoft Corporation for its Windows operating system. The facilities provided by this API are more or less similar, barring a few points. In Windows threading APIs, objects are accessed by their *handle* and the object type is masked. Object types can be threads, synchronization primitives, etc. One can wait for multiple objects of different types, using the same statement and thus remove the additional *join* statements in the Pthreads case. But some would consider this a disadvantage as the code is more ambiguous when used with *handle*.

Listing 8.1: Pthread code for fork-join model.

```
1   #include <pthread.h>
2   #include <stdio.h>
3
4
5   int a = 0;
6   pthread_mutex_t myMutex = PTHREAD_MUTEX_INITIALIZER;
7   void countUp(void *ptr);
8   void countDown(void *ptr);
9
10  int main()
11  {
12          pthread_t thread1, thread2;
13
14          pthread_create( &thread1, (pthread_attr_t *) NULL,
15                          (void *) countUp, (void *) NULL);
16          pthread_create( &thread2, (pthread_attr_t *) NULL,
17                          (void *) countDown, (void *) NULL);
18
19          pthread_join(thread1, (void *) NULL);
20          pthread_join(thread2, (void *) NULL);
21          return 0;
22  }
23
24  void countUp(void *ptr)
25  {
26          for (int i=0, i<5, i++)
27          {       pthread_mutex_lock(&myMutex);
28                  a = a+1;
29                  printf("Thread1: %d\n", a);
30                  pthread_mutex_unlock(&myMutex);
31          }
32  }
33
34
35  void countDown(void *ptr)
36  {
37          for (int i=0, i<5, i++)
38          {       pthread_mutex_lock(&myMutex);
39                  if(a > 0)
40                  {
41                          a = a-1;
42                          printf("Thread2: %d\n", a);
43                  }
44                  pthread_mutex_unlock(&myMutex);
45          }
46  }
```

The arguments about choosing APIs may not be conclusive but there are common issues that require attention while working on this level of abstraction. The highest importance is for the mutual exclusion property required while accessing shared data. There are sections in threads which need sequential update operation on data to maintain data integrity. In the Listing 8.1, the variable a needs to be provided with sufficient protection to avoid conflict between read and write operations of the two functions. Thread APIs have several kinds of objects like mutex, semaphore, critical section etc. which provide mutual exclusion property. These objects ensure that there is a lock placed on the critical piece of data and the key is given to only one thread at a time. A detailed discussion of the data access issues in the threading or transaction-based execution is given in the next section.

8.3 Protections for Data Integrity in a Multi-Threaded Environment

In multi-threaded software, whenever there is sharing of data, out of order update operations on shared memory become a concern. When multiple threads are allowed to write on a memory location, the write operations performed on it should be in order. Even when the order of access is fixed, completion of a write operation must be ensured. There could be read-write conflicts as well, if multiple read operations are performed on a critical section while a write operation is ongoing. The read value from the memory location is now ambiguous, so ordering of write operations is not enough to ensure correct operation. If two threads (Thread A and Thread B) are allowed to enter a critical section, the final value of the shared memory location is unpredictable. If multiple threads are allowed to compete for access to a data point or a memory location, we have a *race condition* delivering unpredictable results. For a deterministic result for each run of the code, the critical sections have to be protected by mutual exclusion primitives.

8.3.1 Mutual Exclusion Primitives for Deterministic Output

The solutions for avoiding corrupted data are based on mutual exclusion property. This strategy is based on giving a single thread access to each critical piece of data. The implementation of such a protocol can be based on a flag-based entry and exit of the critical section. Figure 8.5 illustrates the access of two critical sections of code in separate threads, which also share the same variables. If entry and exit flags are added to the beginning and end of the critical section of each thread as shown in Figure 8.5, we can have synchronized updates of the shared data. The flags are used as a constraint at the entry to the critical section to verify whether any other thread is in the critical

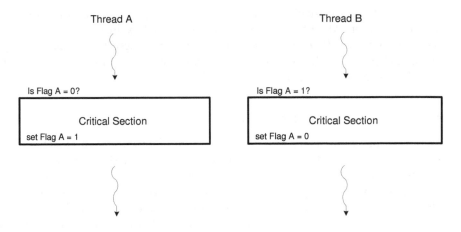

FIGURE 8.5: Scheduling threading structure.

section at that point. A failure of this model is in the protection of the flags that have to be shared amongst the threads. An entry to the critical section does not ensure that the flag set/reset operation has been done in sequential order, and hence the integrity of the flags is questionable. Also a read on the flag should not be processed while a write has been issued on the same flag by another thread. Such protection can be provided only by using *atomic* operations on registers, which sequentialize the write/read operations according to their order of assignment.

In POSIX standard, a *mutex* object can be utilized to perform the atomic operations. A lock and key mechanism is implemented around the protected mutex variable (say z). Each thread will try to obtain the key to access the locked variable. Functions can be called from threads to lock until allowed access($pthread\_mutex\_lock(z)$), or to try lock and return if not allowed access ($pthread\_mutex\_trylock(z)$). An unlock operation ($pthread\_mutex\_unlock(z)$) is performed after the critical section operations are performed. Another synchronization object is *semaphore* (counting mutex) developed by E. W. Dijkstra [27]. A fixed set of threads can enter the critical section and the number of accesses is maintained by the upcount (entry) and downcount (exit) of the semaphore. Here one semaphore is regulating accesses to multiple resources. Along the lines of POSIX synchronization objects, a *critical section* object can be used in Windows threads. There exist other synchronization primitives like *monitor* from Hoare [34] or *event* which are suitable for specific locking and notify situations. They check for new events on a protected variable and notify a set of threads waiting for that particular operation to be completed to resume their individual work. Primitives defined by POSIX/Win32 standards might ensure that the critical section is devoid of any race conditions. But deadlocks (multiple threads waiting for access) or livelocks (multiple threads starved of resources) can still appear during the execution of multi-threaded code if the primitives are not carefully used by the programmer. Algorithms based on these mutual exclusion primitives have been proposed like Peterson's algorithm, Lamport's algorithm, Dekker's algorithm, Bakery algorithm etc. [50] which will assure programming models are free of race conditions, deadlocks, livelocks etc.

8.3.2   Transactional Memory

A transaction can be considered as a collection of a finite set of objects. These objects can be characterized as operations performed on data, tasks scheduled for processors, communication messages between IPs etc. One of the earliest references of this concept is in [43] by Tom Knight, where a **transactional block** was defined as a set of instructions that does not contain any interaction with other blocks or memory accesses in between. A *dependency list* is maintained which contains the list of all memory locations used for the transaction block prior to execution. After the various blocks of instructions are executed in multiple processes, the write operation into memory has to be performed according to their order of memory access. This *confirming* step is a write operation which will modify the content in main memory. So any transaction block which was using the earlier values in these written memory locations for its execution will have to undergo an *abort* operation.

The goal for transactional block concept was to have parallel operations performed on memory, but along with it uphold data integrity. The implementation of such a model was possible only with the use of locks which give atomic access to memory locations. The synchronization involved in these protocols adds considerable overhead in the number of instructions to be executed. To provide lock free synchronization, the **transactional memory** concept was proposed by Herlihy and Moss [33] in 1994. They defined transactional memory as a finite sequence of machine instructions executed by a single process having the properties of serializability and atomicity. *Serializability* of a transaction is the ability to view the set of instructions in a block in a deterministic sequential order, *Atomicity* enables the different blocks to either commit changes to the memory or to abort if any update was performed in between memory accesses. These properties will ensure that the main memory is updated at one go by a block and the pecking order for update operation is maintained. Primitive instructions like *load-transactional* and *store-transactional* were proposed which have become a de facto standard for expressing transactional behavior.

A variant of transactional memory called **software transactional memory** (STM) was proposed by Shavit and Touitou in [49]. They define STM as a shared object which behaves like a memory supporting multiple changes to its location. Ownership of a memory is required to modify its contents and the process which requires ownership is "helped" by other processes to maintain the *wait-free* and *non-blocking* properties in STM. These properties ensure that the threads of execution are not made to wait at any transaction or blocked from accessing a memory location by means of a priority ordering. The ownership transfer was implemented by a compare and swap procedure with the help of a priority queue. The advantages of transactional memory concept include avoiding deadlocks and livelocks with a lock-free mechanism. But priority inversion and overhead of aborted transactions are problems which could result in lower performance or failure of the system.

8.4 Programming Models for Shared Memory and Distributed Memory

Memory models for computing machines are a distinguishing factor for programming in a multiprocessor environment. Multi-core processors are used in a shared memory environment, while multiprocessor systems can work with both memory models. In a shared memory model, each core has the capacity to address the common memory space directly. Distributed memory has parallel machines with exclusive access to its memory module. Such a network of machines is desirable, when the specification is highly parallel and the computation can be localized. Special communication interfaces like message passing interfaces (MPIs) [29] are defined for message passing between cores. The emergence of multi-core systems triggered an interest in the conversion of existing sequential programs into a parallel form for faster execution. The distributed memory models have taken a back seat in handling this task, while synchronization objects defined in POSIX/Windows threads are too low level to tackle this problem. Higher level APIs like OpenMP and thread building blocks (TBBs) have been proposed for this purpose using shared memory based models. Specialized pragmas defined in OpenMp and TBB are associated with loops or any other places which need parallelization. These models of parallel programming are non-invasive, since they can be ignored in an environment that does not support these pragmas.

8.4.1 OpenMP

OpenMP [14] is a set of compiler directives, run-time routines and environment variables used to express parallelism in code. They can be Fortran directives or C/C++ pragmas (*pragma omp*) , which alter the control flow into a fork-join pattern. When encountered with a *parallel* construct followed by an iterative loop, the compiler will create a set of slave threads which divide the iterations among themselves and execute them concurrently. The number of threads can be set by the user by means of a OpenMP directive *omp_set_num_threads*() or by setting a related environment variable *OMP_NUM_THREAD*. The underlying execution model for OpenMP is some implementation which could be an OS process model or a program level threading model (e.g., POSIX threads).

The type and scope of the data structures of each thread are important when used in a multi-threaded context. A variable can be explicitly specified as a *shared* or *private* variable under threaded conditions using OpenMP directives. An OpenMP pragma in C/C++ can specify a set of variables to be shared within a class or structure (default case), which would ensure a single copy is maintained for those variables. For an iterative loop, the iterative index is considered as private for threading purposes. Apart from the index, any variable which will undergo update operation within a loop must also be specified as private to have separate copies for each thread. There is also a reduction operation which functions by making a combination of shared and private variables and makes use of the commutative-associative properties to form intermediate results and thus parallelize the operation. This is different from parallelizing the iterations in a loop as the intermediate results are accumulated to get the final result.

Parallel programming constructs are utilized to increase performance and to ensure correctness of code. There might be statements which should not be executed in parallel or variables whose additional copies should not be made. This can be guaranteed by using *atomic* directive to halt all parallel operation for the concerned statement. *Critical-end* critical directive serves the same purpose for a section of code. There exists *set lock-unset lock* directives similar to Pthread mutex variables for providing exclusive access to variables. In comparison with Pthread constructs, a disadvantage is that the critical section in OpenMP stalls any other critical operation. Even two independent critical sections without common shared variables run in separate threads cannot be executed in parallel. Other mutual exclusion primitives include event synchronization and memory access ordering pragmas. A *barrier* directive can be used to synchronize all threads at a point which acts as a location to halt, join and proceed. The threads which finish execution will wait for others to reach the synchronization point. The *ordered* pragma provides exclusive access to memory by sequentializing a portion of code. This enables the code to perform parallel computation and sequential storing operation.

8.4.2   Thread Building Blocks

Following the lead of OpenMP, new libraries have been proposed for extending parallelization constructs to C++. Thread building blocks (TBBs) [6] represent an effort from Intel Corporation to provide shared memory parallelism in C++ with automatic scheduling of work. It aims to provide better load balancing by using task-based programming instead of lower level threads. The TBB libraries can be used to perform loop parallelization, sorting, scanning etc. which we discuss in this section.

There are two major loop parallelization templates from TBB namely *parallel_for* and *parallel_reduce*. An iterative loop which can be safely parallelized can be done by using *parallel_for* function. The parameters for this function are the datatype, grainsize, number of iterations and the operator function to be parallelized. Grainsize describes the chunk of operations in each parallel processing thread and can be optimized experimentally. The *parallel_reduce* function performs computation in a split-join fashion. A reduction operation is performed by partitioning a long serial operation into smaller independent parts which are merged after the computation. The distribution of computation is shown in Figure 8.6 for both *parallel_for* and *parallel_reduce* functions. The iterations are parallelized in the first case while the sub-blocks are parallelized in the second. Specialized functions like *parallel_scan*, *parallel_sort* etc. are available to exploit concurrency in parallel algorithms. For memory accesses, *containers* or FIFO-like arrangements for multiple threads and standard templates for mutual exclusion are provided.

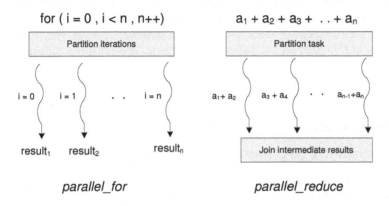

FIGURE 8.6: Parallel functions in thread building blocks.

TBB was designed to remain strictly as a C++ library to support parallelism. Compiler support was required for OpenMP, which is avoided in the case of TBB. TBB also provides nested parallelism support and support for more parallel algorithms. When compared to native threading, TBB influences the scheduling by providing an unfair distribution of processor execution time for each thread. Execution time is allotted based on the load on each thread and thus TBB provides better performance than other shared memory parallelism techniques.

8.4.3   Message Passing Interface

Message passing interface (MPI) is a programming model targeting distributed execution in multiprocessors [29]. The MPI programming model consists of parallel processes communicating with each other in point-to-point fashion. In contrast to the forking of threads in OpenMP, MPI is concurrent from the very beginning. Parallel processes execute in a MIMD-like model and operate on memory with exclusive access. The focus of MPI programming model is on task division, thus reducing the communication between processes.

MPI provides several specialized functions to communicate between processes. The message can be sent or received in a blocking or non-blocking fashion. The message passing functions like *MPI_Send*, *MPI_Recv* will contain parameters which give the starting address, size and type of the data sent/received along with message identifiers and communication handles. The communication handle describes the processes in a group which are allowed to receive the message. Different MPI functions provide facilities to broadcast, distribute or accumulate data within groups. Groups contain an ordered set of processes uniquely defined by their rank. The communication between these processes is termed as *intra-group communication*. It is possible to have message passing between processes that are part of separate groups. In such an *inter-group communication* environment, the identifiers for a process is the communicator (group identifier) and the rank of a process. Apart from the blocking message passing functions, specific synchronization functions (*MPI_Barrier*) are also provided to co-ordinate the communicating processes.

8.5   Parallel Programming on Multiprocessors

As we have discussed before, parallel programming research started with multiprocessor systems. Methodologies applied in that era are the inspiration behind many of the multi-core chips. Some of the significant multi-core processors are CellBE [2] and Sun Niagara [18]. The fundamental difference in architecture between the Niagara processor and the CellBE is in the type of processors employed. Niagara is a homogeneous processor with eight SPARC cores, while CellBE is a heterogeneous processor with IBM PowerPC and several vector processing elements. CellBE has found commercial success in gaming consoles and we will discuss the architecture with an associated programming model of this design in brief.

The Cell broadband engine architecture [2] consists of an IBM PowerPC processor as the power processing element (PPE) with eight vector processors as the synergistic processing elements (SPEs). They are interconnected by an element interconnect bus. The PPE acts as a controller for the SPEs by performing scheduling operations, resource management and other OS services. The Cell processor can support two hardware threads in its PPE and eight hardware threads in its SPEs. But the programming model of the Cell processor is not restricted to a singular methodology. Users are free to create as many software threads and manage the communication between them in shared memory or message passing model. The Cell processor supports OpenMP libraries and is flexible enough to perform multi-threading operations in pipeline, job queue or streaming format. Cell superscalar [20] is one of the applicable programing models for the Cell processor which uses annotations to delegate tasks from the PPE to the SPEs. The PPE contains a master thread which maintains a data dependency graph for the tasks to be performed and a helper thread which schedules tasks for each SPE thread and updates the task dependency graph. The creative freedom present in the applicable programming models for the Cell processor has made it a versatile platform in the multi-core embedded system domain.

8.6 Parallel Programming Using Graphic Processors

Graphics processors have been used to offload vector processing from CPUs for a long time now. Recent advances in gaming technology motivated researchers to look at graphics processors for general purpose computation. The idea is to make use of a large number of multiprocessors in graphics cards to create a massively parallel system for computation. Using general purpose graphics processing units (GPGPUs) for parallel processing delivers a favorable performance-cost metric when compared to the available supercomputing options. The programming of such graphics processors is very different from other embedded systems as they follow a single instruction multiple data pattern. We discuss a leading architecture from NVIDIA Corporation and its programming philosophy in this section.

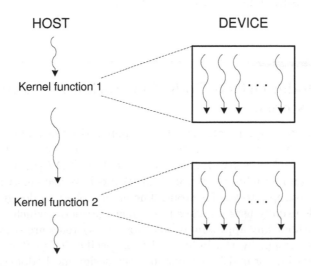

FIGURE 8.7: Program flow in host and device for NVIDIA CUDA.

Compute unified device architecture (CUDA) is a new programming model defined for NVIDIA GPGPUs [13]. In this programming model, the CPU (host) code sets the number of threads to be created in the GPU (device) using a kernel function. Each parallel operation is a kernel function call which halts the host code execution in the CPU and starts a massively parallel operation in the GPU. In the GPU, a set of threads are tied together to form a *warp* and is assigned to a set of processors. The warps assigned to a multiprocessor take turns in execution, memory fetch, etc. Figure 8.7 shows the execution model of the NVIDIA Tesla GPU. The host code is executed in sequential fashion with pauses during the parallel device operation. This shared programming model is scalable for larger applications. There exists a global memory along with shared memory and specialized registers for groups of processing units. Several atomic functions like *atomicAdd()*, *atomicInc*, *atomicAnd* are provided for safe threading operations.

The programming model used in GPUs is of single program multiple data (SPMD) pattern. Streaming data for video rendering was ideal for this model since the same computations were done for multiple pixels. Brook for GPU, a streaming model for GPU general purpose computation similar to CUDA was proposed from Stanford University [17]. The programming language is ANSI C with extensions to declare streams and kernel functions to operate on them. Here the extended C code is transformed to an executable form for graphics processors and no new programming language is required. At a higher level of abstraction is model-driven code generation which has a sound formal basis in its specification format. Streaming model, data flow model, etc. have been used as a references to design high level languages which are transformed into C or RTL using different code generation tools.

8.7   Model-Driven Code Generation for Multi-Core Systems

Model-driven code generation tools have popularized using formal models with sound mathematical bases as the starting point for system design in control systems, embedded software, etc. Tools like LabVIEW [12] from National Instruments and Simulink®[9] from MathWorks have been instrumental in driving these concepts to the forefront. The methodology in designing software in these tools usually uses a higher level language/model which can describe the system without any approximations. Now the systems are translated into the lower level design by the individual tool-specific design flow. In the case of code generation for multi-core systems, the design methodology has to change from the high level specification. The concurrency in the specification needs to be captured correctly at a higher level and it needs to be protected throughout the design flow to generate parallelized code. Event-driven modeling of finite state machines in Simulink Stateflow is a good example of capturing concurrency at a higher level. Individual modules in Stateflow are concurrent and they are composed by means of input/output events. Even then the formalism for Simulink or LabVIEW is not intended for multi-threaded code generation and the major area of focus is different from the multi-core domain. So in this section, we describe StreamIt, a code generation tool with concurrent stream model as a representative of the genre of model-driven tools.

8.7.1 StreamIt

StreamIt [51] is a programming language which is used to model designs which handle streaming flow of data. In this programming model, the smallest basic block is a *filter* which has a single input and output. A filter consists of two special functions, namely *init* and *work*. The function *init* will run at the beginning of the program setting the I/O type, data rate and initialization values while *work* function will run continuously forever. Multiple filter blocks are composed to form structures like pipeline, split-join or feedback loop, which are again single-input, single-output blocks. These StreamIt structures are shown in Figure 8.8. *Pipeline* construct consists of multiple filters in a particular order and it has only an *init* function of its own. *Split-join* construct is used to diverge streams and combine them at a later time. *Feedbackloop* is a combination of a splitter, a joiner with a body and a loop stream. An *initPath* function in the construct will define the initial data and set the delay before joining items in the feedback channel.

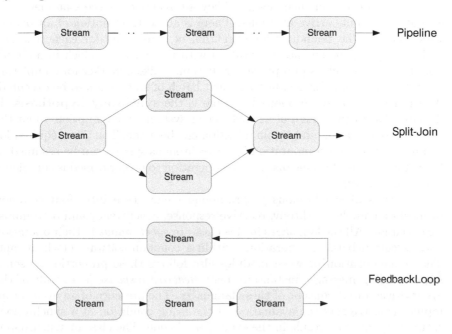

FIGURE 8.8: Stream structures using *filters*.

Dataflow structure makes StreamIt more suitable for multi-core execution. Parallel threads which access a shared memory and communicate using sockets can be generated using a StreamIt compiler. Initially all the threads will be running on the host which compiles the program and will later fork into threads which can execute independent streams. One of the restrictions of StreamIt model, is that filters cannot handle multiple rates. The rate of data flow through a filter remains a constant. StreamIt is more close to synchronous dataflow languages that are capable of providing a deterministic output from a concurrent specification. They are event-driven and perform rigorous formal clock analysis in the backend. The synchronous programming language SIGNAL is multi-rate and hence we will discuss this particular programming model in detail. For a multi-threaded implementation for a safety-critical embedded system, these properties (reactive, deterministic, multi-rate) are attractive which make synchronous programming languages an attractive proposition.

8.8 Synchronous Programming Languages

Synchronous programming languages utilize synchronous execution of code as the central concept in their design. They are reactive, as the statements are executed as events arrive at inputs. There is an abstract notion of an *instant* which defines the boundary for execution of statements for each reaction. This concept of instant has no relation with the hardware clock in a circuit nor the execution clock of a processor. It is more like a marker for completing a set of actions and for deciding the next batch of statements to be executed. At the heart of a synchronous language is the **synchrony hypothesis**. It declares that in the design of a synchronous system, an assumption about the time for computation and communication can be made. The time required for communication and computation in a synchronous system can be assumed to be instantaneous. This means that the operation to be performed is completed within an *instant*.

The class of synchronous programming languages exhibit four common properties, namely synchrony, reactive response, concurrency and deterministic execution. All the languages in this class are *synchronous* in their operation and execute a batch of operations within a common software clock instant. The communication between modules also follows these properties by sending or reading messages instantaneously. *Reactive response* is a result of the event driven input concept of these languages. The presence of an event at an input signal triggers the evaluation of the firing condition of a synchronous statement, and may result in the execution of code. The class of synchronous programming languages has the ability to capture *concurrency* at a high level. The execution of modules or statements can be specified independently of each other. If unrelated signals are triggering mutually exclusive sets of statements, the lower level code will be executed completely in parallel. This might not be true in the case of compilers which generate sequential code. Finally, *deterministic execution* of synchronous programming languages can be guaranteed since the computation and communication are to be completed before the next instant. Given the set of input events and the synchronous program, the output of each module can be predicted for every instant.

Several proposed synchronous programming languages encompass these properties for synchrony [32]. They differ in terms of their applications, compilation schemes, specification (textual, visual) or code generation methods (sequential, parallel). Some of them have been commercialized as software tools and have found acceptance in safety-critical fields like aviation, power plants etc. This chapter will cover **Esterel** [22], **LUSTRE** [31] and **SIGNAL** [30] languages in detail in the next subsections. Some other synchronous programming languages are briefly introduced here for the readers.

Argos is a automata-based synchronous language developed at IMAG (Grenoble) which uses the graphs to describe reactive systems [47]. In Argos, a process is expressed as an automaton using graphical representation. Parallel composition of processes and hierarchical decomposition of the processes are used to construct large systems.

ChucK is a programming language for concurrent real-time audio synthesis [52]. The backbone of this language is a highly precise timing/concurrency programming model which can synthesize complex programs with determinism, concurrency and multiple, simultaneous, dynamic control rates. A unique feature of the language is *on the fly programming* which is an ability to dynamically modify code when the program is running. This language is not primarily optimized for raw performance instead it gives more priority to readability and flexibility.

SOL (Secure Operations Language) developed jointly by the United States Naval Research Laboratory and Utah State University is a domain-specific synchronous programming language for implementing reactive systems [24]. SOL is a secure language which can enforce safety and security policies for protection. These security policies can be expresses as *enforcement automata* and the parts of the SOL program which do not abide by the policies are terminated.

LEA is a multi-paradigm language for programming synchronous reactive systems specified in Lustre, Esterel or Argos. The synchronous specification in any of the three languages is translated into common format *Boolean automata*, and thus the integration of modules specified in different languages is performed. Synchronie WorkBench (SWB) is the integrated development environment for specifying, compiling, verifying and generating code for the synchronous languages [35].

8.9 Imperative Synchronous Language: Esterel

Esterel is an imperative synchronous programming language for the development of complex reactive systems [22]. The development of the language started in the early 1980s as a project conducted at INRIA and ENSMP. Esterel Technologies provides a development environment called Esterel Studio [3] based on the Esterel language. Esterel Studio takes Esterel specification as input and generates C code or hardware (RTL) implementations. In this section, we briefly introduce the basic concepts of Esterel and then move on to its programming models for multi-core/multiprocessor implementations.

8.9.1 Basic Concepts

In Esterel, there are two types of basic objects: signals and variables. **Signals** are the means for communication and can be used as inputs or outputs for the interface or as local signals. There are two parts to a signal, namely the status and the value. The *status* denotes whether the signal is present or absent at a given instant and on presence, *value* provides the data contained in the signal. The *value* attribute of a signal is permanent and if the signal is absent, it will retain the information from previous instant. Esterel assumes instantaneous broadcasting of signals. Once a signal A is **emit** by a statement, the statements which are "listening" to this signal will be active. The scope of a signal is valid all through the module it is defined in and can be passed to another module for computation. **Variable** is local to the module it is defined in and unlike the *signal*, can be updated several times within an instant.

An Esterel program consists of modules, which in turn are made of declarations and statements. The declarations are used to assign data types and initial values (optional) for signals and variables. Statements consist of expressions which are built from variables, signals, constants, functions, etc. The expressions in Esterel are of three basic types, namely data, signal and delay. Data expressions are computations performed using functions, variables, constants or current value of a signal (denoted by '$?A$'). Signal expressions are Boolean computations performed on the *status* of a signal. Logical primitives like **and**, **or**, **not** are used in these expressions to obtain a combinational output (e.g., a **and not** b). Delay expressions are used in temporal statements along with primitives like **await**, **every**, etc. to test for presence or to assign the statements to be executed. For example, **present** A **then** $< bodyA >$ **else** $< bodynotA >$ **end** checks for the presence of A and selects between two sets of statements $bodyA$ and $bodynotA$. Esterel expressions are converted to finite state automata with the statements as datapath and conditions as guards. The finite state machine programming model is used as the underlying formalism to convert Esterel expressions to RTL or C code during synthesis.

8.9.2 Multi-Core Implementations and Their Compilation Schemes

Esterel expressions are converted into a finite automaton and synthesis is performed to generate sequential code [22]. An input automaton at state P when in the presence of an input event i, generates an output event o and moves into a derivative state P'. In this manner, a finite state machine (FSM) can be formed which produces a deterministic sequential output from a concurrent specification. The datapath of the FSM at each state will include the code that has to be executed at each instant. Esterel compiler can generate C code or RTL from this finite state automata.

The underlying concurrency in the specification makes Esterel a good candidate for distributed implementation. A work on automatic distribution of synchronous programs proposed a common algorithm for conversion of an **object code** (OC) into a distributed network of processors [26]. Esterel, LUSTRE and Argos compilers can output code in this common format. The distribution method from the OC form is as follows:

1. The centralized object code is duplicated for each location.
2. Decision is made on mapping each instruction to a unique location and copies are removed from the rest of the locations.
3. Analysis is performed to find the communication required between locations to maintain the data dependency between instructions.
4. New instructions are inserted (put, get) to pass the variables that were computed in a different location.

An optimization can be performed to reduce the redundant code in the network. A sample object code is shown in Listing 8.2 and its distributed implementation for two locations is shown in Figure 8.9. The code is first duplicated on both locations and then the body of the code is removed from one of them. Later the communication instructions (put(a,0), get(1)) are placed in the locations as required. In Figure 8.9, on a *true* result on the 'If a' condition, *body*1 is executed on Loc0 and Loc1 remains idle. On a *false* result, Loc1 computes *body*2 and sends the value of a to Loc0. In Loc0, a **get** operation is performed to update the latest value of a and then *body*3 is executed.

Listing 8.2: Object Code for an if-else condition.

```
 1   Location      State 0
 2   0             put_void(1);
 3   1             put(0,a);
 4   0,1           if (a) then
 5   1                 put(0,a);
 6   0                 body1
 7   0                 output(b);
 8   0,1           else
 9   1                 body2
10   1                 put(0,a);
11   0                 body3
12   0                 output(b);
13   0,1           end if
14   0,1           go to State1
```

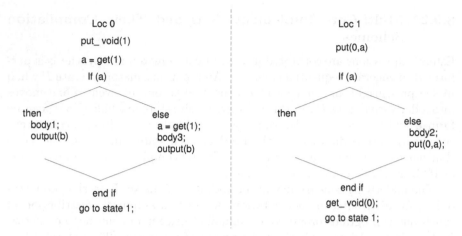

FIGURE 8.9: OC program in Listing 8.2 distributed into two locations.

The Columbia Esterel Compiler [28] has implemented a few code generation techniques to form C code from Esterel. One method divides the code into atomic tasks and performs aggressive scheduling operations. Another method is to form a linked list of the tasks by finding their dependencies. Here also the focus is on fine grained parallelism as in OC method [26]. A distributed implementation on multiprocessors [53] uses a graph code format proposed in [48] to represent parallelism in Esterel. Here each thread is a distinct automaton (or a reactive sub-machine). Instead of scheduling tasks during runtime as in other techniques, each sub-machine is assigned to a processor core and they are combined together to form the main machine which represents the whole Esterel code.

8.10   Declarative Synchronous Language: LUSTRE

LUSTRE is a declarative synchronous language based on a data flow model [31]. The data flow approach allows the modeling to be functional and parallel, which helps in verification and safe transformation. The language was developed by Verimag and it is the core language behind the tool SCADE from Esterel Technologies [25]. The data flow concept behind LUSTRE enables easier verification and model checking using the tool Lesar and hence is popular for modeling safety critical applications like avionics, nuclear plants, etc.

8.10.1 Basic Concepts

In LUSTRE, a variable is an infinite stream of values or a *flow*. Each variable is associated with its *clock* which defines the presence or absence of the variable at an instant. The statements in LUSTRE are made of data flow equations, which result in the clock equations of the respective variables as well. There are four temporal operators in LUSTRE, namely Pre, ->(followed by), when and current.

1) $Pre(e)$ provides the previous value in the flow of the event e.

2) x -> y orders sequence x followed by y.

3) $z = x$ when y is a sampler which passes value of x to the output z when the Boolean y input is true.

4) $current(z)$ is used with $z = x$ when y and it memorizes the last value of x for each clock instance of y.

Apart from the equations, there can be assertions in a LUSTRE program. They are used to specify the occurrence or non-occurrence of two variables at the same time or any other property of the design. In the LUSTRE compiler, *clock calculus* is performed to find the clock hierarchy of the variables. A finite automaton is built from the state variables in a similar manner as in Esterel and code generation is performed to obtain the C or RTL.

8.10.2 Multi-Core Implementations from LUSTRE Specifications

The LUSTRE compiler can also generate output in object code form [26] which can be used for distributed implementation as described in the Section 8.9.2. Another work on multiprocessor implementation is based on time triggered architectures (TTAs). SCADE can be used to map LUSTRE specifications on the synchronous bus by having some extensions on the LUSTRE code in the form of annotations [25]. Code distribution annotations are used to assign parts of LUSTRE program to unique locations in the distributed platform. Execution time, period and deadlines can also be specified along with the code. The methodology for implementation of LUSTRE program in a TTA is shown in Figure 8.10. The LUSTRE specification given to the analyzer which builds a partial order of tasks with the help of the deadline and execution time annotations. The timing details are used by the scheduler to solve a multi-period, multiprocessor scheduling problem. The bus and processor schedules for a solution to this problem are given to the integrator block. Integrator obtains the different LUSTRE modules from analyzer and generates a glue code to interface these modules.

In LUSTRE and Esterel, the parallel implementations have focused on locating the computation on the platform rather than identifying streams of data to be assigned as tasks. The textual representation and lack of visual means to project the specification is a handicap with Esterel. The data flow representation in LUSTRE does address this problem, but the distributed im-

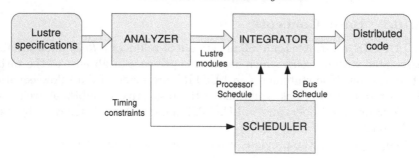

FIGURE 8.10: LUSTRE to TTA implementation flow.

plementation methods remain the same. Both languages try to convert the
automata generated from the respective specifications into an intermediate
form ready for deployment in a distributed system. Within the family of synchronous languages, a new formalism SIGNAL has tried to address multi-
threading for multi-core aspect in a different manner. In structure, SIGNAL
is closer to LUSTRE, but better suited for multi-threaded programming. In
the next section, the SIGNAL language, semantics and the multi-threading
methodologies proposed in literature are discussed in detail.

8.11   Multi-Rate Synchronous Language: SIGNAL

SIGNAL is a declarative synchronous language that is multi-rate [30]. SIG-
NAL captures computation by data flow relations and by modularization of
processes. The variables in this language are called as signals and they are
multi-rate. This means that two signals can be of different rates and can remain unrelated throughout the program. This is a significant departure from
LUSTRE data flow specifications which define a global clock which is synchronous with every clock in the code. SIGNAL language and its Polychrony
compiler were developed by IRISA, France.

8.11.1   Basic Concepts

The SIGNAL language consists of statements written inside processes, which
can be combined together. A signal x is tied to its clock $\hat{x}$ which defines the
rate at which the signal gets updated. A signal can be of different data types
like Boolean, integer, etc. The statements inside a process can be assignment
equations or clock equations. If there is no data dependency between the input
signals of one statement with the output signal of another statement, they are
concurrent within the process. In contrast to Esterel, no two signals can be

repeatedly assigned within a process. The assignment statements will consist of either function calls which are defined by other processes or any of the four primitive SIGNAL operators. They are as follows:

The **function** operator f when applied on a set of signals $x_1, x_2, .., x_n$ will produce an event on the output signal y and is represented in SIGNAL as :

$$y := f(x_1, ..., x_n) \qquad (8.1)$$

Along with the function operator, the clocking requirements for the input signals are specified. To evaluate an operation on n inputs, all n inputs need to be present together and this equates the rates of y with each of the input signals.

The **sampler** operator when is used to check the output of an input signal at the true occurrence of another input signal.

$$y := x \text{ when } z \qquad (8.2)$$

Here z is a Boolean signal whose true occurrence passes the value of x to y. The true occurrence of z is represented as $\widehat{[z]}$. The clock relation of y is defined as the intersection of the clocks of x and z.

The **merge** operator in SIGNAL uses default primitive to select between two inputs x and z to be sent as the output, with a higher priority to the first input.

$$y := x \text{ default } z \qquad (8.3)$$

Here the input x is passed to y whenever $\hat{x}$ is true, otherwise z is passed on whenever $\hat{z}$ is true. So the clock of y is the union of the clocks of x and z.

The **delay** operator in SIGNAL sends a previous value of the input to the output with an initial value k as the first output.

$$y := x\$ \text{ init } k \qquad (8.4)$$

Here previous value of x, denoted by $x\$$ is sent to y with initial value of k, a constant. The clock of signals y and x are equated by this primitive. The clock equations of the SIGNAL operators are summarized in Table 8.1.

8.11.2   Characterization and Compilation of SIGNAL

Unlike the synchronous languages described above, SIGNAL specification does not require every signal in a program to be working at a clock that is a subset of the global clock. The multi-rate specification demands an independent clock structure between unrelated signals. But in the current Polychrony compiler, the global clock is enforced by defining a global clock based on the fastest clock in the program. **Endochrony** describes the property of a SIGNAL code to

TABLE 8.1: SIGNAL Operators and Clock Relations

SIGNAL operator	SIGNAL expression	Clock relation
Function	$y = f(x_1, x_2, \ldots, x_n)$	$\widehat{y} = \widehat{x_1} = \ldots = \widehat{x_n}$
Sampler	$y = x \; when \; z$	$\widehat{y} = \widehat{x} \cap [z]$
Merge	$y = x \; default \; z$	$\widehat{y} = \widehat{x} \cup \widehat{z}$
Delay	$y = x \; \$ \; init \; k$	$\widehat{y} = \widehat{x}$

construct a clock hierarchy where there exists a clock from which the signals of the program can be derived. This property would mean that a static schedule can be found for the computations in the program. Formal definition about this property can be found in [21] and the examples which explain these characterizations can be found in [41]. The current version of Polychrony compiler requires endochrony as a sufficient condition for a SIGNAL program to be transformed into sequential C code.

In a multi-rate SIGNAL code, a process is said to be **weakly endochronous**, if it satisfies the 'diamond property' or in other words, if the computation is confluent. Confluence in the SIGNAL context means that irrespective of the order of computation, the final output of the process remains the same. An example of a weakly endochronous SIGNAL code is shown in Listing 8.3:

Listing 8.3: Weakly endochronous SIGNAL program.

```
1   process wendo = (? event x, y;
2                     ! boolean a, b;)
3   (| ia := 1 when x|
4    | ib := 2 when y|
5    | a := ia$ init 0|
6    | b := ib$ init 0|
7   )
```

Here the computations of a and b are independent of each other and they are truly concurrent. Such a piece of code need not be restrained by a global clock connecting the inputs x and y. The diamond property present in this code is shown in Figure 8.11. There are three different orders of execution, with event x happening before y as shown in the top path, y before x shown in the bottom path and the synchronous event of x and y shown in the middle path. When the order of execution is not synchronous, an absent value is the intermediate output event. These different cases are among the possible behaviors for a multi-threaded implementation in C.

8.11.3   SIGNAL Implementations on Distributed Systems

SIGNAL, due to its multi-rate formalism, was initially used in prototyping for real time multiprocessor systems. An early work on clustering and schedul-

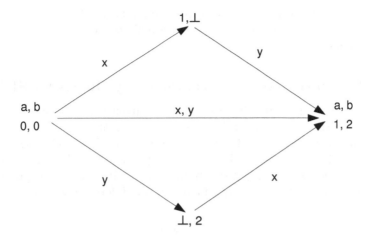

FIGURE 8.11: Weakly endochronous program with diamond property.

ing SIGNAL programs [46] discusses combining SIGNAL with the SYNDEX (SYNchronous Distributive EXecutive) CAD tool. SYNDEX can provide rapid prototyping and optimization of real time embedded applications on multiprocessors and it is based on a **potential parallelism** theory. In this method, parallelism will be exploited only if the hardware resources for parallel execution are available. SYNDEX communicates with its environment using operators like **sensors** and **actuators** and hence requires conversion of SIGNAL operators to SYNDEX form. A SIGNAL-SYNDEX translation strategy is defined by using an intermediate representation compatible with both languages, the directed acyclic graph. Directed acyclic graphs are built by considering nodes as tasks and the precedences as the edges between the nodes. A synchronous flow graph is a five-tuple with nodes, clocking constraints, precedence constraints, etc., as its elements. Once the equivalent graph for SIGNAL is constructed in SYNDEX, a clustering and scheduling strategy is applied to obtain the optimized real-time mapping onto a distributed system.

A clustering phase is used to increase granularity, thus reducing the complexity of the scheduling problem into multiple processors. Clustering can be of two types: linear and convex. In linear clustering, there is a pre-order between the nodes and sequentially executable nodes are merged. In convex clustering, a macro actor is formed with a set of nodes, and the triggering of execution for the macro actor is combined. For the macro actor, once all the inputs are available, computation and emitting of the outputs can be performed at once. Compositional deadlock consistency is a qualitative criterion defined in the framework for combining both linear as well as convex clustering. After the clustering phase is done, mapping of u clusters onto p processors ($u \geq p$) is done using the SYNDEX tool. After the virtual processors are mapped to physical processors, clusters are formed within each processing element and an efficient static schedule is found for each cluster. Meanwhile the resultant sequence in each cluster is dynamically scheduled according to the arrival of input events.

8.11.4   Multi-Threaded Programming Models for SIGNAL

Multi-threaded programming requires concurrency in the specification language. Deterministic output is a property present in SIGNAL that would maintain the equivalence of threaded implementation against the specification. There have been several strategies applied to the SIGNAL code conversion process for generating multi-threaded code. In general, the granularity of the threads seems to be a major factor in the different strategies.

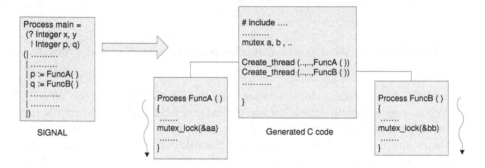

FIGURE 8.12: Process-based threading model.

A coarse grained multi-threaded code generation was proposed in [42] for SIGNAL. The key idea here is to utilize the modularity of SIGNAL processes for separating the threads. A SIGNAL program consists of concurrent statements, some of which are processes that are parallel themselves. Hence the SIGNAL top level process can be implemented as a master thread which forks and joins several worker threads. This **process-based multi-threading model** for multi-cores is shown in Figure 8.12. Here a SIGNAL description with the equivalent C code is shown side-by-side, with the *main* process mapped as the master thread. The master thread forks different worker threads like *FuncA* and *FuncB* for the respective SIGNAL sub-processes. The master thread contains a glue logic which holds together the different processes and protects the reads and writes into the shared variables. This strategy is thread-safe with respect to writes, since according to SIGNAL semantics, no signal can be assigned twice within a SIGNAL process. An added advantage in this model is the flexibility in assigning the threads to different cores. There are no additional instructions required due to the SIGNAL specification in contrast to the other distributed implementations. The communication between cores will be defined by the input and output parameters of each SIGNAL process. A drawback of this strategy is that the concurrency is still not fully exploited by the multi-threaded code. As the code grows, the number of threads do not scale proportionally and will not be able to benefit from the parallelism. The sequential execution of sub-processes is an under-utilization of the parallelization opportunities in SIGNAL.

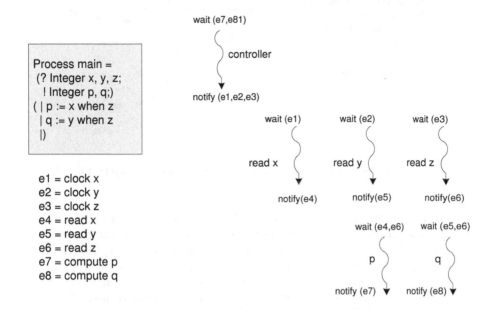

FIGURE 8.13: Fine grained thread structure of polychrony.

The current Polychrony compiler for SIGNAL from IRISA has implemented a multi-threaded code generation scheme. The strategy here is to use semaphores and event-notify schemes to synchronize the communication between threads. Each concurrent statement in SIGNAL is translated into a thread with a *wait* for every input at the beginning and a *notify* for every output at the end. This **micro-threading model** for a SIGNAL code is shown in Figure 8.13. A *controller* ticks according to the endochronous SIGNAL global clock which is a superset of all the input events. The *controller* thread notifies the read operation for the particular input and the respective threads associated with the inputs are triggered. For example, p is computed using inputs x, z and the computation will be triggered by events $e4$ and $e6$. The semaphore wait and notify statements provide the synchronization between the threads. The multi-threading model of Polychrony is modeled to be reactive to the input and will aggressively schedule computations whenever they are available. But at the same time, the fine grained nature of the tool results in more communication and less computation for a small task. When applied to larger SIGNAL programs, the number of threads increases exponentially since each concurrent statement in the code is forked out as a thread.

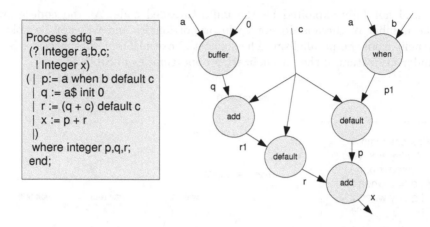

```
Process sdfg =
(? Integer a,b,c;
 ! Integer x)
( |  p:= a when b default c
 |  q := a$ init 0
 |  r := (q + c) default c
 |  x := p + r
 |)
 where integer p,q,r;
end;
```

FIGURE 8.14: SDFG-based multi-threading for SIGNAL.

From the previous two strategies discussed above we can conclude that the middle ground in the granularity of threads should be the general solution to multi-threaded code generation from SIGNAL. Even though this is a subjective answer, depending on the platform and resources available for implementation, algorithms which can fine tune the trade-off to this target have been proposed. An algorithm for constructing **synchronous data flow graphs (SDFG) for multi-threaded code generation** from SIGNAL proposed in [41] aims to break down complex expressions in SIGNAL and find the right amount of computation for each thread. A SIGNAL program consists of complex statements like $x := a$ when b default c is broken into normalized statements $x1 := a$ when b and $x := x1$ default c. From the normalized SIGNAL program, an SDFG is built as a dependence graph based on the flow of data. Each node is a normalized statement and each edge is the resultant clock relation of the data sent between nodes. Figure 8.14 shows the SDFG for a sample SIGNAL code. The normalization operation is visible for $p := a$ when b default c (intermediate node output $p1$) and also for $(q + c)$ default c. The SDFG is analyzed for weak endochrony and nodes are grouped for forming threads. An aggressive clustering of nodes can form threads of execution which are parallel. From Figure 8.14 it can be observed that the nodes leading to the output node 'add' form two chains that are parallel. Here the threading methodology tries to combine the benefits of clustering from distributed implementation strategy and the data flow from the Polychrony tool strategy.

8.12   Programming Models for Real-Time Software

Real-time applications of embedded systems in the fields of aviation, medicine etc. are highly safety-critical and failure of embedded systems in these fields could be fatal. Hence these devices are designed with models which, in case of an error, will try to avoid system failure. Programming models for real-time software have conventionally focused on task handling, resource allocation and job scheduling. We first introduce the legacy real-time scheduling algorithms and move onto the parallel implementations from instruction level to multiprocessor level. Synchronous real-time implementations are given special attention to drive the point about the importance of deterministic software synthesis for multi-core systems.

Earliest deadline first (EDF) is an intuitive job scheduling algorithm that has proven optimal for uniprocessor systems. Optimality in this context means that a set of tasks that cannot be scheduled using EDF cannot be scheduled using any other algorithm. Rate monotonic algorithm (RMA) was another job scheduling technique which gave priority to the period of the task to be scheduled. The logic behind scheduling a task with shorter period is that the next instance of the same task can add up to the pending tasks in queue. Liu and Leyland have discussed the optimality of scheduling for uniprocessors and have derived a sufficiency condition for schedulability of tasks using RMA [45].

In the current superscalar processors, simultaneous multi-threading (SMT) has gained importance in the design of real-time systems. *Simultaneous multi-threading* is the technique employed in superscalar architectures, where instructions from multiple threads can be issued per cycle. This is opposite to *temporal multi-threading* which has only one thread of execution at a time and a context switch is performed for execution of a different thread. The issues to be noticed for real-time scheduling in SMT processors is determining the tasks that need to be scheduled together (co-schedule selection) and the partition of resources among co-scheduled tasks. More information about the relative performance of popular algorithms is found in [38]. Moving from instruction level parallelism to a higher level of abstraction, programming models tend to concentrate on efficient job scheduling more than penalty due to context switches. Proportionate-fair (Pfair) scheduling [36] is a synchronized scheduling method for symmetric multiprocessors (SMP). It uses a weighted round-robin technique to calculate the utlization of processors and thus eventually achieve the optimal schedule. A staggered approach to distribute work was adopted to reduce bus contention at synchronization points. At the operating system level, real-time scheduling for embedded systems is a trade-off between building new RTOS for specific applications vis a vis customizing the commercial products in the market. Commercial RTOS products like VxWorks from Wind River Systems [19] and Nucleus from Mentor Graphics [10] offer implementations for multi-core systems.

8.12.1   Real-Time Extensions to Synchronous Languages

Synchronous programming languages have been proposed for real-time applications in the embedded world. The multi-core implementations for these languages and the real-time extensions in the industry have not exactly merged together as of now. But multiple synchronous languages have had success in incorporating real-time features in their software development tools and compilers. SCADE suite from Esterel Technologies, based on LUSTRE, has a timing and stack verifier which can estimate the worst case execution time and stack size on the MPC55xx Family embedded processor from Freescale Semiconductor. Esterel and SIGNAL have a few model checking and code generation tools with real-time characteristics embedded into them.

A software tool based on Esterel called TAXYS has been proposed to capture the temporal properties in an embedded environment [23]. The TAXYS architecture as shown in Figure 8.15 consists of two basic blocks, the external event handler (EEH) and the polling execution structure (PES). The function of the EEH is to accept the stimuli from the environment and store them in a FIFO queue in order. The Esterel code which is extended with pragmas to specify temporal constraints is compiled to generate the PES intermediate C code. When stimulated, a REACT procedure is called by the PES code, which executes *halt point functions*. A halt point is a control point in the Esterel code like `await`, `trap` etc., which on activation by stimuli executes an associated C code. For model checking purposes, separate tools have been made to model the environment, event handler and intermediate PES description.

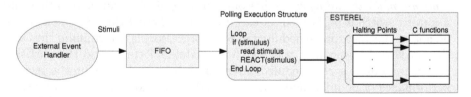

FIGURE 8.15: TAXYS tool structure with event handling and code generation [23].

Multi-rate reactive systems have been extended to perform real time scheduling using the EDF scheduling policy in [37]. Synchronous real-time language semantics close to SIGNAL is proposed and additional parameters for execution time, deadline etc. are added to its syntax. In the modified language, the clock of a signal can be increased or decreased by a constant value. The task precedence graph representing the data dependencies between statements in a synchronous program is constructed first and then the rate adjustments are performed. Now the EDF scheduling policy is enforced while ensuring the deadlines are not missed. For example, consider two tasks A and B with a constant period T. In this extended language, if we increase the rate of task A by two, its period shrinks by half and there are two instances of A to be scheduled for every task B. Figure 8.16 case (a) demonstrates this increased rate example, while case (b) shows a rate decrement example for tasks C and D. Here we can observe that the first instance of Task A has to finish execution before Task B is scheduled due to possible data dependency, but Task A has to be scheduled early enough to ensure Task B meets its deadline. Conversely, in case (b) the Task C is scheduled early enough to ensure two instances of Task D meet their respective deadlines.

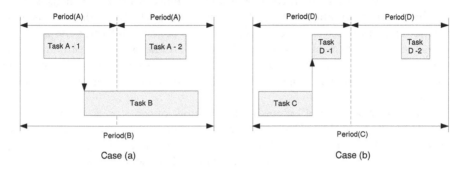

FIGURE 8.16: Task precedence in a multi-rate real time application [37].

The real-time extension proposed in [37] is under implementation in a framework similar to SIGNAL. The new formalism is called MRICDF or multi-rate instantaneous channel connected data flow [40]. This is an actor-based formalism with primitives having similar capabilities as SIGNAL or LUSTRE. EmCodeSyn [39] is a software tool to model MRICDF specifications. Currently sequential C code is generated from EmCodeSyn from MRICDF specification. Multi-threaded code generation based on SDFG-based threading strategy and real time extensions discussed in the chapter are among the goals of the EmCodeSyn project. Multi-threaded code generation is an important addition to any real time software tool targeting the multi-core market but it is desirable only if correctness and determinism are assured to the user. Amidst performance gains brought out by the revolution in multi-core technology, safety and deterministic execution have not lost their importance. And quite rightly so.

8.13 Future Directions for Multi-Core Programming

Multi-core processors seem to be the way of the future, and programming models for exploiting the parallelism available in such processors is appearing. There needs to be more academic debate on the choice of the right programming models for multi-core processors. Adapting existing parallelization techniques which evolved with the von Neumann sequential execution model in mind may not be the right answer to such debate. It is conceivable that a new innovative model of computation will emerge for multi-core programming.

In the absence of a real alternative, we tried to cover many of the significant parallel/concurrent programming libraries, APIs and tools existing today (in industry and academia). Intel Corporation has been trying to popularize the use of these APIs (like TBBs [6]) and also to help the programmer write correct multi-threaded code using software such as Intel Thread Checker [7]. Other tools including the Intel VTune Performance Analyzer [8] help improve application performance by identifying the bottlenecks through analysis of information from the system to source level. We believe that these are attempts to use the existing tools and technologies to handle problems of adapting to the multi-core domain. However, chances are that such approaches might be insufficient for efficient usage of the resources on-chip waiting to be exploited during execution.

Whether the trend of increasing the number of cores on a chip will be sustained has to be seen and hence we might have to shift again from multi-core technology. Industry experts have different opinions as to whether homogeneous or heterogeneous cores on chips will be beneficial in the long run [16]. In the midst of these undecided issues, there is still consensus on one topic: The future software programming models will be parallel.

Review Questions

[Q_1] What is meant by a programming model? How does one differentiate between a programming model, a model of computation, and a programming language?

[Q_2] Why is it important to study the programming models for programming multiprocessor architectures in the context of multi-core programming?

[Q_3] Abstraction is a key concept in computing. A programming model is an abstraction of the underlying execution engine. Can one consider multiple abstraction layers and multiple programming models for the same architecture?

[Q_4] Distinguish between user and kernel threads. Threads can be of different kinds, cooperative threads versus preemptive threads. Cooperative threads are like coroutines, and usually not scheduled by the operating system. Why would one use a cooperative threading model? Why are preemptive threading models more relevant for programming multi-core architectures?

[Q.5] Threading is often used on single core machines to hide the latency of input/output or memory access activities, and keep a CPU utilized. However, such usage of threads is different from cases when one has multiple processor cores and can use parallelism. Distinguish between concurrent programming and parallel programming models along these lines.

[Q.6] How threads interact with each other distinguishes between work distribution, pipeline, master/slave, and other models. Think of applications where each of these models would be an appropriate threading structure.

[Q.7] Write a multi-threaded code using POSIX primitives which can perform add, subtract, multiply and divide operations on two input operands. The order of the operations is random and the programming model for threading is the work distribution model. Ensure data access is protected using synchronization primitives.

[Q.8] Write sample programs in C for performing the add, subtract and multiply operations on a streaming input data using the pipeline model.

[Q.9] Explain the need for mutual exclusion primitives. Why would a two flag arrangement outside a critical section fail in protecting data?

[Q.10] Distinguish between the mutex and semaphore mutual exclusion primitives.

[Q.11] What is transactional memory? What are the major properties of transactional memory?

[Q.12] Explain how software transactional memory (STM) helps in avoiding deadlocks and livelocks.

[Q.13] Name a point-to-point communication based programming model for multiprocessors. Contrast this model to other shared memory models like OpenMP.

[Q.14] Explain how loop parallelization is obtained using *parallel_for* and *parallel_reduce* functions in thread building blocks.

[Q.15] Explain the difference between heterogeneous and homogeneous multiprocessor programming models.

[Q.16] What are general purpose graphics processing units? Why are they gaining importance for high performance computing? Explain the flow of execution in the CUDA programming model.

[Q.17] What are model-driven code generation techniques? Explain how StreamIt constructs are used to perform parallel computation.

[Q.18] Explain synchrony hypothesis. List a few relevant synchronous programming languages. What are the properties of synchronous languages which appeal to concurrent programming?

[Q.19] List the steps for converting an object code [26] into its distributed form. Show by means of a diagram how a sample *"if elseif end"* program can be allocated into two memory locations.

[Q.20] Explain the process of converting LUSTRE specifications into time triggered architectures with the help of a block diagram.

[Q.21] Compare the similarities and differences between the SIGNAL language and the Esterel and LUSTRE languages.

[Q.22] What characteristics of the SIGNAL programming model make it a good candidate perhaps for multi-core programming?

[Q.23] Explain distributed implementation of SIGNAL with SYNDEX CAD tool. What is the difference between convex and linear clustering?

[Q.24] Explain the process-based threading model and the micro-threading model for SIGNAL. What is the importance of granularity of computation for parallelization?

[Q.25] Consider a sample program in SIGNAL with the basic primitives. Draw its equivalent synchronous data flow graph for multi-threading. Explain how parallelization can be applied from the SDFG.

[Q.26] What are the different scheduling algorithms applicable for multi-core domains?

[Q.27] Explain the TAXYS tool structure.

[Q.28] Write a short paragraph on how you think multi-core programming models are going to evolve in the near future.

Bibliography

[1] ARM Cortex-A9 MPCore - Multicore processor.
http://www.arm.com/products/CPUs/ARMCortex-A9_MPCore.html.

[2] CellBE: Cell Broadband Engine Architecture (CBEA).
http://www.research.ibm.com/cell/.

[3] Esterel-Technologies, Esterel Studio EDA Tool.
http://www.esterel-technologies.com/.

[4] IEEE POSIX standardization authority.
http://standards.ieee.org/regauth/posix/.

[5] Intel Hyper Threading Technology.
http://www.intel.com/technology/platform-technology/hyper-threading/.

[6] Intel Thread Building Blocks.
http://www.threadingbuildingblocks.org/.

[7] Intel Thread Checker.
http://software.intel.com/en-us/intel-thread-checker/.

[8] Intel VTune Performance Analyzer.
http://software.intel.com/en-us/intel-vtune/.

[9] MathWorks SIMULINK®.
http://www.mathworks.com/products/simulink/.

[10] Mentor Graphics: NUCLEUS RTOS.
http://www.mentor.com/products/embedded_software/nucleus_rtos/.

[11] Microsoft Windows Threads.
http://msdn.microsoft.com/.

[12] National Instruments LabVIEW.
http://www.ni.com/labview/.

[13] NVIDIA Compute Unified Device Architecture.
www.nvidia.com/cuda.

[14] OpenMP API specification for parallel programming.
http://openmp.org/wp/.

[15] Renesas SH2A-DUAL SuperH Multi-Core Microcontrollers.
http://www.renesas.com/.

[16] Rick Merrit, CPU designers debate multi-core future. http://www.eetimes.com/showArticle.jhtml?articleID=206105179.

[17] Stanford University graphics Lab, BrookGPU. http://graphics.stanford.edu/projects/brookgpu/index.html.

[18] Sun Niagara Processor. http://www.sun.com/processors/niagara/.

[19] Windriver VxWorks. http://www.windriver.com/products/vxworks/.

[20] P. Bellens, J. M. Perez, R. M. Badia, and J. Labarta. CellSs: a programming model for the cell BE architecture. *In Proceedings of the ACM/IEEE Conference on Supercomputing,* 2006.

[21] A. Benveniste, B. Caillaud, and P. L. Guernic. From synchrony to asynchrony. *In Proceedings of the 10th International Conference on Concurrency Theory,* Springer-Verlag, London, 1664:162–177, 1999.

[22] G. Berry and G. Gonthier. The ESTEREL synchronous programming language: design, semantics, implementation. *Sci. Comput. Program,* 19(2):87–152, 1992.

[23] V. Bertin, M. Poize, J. Pulou, and J. Sifakis. Towards validated real-time software. *Proc. of 12th Euromicro Conference on Real-Time Systems,* pages 157–164, 2000.

[24] R. Bharadwaj. SOL: A verifiable synchronous language for reactive systems. *Electronic Notes in Theoretical Computer Science,* 65(5), 2002.

[25] P. Caspi, A. Curic, A. Maignan, C. Sofronis, S. Tripakis, and P. Niebert. From simulink to SCADE/LUSTRE to TTA: a layered approach for distributed embedded applications. *SIGPLAN Not.,* 38(7):153–162, July 2003.

[26] P. Caspi, A. Girault, and D. Pilaud. Automatic distribution of reactive systems for asynchronous networks of processors. *IEEE Transactions on Software Engineering,* 25(3):416–427, May 1999.

[27] E. W. Dijkstra. Cooperating sequential processes. *Communications of the ACM,* 26(1):100–106, Jan. 1983.

[28] S. A. Edwards and J. Zeng. Code Generation in the Columbia Esterel Compiler. *EURASIP Journal on Embedded Systems,* pages 1–31, 2007.

[29] W. Gropp, E. Lusk, N. Doss, and A. Skjellum. A high-performance, portable implementation of the message passing interface standard. *J. Parallel Computing,* 22(6):789–828, Set. 1996.

[30] P. L. Guernic, T. Gautier, M. L. Borgne, and C. L. Maire. Programming real-time applications with SIGNAL. *Proceedings of the IEEE*, 79(9):1321–1336, 1991.

[31] N. Halbwachs, P. Caspi, P. Raymond, and D. Pilaud. The synchronous data flow programming language LUSTRE. *Proceedings of the IEEE*, 79(9):1305–1320, Sept. 1991.

[32] Nicolas Halbwachs. *Synchronous Programming of Reactive Systems*. Kluwer Academic Publishers, Netherlands, 1993.

[33] M. Herlihy and J. E. Moss. Transactional memory: architectural support for lock-free data structures. *SIGARCH Comput. Archit. News*, 21(2):289–300, May 1993.

[34] C. A. Hoare. Monitors: an operating system structuring concept. *Communications of the ACM*, 17(10):549–557, Oct. 1974.

[35] L. Holenderski and A. Poign. The multi-paradigm synchronous programming language LEA. *In Proceedings of the Intl. Workshop on Formal Techniques for Hardware and Hardware-like Systems*, 1998.

[36] P. Holman and J. H. Anderson. Adapting Pfair scheduling for symmetric multiprocessors. *J. Embedded Comput.*, 1(4):543–564, Dec. 2005.

[37] D. Lesens J. Forget, F. Boniol and C. Pagetti. Multi-periodic synchronous data-flow language. *In 11th IEEE High Assurance Systems Engineering Symposium (HASE08)*, Dec. 2008.

[38] R. Jain, C. J. Hughes, and S. V. Adve. Soft real-time scheduling on simultaneous multithreaded processors. *In Proceedings of the 23rd IEEE Real-Time Systems Symposium, Washington DC*, Dec. 2002.

[39] B. A. Jose, J. Pribble, L. Stewart, and S. K. Shukla. EmCodeSyn: A visual framework for multi-rate data flow specifications and code synthesis for embedded application. *12th Forum on Specification and Design Languages (FDL'09)*, Sept. 2009.

[40] B. A. Jose and S. K. Shukla. MRICDF: A new polychronous model of computation for reactive embedded software. *FERMAT Technical Report 2008-05*, 2008.

[41] B. A. Jose, S. K. Shukla, H. D. Patel, and J. Talpin. On the multi-threaded software synthesis from polychronous specifications. *Formal Models and Methods in Co-Design (MEMOCODE), Anaheim, California*, pages 129–138, Jun. 2008.

[42] Bijoy A. Jose, Hiren D. Patel, Sandeep K. Shukla, and Jean-Pierre Talpin. Generating multi-threaded code from polychronous specifications. *Synchronous Languages, Applications, and Programming (SLAP'08), Budapest, Hungary*, Apr. 2008.

[43] T. Knight. An architecture for mostly functional languages. *In Proceedings of the ACM Conference on LISP and Functional Programming*, pages 105–112, 1986.

[44] Edward A. Lee. The problem with threads. *Computer*, 39(5):33–42, May 2006.

[45] C. L. Liu and J. W. Leyland. Scheduling algorithms for multiprogramming in a hard real-time environment. *Journal of the ACM*, pages 46–61, Jan. 1973.

[46] O. Maffeis and P. L. Guernic. Distributed implementation of SIGNAL: scheduling and graph clustering. *In Proceedings of the 3rd International Symposium on Formal Techniques in Real-Time and Fault-Tolerant Systems*, Springer-Verlag, London, 863:547–566, Sept. 1994.

[47] F. Maraninchi and Y. Rémond. Argos: an automaton-based synchronous language. *Elsevier Computer Languages*, 27(1):61–92, 2001.

[48] D. Potop-Butucaru. Optimizations for faster simulation of Esterel programs. Ph.D. thesis, Ecole des Mines, 2002.

[49] N. Shavit and D. Touitou. Software transactional memory. *In Proceedings of the 14th Annual ACM Symposium on Principles of Distributed Computing*, pages 204–213, Aug. 1995.

[50] Gadi Taubenfeld. *Synchronization Algorithms and Concurrent Programming*. Pearson Education Limited, England, 2006.

[51] W. Thies, M. Karczmarek, and S. P. Amarasinghe. StreamIt: A language for streaming applications. *In Proceedings of the 11th International Conference on Compiler Construction*, Springer-Verlag, London, 2304:179–196, Apr. 2002.

[52] G. Wang and P. Cook. ChucK: a programming language for on-the-fly, real-time audio synthesis and multimedia. *In Proceedings of the 12th Annual ACM International Conference on Multimedia*, pages 812–815, 2004.

[53] L. H. Yoong, P. Roop, Z. Salcic, and F. Gruian. Compiling Esterel for distributed execution. *In Proceedings of Synchronous Languages, Applications, and Programming (SLAP)*, 2006.

9

Operating System Support for Multi-Core Systems-on-Chips

Xavier Guérin and Frédéric Pétrot

TIMA Laboratory
Grenoble, France
xavier.guerin@imag.fr

CONTENTS

9.1	Introduction		310
9.2	Ideal Software Organization		311
9.3	Programming Challenges		313
9.4	General Approach		314
	9.4.1	Board Support Package	314
		9.4.1.1 Software Organization	315
		9.4.1.2 Programming Model	315
		9.4.1.3 Existing Works	317
	9.4.2	General Purpose Operating System	317
		9.4.2.1 Software Organization	318
		9.4.2.2 Programming Model	318
		9.4.2.3 Existing Works	320
9.5	Real-Time and Component-Based Operating System Models		322
	9.5.1	Automated Application Code Generation and RTOS Modeling	322
		9.5.1.1 Software Organization	323
		9.5.1.2 Programming Model	323
		9.5.1.3 Existing Works	324
	9.5.2	Component-Based Operating System	326
		9.5.2.1 Software Organization	327
		9.5.2.2 Programming Model	328
		9.5.2.3 Existing Works	328
9.6	Pros and Cons		329

9.7   Conclusions   330
Review Questions and Answers   332
Bibliography   333

9.1 Introduction

Most of the modern embedded applications include complex data-crunching algorithms that are highly demanding in terms of memory resources and processing power. Today, not only must embedded appliances be low-cost and energy-saving, but they must also provide cutting-edge performance to application designers.

In addition, as multimedia and telecommunication standards evolve quickly, having a pure hardware approach is no longer considered viable. Consequently, hardware platforms based on a multi-core SoC (MC-SoC) embedding several heterogeneous cores have become the preferred choices over solutions composed of one or several general-purpose processors, which are common to computer science experts but currently not suited to meet the power/performance challenges of portable appliances.

A heterogeneous MC-SoC (HMC-SoC) is generally composed of small amounts of on-chip memory, several hardware devices, and heterogeneous programmable cores (Figure 9.1). An application can be split into several parts that can benefit from the different abilities of these cores. Hence, an application running on a HMC-SoC can reach the same performance level as if running on a traditional platform but with lower operating frequency and voltage, therefore keeping the electrical consumption and the production costs low.

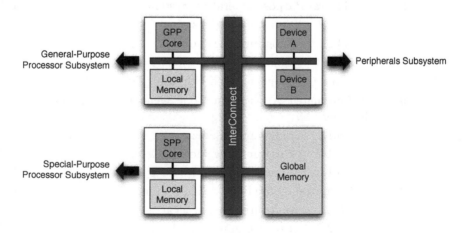

FIGURE 9.1: Example of HMC-SoC.

The major drawback of HMC-SoCs resides in their programming. Due to the heterogeneity of their cores, the following obstacles are to be expected:

- **Architectural differences:** the cores can have different instruction sets, different word representations (16-bit versus 32-bit), and/or different *endianness*.

- **Non-uniform ways to access memory:** the part of the memory accessible from each core may not be the same or may not be accessible the same way (e.g., different latencies, data bursts, etc.).

- **Application distribution:** the application has to be split between each core, and the computations and communications should be carefully designed in order to benefit from the parallelism.

As a consequence, and contrary to homogeneous configurations, HMC-SoCs cannot be efficiently operated with generic software solutions. The difficulties mentioned above need to be overcome by using an application-programming environment, specific to the hardware platform. This kind of environment usually contains several compilation tools, software libraries that provide support for distributing the application, and mechanisms to provide an abstract view of the underlying hardware.

In this chapter, we present the approaches used by the existing application programming environment. They are organized in two categories: the general approaches and the model-based approaches. The former category will deal with board support packages (BSPs) and general-purpose operating systems (GPOSs). The latter category will describe the automatic application generation with real-time operating system modeling and component-based operating systems.

9.2 Ideal Software Organization

In this section, we present the software organization that will be used as a reference throughout this chapter. It can be seen as a cross-section of a software binary executed on a HMC-SoC hardware platform. It is composed of several layers: the application layer, the operating system layer, and the hardware abstraction layer (Figure 9.2).

This particular layout corresponds to an ideal organization of the key software roles (the application, the system functions, and the hardware dependencies), that offers maximum flexibility and portability, since the application is not bound to any particular operating system, nor is the operating system dedicated to a particular processor/chipset. Moreover, it also supposes that the layers on which the application depends can be tailored to its needs in order to reduce the final memory footprint.

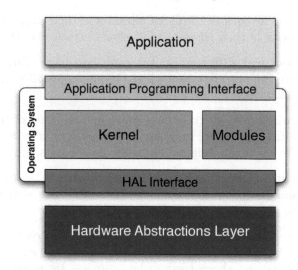

FIGURE 9.2: Ideal software organization.

The **application** layer contains an executable version of the application's algorithm. Its implementation depends on the design choices made by the application developer and should not be constrained by the underlying operating system. It uses external software libraries or language-related functions to access the operating system services, specific communication, and workload distribution interfaces.

The **operating system** layer provides high-level services to access and multiplex the hardware on which it is running. Such services can be (and are not limited to) multi-threading, multi-processing, inputs/outputs (I/O) and file management, and dynamic memory. It can be as small as a simple scheduler or as big as a full-fledged kernel, depending on the needs of the application. It relies on the hardware abstractions interface to perform hardware-dependent operations, hence ensuring that its implementation is not specific to a particular hardware platform or processor.

The **hardware abstractions** layer contains several functions that perform the most common hardware-dependent operations required by an operating system. These functions deal with execution contexts, interrupts and exceptions, multi-processor configurations, low-level I/Os, and so on. This layer usually does not have any external dependencies.

In the following sections, we will use this organization to draw the blueprints of the software organization resulting from each application development method we present. By doing so, we hope to highlight their main differences, their advantages, and the constraints they imply.

9.3 Programming Challenges

Programming a MC-SoC is very different from programming a typical uniprocessor machine. This is true at several stages in the design of an application that targets this kind of hardware. In this section, we explain the major challenges prompted by these hardware platforms.

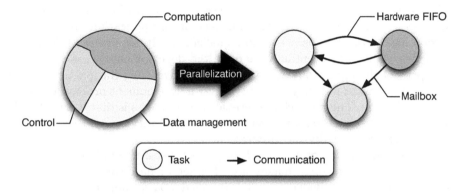

FIGURE 9.3: Parallelization of an application.

At least two characteristics of MC-SoCs have a drastic influence on the design of an application: its multi-core nature and its (possible) heterogeneity. The multi-core characteristic implies that the application must be statically parallelized from one to multiple computation tasks in order to take advantage of the multiple cores. When an application is parallelized (Figure 9.3), the following questions have to be answered:

- **How to balance the computing needs?** Each task must be well defined and its role must be clearly identified.

- **How will the tasks communicate?** The hardware and software communications have to be known and wisely allocated. The software communication primitives have to be wisely chosen and their hardware supports carefully implemented.

The parallelization of an application is a delicate operation: badly defined tasks or unwisely selected communications will lead to poor overall performance.

The heterogeneous characteristic is probably the most complicated. The difference in the processors' architectures implies that a different method is required for each core to execute the software. The principal consequence of this statement is that the communications between tasks mapped on two different cores must be allocated to channels shared by two different control entities. Such channels are not easy to implement, since they require additional synchronization points to perform correctly.

9.4 General Approach

Compared to the hardware-oriented design approach of consumer electronic devices used in the past, the development of an application on a heterogeneous, multi-core SoC is a lot more challenging. It requires additional time and new skills that eventually increase the developmental costs, increasing the price of the final product. To stay competitive, the principal actors of the embedded devices industry generally prefer a development process focused on the application. The main characteristic of this process is that it heavily relies on existing system tools to provide the low-level services and the hardware adaptations required by the application.

Depending on the complexity of the project, two different kinds of approach are considered. The first makes use of a set of hardware-specific, vendor-specific functions provided in what is called board support packages. The second relies on the services provided by general-purpose operating systems. In the following subsections we, for each approach, describe its software organization. Then, we detail its compatible programming model. Finally, we present some existing works based on these approaches.

9.4.1 Board Support Package

Board support packages are software libraries that are provided with the hardware by the vendor. They can be used through their own application programming interfaces (APIs) and are bound to specific hardware platforms. Each hardware device present on the platforms can be configured and accessed with its own set of functions. Contrary to operating systems, BSPs do not provide any kind of system management. In addition, their thoroughness and quality vary widely from one vendor to another.

Two kinds of BSPs are available: general-purpose BSPs that export their own specific API, and OS-specific BSPs designed to extend the functionalities of an existing general-purpose operating system. In this section, we focus on BSPs of the first type since BSPs of the second type are barely usable outside their target OSs.

BSPs are directly used when nothing more complex is required or affordable. This is usually the case in the following situations:

- **Small or time-constrained application:** the application is too simple to require the use of an operating system or its real-time requirements are too high to allow unpredictable behaviors.

- **Limited hardware resources:** the targeted hardware has a very limited memory size or contains only micro-controllers (μCs) or digital signal processors (DSPs), not compatible with generic programing models.

- **Limited human and financial resources:** the use of a more complete software environment is not affordable in terms of development costs and/or work force.

While the direct use of a BSP may not prevent long-lasting headaches in the last scenario, it can prove to be really useful in cases where the full control over the software — concerning its performances or its final size — is required.

9.4.1.1   Software Organization

A BSP takes the form of several software libraries which provide functions to access and control the hardware devices present on a SoC. It also provides bootstrap codes and memory maps for each of the processors. These libraries are specific to one type of processor and to one type of SoC. Hence, one specific version of a BSP is necessary for each type of processor present on a hardware platform. This version of a BSP thus cannot be used with other SoCs.

FIGURE 9.4: BSP-based software organization.

Applications that directly use BSPs are not usually designed to be reused on different hardware platforms or processors. Each part of an application is dedicated to run one processor of a specific platform and consequently makes intensive use of the processor's assembly language and direct knowledge of the memory and peripherals organization (Figure 9.4). Moreover, since the application uses the BSP's interface, they both must use the same or a compatible programming language.

9.4.1.2   Programming Model

To develop an application that directly interacts with a BSP, a software designer first needs to manually split the application in parts which will be executed on different processors of the platform. This process is by far one of the most complicated, since the full algorithm needs to be thoroughly analyzed in order to achieve the best partition. However, fair results are usually obtained on small applications. The next step is to either use an integrated development environment (IDE) or manually use the tools and libraries pro-

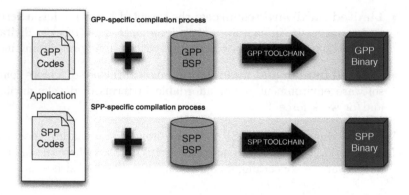

FIGURE 9.5: BSP-based application development.

vided by the hardware vendor to produce the software binary of each processor (Figure 9.5).

Once the application is adapted to the BSP and compiles properly, it needs to be debugged. In this configuration, the debugging of an application is done directly on the hardware using boundary-scan emulators which are connected on the test access port (TAP) of the processor (if one is available) and an external debugger[19][36].

Once the application is developed, the binaries produced and validated, the booting sequence must be configured. Although it closely depends on the hardware architecture, two methods can be distinguished: a) each binary is placed on a read-only memory (ROM) or an electrically erasable programmable read-only memory (EEPROM) specific to each processor, or b) all the binaries are placed on the same ROM device.

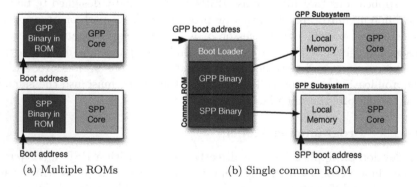

(a) Multiple ROMs   (b) Single common ROM

FIGURE 9.6: BSP-based boot-up sequence strategies.

In method a), as depicted in Figure 9.6a, all the processors are independent from each other and autonomously start at the address on which their local ROM is mapped. In method b), as depicted in Figure 9.6b, one processor is designated to boot first while the others are put in an idle mode. This processor is responsible for dispatching the binaries and starting the remaining processors.

9.4.1.3 Existing Works

Nowadays, each board is shipped with a more or less functional board support package. Hence, examples of BSPs are numerous. A few examples of both types are provided below.

On the one hand, vendors such as Altera [2], Xilinx [39], Tensilica [32], etc., provide general-purpose BSPs coupled with IDEs dedicated to the development of applications for their hardware platforms. They also provide software libraries containing standard C functions, network management functions, and basic thread management functions.

On the other hand, vendors such as Texas Instruments [33], Atmel [3], Renesas [29], etc., provide OS-specific BSPs for systems such as Windows Mobile or Linux. These BSPs are not supposed to be used outside of their specific operating system targets, and they are usually prepackaged to be directly installed in the OS's development environment.

9.4.2 General Purpose Operating System

A general-purpose operating system is a full-featured operating system designed to provide a wide range of services to all types of applications. These services usually are (but are not limited to) multiprocessing, multi-threading, memory protection, and network support. Their point in common is that they are not specifically designed to operate an HMC-SoC, but they are only adapted from other computing domains such as desktop solutions or uniprocessor embedded solutions in order to provide a development environment similar to what software developers are generally used to.

General-purpose OSs are used when hardware resources are sufficient and when a more specific system solution is not required. This is usually the case in the following situations:

- **Portability or limited knowledge of the target hardware:** the application needs to be adapted to multiple hardware targets. Gaining a perfect knowledge of each target is not feasible.

- **Application complex but not critical:** the application requires high-level system services such as thread management or file access and does not have particular performance constraints.

- **Limited time resources:** for different reasons the development time of the application is limited. Hence, additional developments required by the main application must be kept to a minimum.

One of the principal advantages for using a general-purpose operating system is the availability of a large number of resources from its community such as external support, existing hardware drivers, etc., that can greatly accelerate the software development process. The other advantage is the availability of a well-established development environment containing many libraries for application support and several tools such as compilers, profilers, and advanced debuggers.

9.4.2.1   Software Organization

A general-purpose operating system is a stand-alone software binary, running in supervisor mode, that provides services to applications, running in user mode, through system calls. It usually requires hardware support for atomic operations and virtual memory management, and consequently is dedicated to run on a general-purpose processor (Figure 9.7).

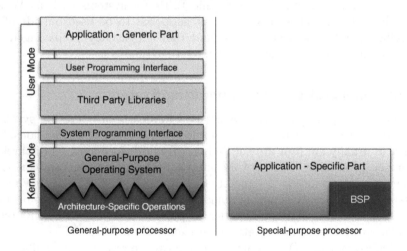

FIGURE 9.7: Software organization of a GPOS-based application.

Another consequence of these requirements is that the other processors of the hardware platform are seen simply as hardware devices that can only be accessed (hence programmed) through device drivers of the GPOS. This particularity radically changes the programming model of the application as compared to the programming model of the previous approach. This point is discussed in the next section.

9.4.2.2   Programming Model

In the GPOS-based approach the hardware platform is assimilated to a standard hardware configuration, where the general-purpose processor (GPP) is seen as the master processor and the specific-purpose processors (SPPs) are seen as co-processors dedicated to specific tasks such as video or audio decoding. Strong hypotheses are made concerning the GPP capabilities for the GPOS to run correctly:

1. It is supposed to be the only processor to exert complete control over the entire hardware platform.

2. It has no limitation in terms of addressing space.

3. It can decide whether or not a SPP can be started.

4. It is the only processor to have a memory management unit (MMU).

The parts of the application dedicated to run on the GPP are developed using toolchains specific to the processor and to the GPOS. They cannot access the peripherals directly and, although the use of assembly code is allowed, only mnemonics available in user mode can be used. The exchange of data with SPPs is executed using the corresponding device drivers of the GPOS. The parts of the application dedicated to run on the SPPs are mainly developed using a BSP-based method as explained in the previous section (Figure 9.8).

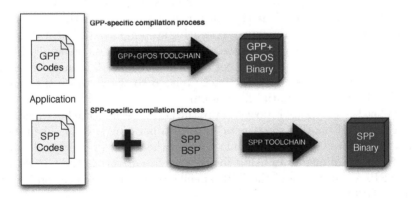

FIGURE 9.8: GPOS-based application development.

The debugging of an application that partially relies on a GPOS is slightly more difficult than when only BSPs are used. Although the method is the same for the parts of the application running on the SPPs as when a BSP is directly used, two methods for the parts running on the GPP are available: one can use either an external debugger connected to the TAP port of the platform or an internal debugger running on the GPOS.

However, none of these approaches is truly efficient. In the first approach, the external debugger must be able to load and boot the kernel of the GPOS and, in that case, not only the application is debugged but the whole operating system as well, increasing the complexity of the operation by a hundredfold.

In the second approach, only the application is debugged. However, if something corrupts the kernel of the GPOS (such as a bug in a driver or a

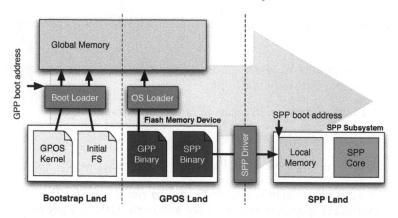

FIGURE 9.9: GPOS-based boot-up sequence.

failure from one of the SPPs) then the whole operating system will crash, including the debugger and the application being debugged.

The boot-up sequence of the GPOS-based approach heavily relies on the hypothesis that the GPP is the master of the board and the SPPs have not started until the general-purpose operating system initiates them (Figure 9.9). The binaries, including a boot loader, the GPOS, and its (generally huge) initial file system, can be placed either on an internal ROM or on an internal flash memory device. This choice is closely related to the memory space required by the GPOS and its initial file system.

When the hardware is powered up, the GPP executes the boot loader which is in charge of booting the general-purpose operating system. Then, once the GPOS is booted, the SPP-specific parts of the application are uploaded onto the SPPs' local memories using the SPP device drivers of the GPOS. Finally, the SPPs are started and the GPP-specific part of the application is executed on the GPOS.

9.4.2.3 Existing Works

In this section, we give a short presentation of the most used GPOS solutions in the embedded system industries [1]: VxWorks, Windows CE, QNX, eCos, and Linux. If not specified otherwise, the real-time attribute means that the operating system has soft real-time scheduling and time-determined interrupt handling capabilities.

VxWorks [37] is a real-time, closed-source operating system developed and commercialized by Wind River Systems. It has been specifically designed to run on embedded systems. It runs on most of the processors that can be found on embedded hardware platforms (MIPS, PowerPC, x86, ARM, etc.) and its micro-kernel supports most of the modern operating system services (multi-tasking, memory protection, SMP support, etc.). Applications targeting this operating system can be developed using the Workbench IDE. VxWorks has been used in projects such as the Honda Robot ASIMO, the Apache Longbow helicopter, and the Xerox Phaser printer.

Windows CE [24] is Microsoft's closed-source, real-time operating system for embedded systems. It is supported on the MIPS, ARM, x86, and Hitachi SuperH processor families. Its hybrid kernel implements most of the modern system services. Applications targeting this operating system can be developed using Microsoft Visual Studio or Embedded Visual C++. Windows CE has been used on devices such as the Sega Dreamcast or the Micros Fidelio point of sales terminals.

QNX [28] is a micro-kernel based, closed-source, UNIX-like operating system designed for embedded systems developed and commercialized by QNX Software System. It is supported on the x86, MIPS, PowerPC, SH-4, and ARM processor families. Its kernel implements all the modern operating system services and supports all current POSIX API. It is known for its stability, its performance, and its modularity. Applications targeting this operating system can be developed using the Momentics IDE, based on the Eclipse framework.

eCos [22][12] is a real-time, open-source, royalty-free operating system specifically designed for embedded systems initially developed by Cygnus Solutions. It mainly targets applications that require only one process with multiple threads. It is supported on a large variety of processor families, including (but not limited to) MIPS, PowerPC, Nios, ARM, Motorola 68000, SPARC, and x86. It includes a compatibility layer for the POSIX API. Applications targeting this operating system can be developed using specific cross-compilation toolchains.

Symbian-OS [31] is a general-purpose operating system developed by Symbian Ltd. and designed exclusively for mobile devices. Based on a micro-kernel architecture it runs exclusively on ARM processors, but unofficial ports on the x86 architecture are known to exist. Applications targeting this operating system can be developed using an SDK based either on Eclipse or CodeWarrior.

μC-OS/II [23] is a real-time, multi-tasking kernel-based operating system developed by Micrium. Its primary targets are embedded systems. It supports many processors (such as ARM7TDMI, ARM926EJ-S, Atmel AT91SAM family, IBM PowerPC 430, ...), and is suitable for use in safety critical systems such as transportation or nuclear installations.

Mutek [26] is an academic OS kernel based on a lightweight implementation of the POSIX Threads API. It supports several processor architectures such as MIPS, ARM, and PowerPC, and application written using the *pthread* API can be directly cross-compiled for one of these architectures using Mutek's API.

Linux [21] is an open-source, royalty-free, monolithic kernel first developed by Linus Torvald and now developed and maintained by a consortium of developers worldwide. It was not designed to be run on an embedded device at first but due to its freedom of use, its compatibility with the POSIX interface, and its large set of services it has become a widely adopted solution. It supports a very large range of processors and hardware architectures, and it benefits from an active community of developers. Soft real-time (PREEMPT-RT), hard real-time (Xenomai [38]), and security (SE-linux [34]) extensions can be added to the mainline kernel. Applications targeting operating systems based on this kernel can be developed using specific cross-compilation tool chains.

9.5   Real-Time and Component-Based Operating System Models

In the previous section, we saw that an ideal application programming environment should be able to abstract the operating system and hardware details for the application programmer and still produce memory- and speed-optimized binary code for the targeted platform. It should also provide mechanisms to help the application designer distribute his application on the available processors.

There are two approaches that propose these kinds of services. The first takes the form of a design environment that allows the application designer to describe in an abstract way the application and its software dependencies, while automatically generating the code for both. The second takes the form of a programming model where the application developer would describe the OS dependences into the application's code. Then, a set of tools would analyze these descriptions and generate the binaries accordingly.

These two solutions generally produce small binary executables that are comparable to those produced by a BSP-based development approach. They particularly share the same debugging methods and boot-up sequences. Hence, in the following sections we will focus on software organizations and programming models, which are the innovations proposed by those two solutions.

9.5.1   Automated Application Code Generation and RTOS Modeling

Automated application code generation stems from the system level design approach, where the implementation of an application is decoupled from its specification. Formal mathematical models of computation are used instead of standard programming language to describe the application's behavior. Software dependences such as specific libraries of real-time operating system (RTOS) functionalities can also be modeled. This allows the software designer to perform fast functional simulations and validate the application early in the development process.

This solution is generally used when the application's behavior needs to be thoroughly verified and the validation of the software needs to be fast and accurate. This is usually the case in the following situations:

- **Safety-critical applications:** the application will be embedded in high-risk environments such as cars, planes, or nuclear power plants.

- **Time-critical applications:** each part of the application needs to be accurately timed in every possible execution case.

It is also used in industries that must rely on external libraries that have earned international certifications, such as DO-178B/EUROCAE ED-12B [30] for avionics or SIL3/SIL4 IEC 61508 [18] for transportation and nuclear systems.

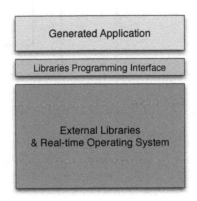

FIGURE 9.10: Software organization of a generated application.

9.5.1.1 Software Organization

The software organization of this approach is composed merely of two parts: the application generated from the high-level model and the external libraries (Figure 9.10). The programming interface exported by those libraries, is completely opaque to the developer and is automatically used by the code generator.

In addition, what is done in the external libraries and how it is done are usually not documented. This is not a problem since, as evoked above, the behavior of each function contained in these libraries has previously been validated and certified. This characteristic is generally what truly matters regarding the external dependencies of such approaches.

9.5.1.2 Programming Model

The development of an application that uses this approach starts with the description of the application's algorithm in a particular model of computation (Figure 9.11). This model of computation must fit the computational domain of the algorithm and it must be supported by the code generation tool. The most widely used models of computation are: synchronous data flow (SDF),

control data flow (CDF), synchronous and control data flow (SCDF), final
state machine (FSM), Kahn process network (KPN), and Petri nets (PN).

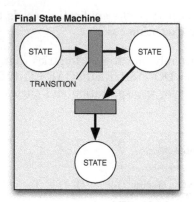
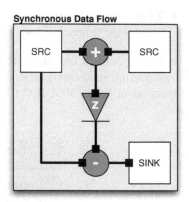

FIGURE 9.11: Examples of computations models.

Different execution models may also be available. The execution can be
time-triggered, resulting in the repeated execution of the whole algorithm at
any given frequency. Or it can be event-triggered, causing the algorithm to be
executed as a reaction to external events.

Next, the model can be organized in a hierarchical task graph (HTG),
where different parts of the application are encapsulated into tasks. Then,
real-time operating system elements can be added to the model and connected
to the tasks in order to extract information and timings about the behavior
of the whole software organization early in the development process (Figure
9.12). These elements can be real-time schedulers, interrupt managers, or in-
put/output managers [25]. They can also be adjusted (e.g., the scheduling
policy can be changed) to fit the application requirements.

Finally, the code of the application is generated in a language compatible with the code generator and the designer's choice. Operating system and
communication elements are considered as external libraries, in which the programming interface is known by the code generator. The tasks defined in the
model are encapsulated in execution threads compatible with the supported
operating system. Communications between the tasks are performed using
functions from one of the external communication libraries.

9.5.1.3   Existing Works

The software development process of some existing works starts from the same
kind of functional model, although their operating modes eventually differ.
Hence, we regrouped them in function of their input model in order to show
the possibilities of what can be done with a same application programming
paradigm.

SPADE [20], Sesame [27], Artemis [9], and Srijan [11] start with functional

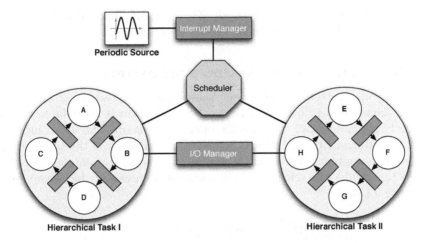

FIGURE 9.12: Tasks graph with RTOS elements.

models in the form of KPN. These approaches are able to refine automatically the software from a coarse-grained KPN, but they require the designer to determine the granularity of processes, to specify manually the behavior of the tasks, and to express explicitly the communication between tasks using communication primitives.

Ptolemy [6], Metropolis [4], and SpecC [7] are high-level design frameworks for system-level specification, simulation, analysis and synthesis. Ptolemy is a well-known development environment for high-level system specification and simulation that supports multiple models of computation. Metropolis enables the representation of design constraints in the system model. The meta-model serves as input for all the tools built in Metropolis. The meta-model files are parsed and developed into an abstract syntax tree (AST) by the Metropolis front-end. Tools are written as back-ends that operate on the AST, and can either output results or modify the meta-model code.

MATCH [5] uses MATLAB® descriptions, partitions them automatically, and generates software codes for heterogeneous multiprocessor architectures. However, MATCH assumes that the target system consists of commercial off-the-shelf (COTS) processors, DSPs, FPGAs, and relatively fixed communication architectures such as ethernet and VME buses. Thereby MATCH does not support software adaptations for different processors and protocols.

Real-Time Workshop (RTW) [35], dSpace [10], and LESCEA [17] use a Simulink® model as input to generate software code. RTW generates only single-threaded software code as output. dSpace can generate software codes for multi-processor systems from a specific Simulink model. However, the generated software codes are targeted to a specific architecture consisting of several COTS processor boards. Its main purpose is high-speed simulation of control-intensive applications. LESCEA can also generate multi-threaded software code for multiprocessor systems. The main difference with dSpace is that it is not limited to any particular type of architecture.

9.5.2   Component-Based Operating System

This approach, using a component-based operating system, is radically different from the previous ones. Here, the application is not adapted (neither directly nor indirectly) to a specific operating system. The operating system and the external libraries adapt themselves to the needs of the application. To do so, this approach introduces a new software organization and a new set of tools that enable the selection of the components necessary for the application (Figure 9.13).

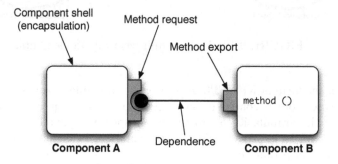

FIGURE 9.13: Component architecture.

This approach is generally used when a high level of portability, reconfigurability, and adaptation is required. This is usually the case in the following situations:

- **Limited physical access to the hardware:** remotely-managed hardware devices require mechanisms to disable, update, and restart software services.

- **Self-manageable and fault-tolerant:** when a problem occurs, the system must be able to identify the source of the problem, disconnect the faulty piece of software, and send an administrative alert.

- **Deployment capabilities:** the same software needs to be repeatedly deployed on a large set of identical devices.

This approach also has good properties for memory constrained heterogeneous embedded systems: since the software required by the application is tailored to its needs, a unified programming interface and a minimal memory usage can be guaranteed to the application designer.

9.5.2.1 Software Organization

A software component is a set of functionalities encapsulated in a shell that contains the component's interface (the signature of the methods it exports) and its dependencies on other components. While the behavior of each method is explicit, its implementation is completely opaque to the user (Figure 9.14).

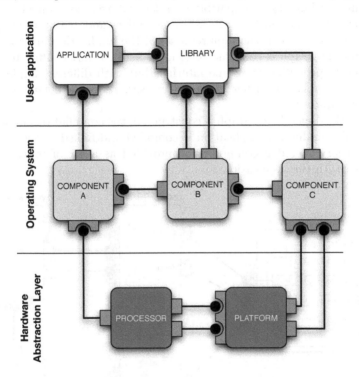

FIGURE 9.14: Component-based OS software organization.

The component's interface is usually described using an interface description language (IDL), while its dependences and method requests are gathered in a separate file whose format depends on the tools used to resolve the dependences. One important point is that the access to an exported method may be more complicated than a simple function call. A specific *communication* component can be inserted between the *caller* and the *callee* and perform a more complex procedure call that may involve other components.

With this approach, the operating system's programming interface is the union of the components' interfaces that compose it, which means that it may vary a lot between different configurations of the OS. This is not a problem since it corresponds exactly to the needs of the application. Each component that requires a hardware-dependent function relies on the hardware abstractions corresponding to the targeted processor and platform.

9.5.2.2 Programming Model

There are two ways to start the development of an application using this approach. Either the application is described as a component itself using a paradigm compatible with the one of the other components and with the dependence resolution tools, or the application can be directly written in a programming language compatible with the external components and its dependencies that are gathered in a format understandable by the dependence resolution tools. Then, the requirements of the application are analyzed and a dependency graph is constructed (Figure 9.15). If any conflict between two components providing the same interface but with different implementation occurs, the tools can either prompt the user, choose a default solution, or abort the process.

Once the dependency graph is constructed, the compilation environment is generated. Finally, the application is compiled and linked to the components into a binary file. If a component is provided as a set of source files, it is compiled with the application.

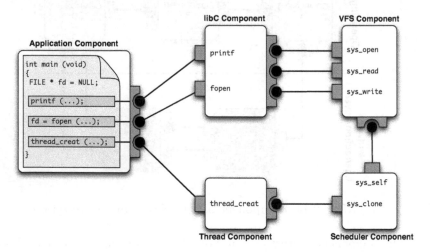

FIGURE 9.15: Example of a dependency graph.

9.5.2.3 Existing Works

Since the adoption of component-based software development approaches is rather new in the embedded software world, some of the works presented below were not specifically designed to operate HMC-SoCs. However, each of them, in its own domain, obtains interesting results that illustrates the benefits of the approach.

Choices [8] is written as an object-oriented operating system in C++. As an object-oriented operating system, its architecture is organized into frameworks of objects that are hierarchically classified by function and performance.

The operating system is customized by replacing sub-frameworks and objects. The application interface is a collection of kernel objects exported through the application/kernel protection layer. Kernel and application objects are examined through application browsers. Choices runs on bare hardware on desktop computers, distributed and parallel computers, and small mobile devices. Choices is supported on the SPARC, x86, and ARM processor architectures.

OSKit [14] is a framework and a set of 34 operating system components' libraries. OSKit's goal is the manipulation of operating system elements in a standard software development cycle by providing a modular way to combine predefined OS components.

Pebble [15] is a toolkit for generating specialized operating systems to fit particular application domains. It is intended for high-end embedded applications, which require performance near to the bare machine, protection, and modularity. Pebble consists of a tiny nucleus that manages context switches, protection domains, and trap vectors. Pebble also provides a set of run-time replaceable components and implements efficient cross-domain communication between components via portal calls. Higher level abstractions, such as thread and IPC are implemented by server components, and run in separate protection domains under hardware memory management.

THINK [13] is a software framework for implementing operating system kernels from components of arbitrary sizes. A unique feature of THINK is that it provides a uniform and highly flexible binding model to help OS architects assemble operating system components in varied ways. An OS architect can build an OS kernel from components using THINK without being forced into a predefined kernel design (e.g., exo-kernel, micro-kernel, or classical OS kernel).

APES [16] is a component-based system framework specially designed to fully take advantage of heterogeneous, embedded hardware architectures. It includes several components such as processor support, thread libraries, and C libraries. It also includes a set of micro-kernel components that provides services such as task management, memory management, and I/O management. It currently supports several RISC processors (ARM, SPARC, MIPS, Xilinx Microblaze, ...) and DSP (Atmel mAgicV) processors.

9.6 Pros and Cons

What solution is best suited to a particular kind of project? Well, there are no straight answers and the choice closely depends on the size and the needs of the application, as well as the complexity of the targeted HMC-SoC. Table 9.1 recapitulates the characteristics of each solution.

Small projects (such as embedded audio players or gaming devices) with a long life cycle will usually use a developmental approach based on BSPs, because it offers complete control over each element of the software and it provides only the strict minimum to the application. If the need arises, a software development kit (SDK) containing high-level functions to manipulate specific parts of the hardware can also be provided.

Safety- or time-critical projects (such as software final state machines (FSMs) or event-driven systems that are developed in the automotive or the avionic industries) should use an environment that is able to generate validated applications from mathematical models using certified libraries. Each part of the application can be precisely time-bounded, and the generated code is usually guaranteed to have the same behavior as the initial model. However, complex services such as network stacks or virtual memory management should not be considered or expected. Fortunately, they are not of critical importance in this kind of project.

TABLE 9.1: Solution Pros and Cons: $*$ = *low*, $***$ = *high*

	BSP	GPOS	Application generation	Component OS
Application development	*	**	* * *	**
Application portability/reuse	*	**	*	* * *
Application debug	**	*	* * *	**
Devices support	*	* * *	**	* * *
Hardware optimization	* * *	*	**	**

Set-top boxes, routers, and multimedia platforms are the kinds of applications that can benefit from general-purpose or component-based operating systems. They are not developed specifically for a hardware platform, they require strong operating system support, and they are not too demanding in terms of response time. GPOS-based development approaches offer good programming environments and ensure the compatibility of the developed applications with all the hardware architectures supported by the GPOS. Unfortunately, taking full advantage of the specificities of HMC-SoC platforms using this approach requires additional mechanisms that are not yet available. Component-based development approaches offer a better use of the underlying hardware for an equivalent set of services. However, their programming model is radically different from GPOS-based solution, which can be seen as the "show-stopper" by software engineers.

9.7 Conclusions

In this chapter, we presented the approaches used by the existing application programming environments which target heterogeneous, multi-core systems-on-chips.

The BSP-based development solution is suited for the development of small-budget or time-critical applications. The application, manually distributed on each processor of the hardware platform, directly makes use of the interface exported by the BSPs and is available for each processor. Likewise, each part of the application makes use of the assembly language of the processor it belongs to, so as to improve its overall performance. If this approach allows fast development cycles and brings high performance to small applications on average-sized hardware platforms, it does not fit the development of large applications on more complex HMC-SoCs.

The GPOS-based development solution is best suited to the development of complex but not critical applications. It increases their portability by providing a stable API on all the processors and hardware architectures supported by the OS. It also reduces development times when multiple hardware platforms are targeted or when the application already exists and makes use of an API supported by the OS. The application still must be manually distributed over the processors and adjusted to benefit from their specificities.

The part of the application dedicated to run on the GPOS makes use of its API and is loaded by the OS as one of its processes, while the parts of the application dedicated to run on the SPPs directly use their BSP interfaces and are loaded by the GPOS through specific device drivers. If this approach brings flexibility to the development cycle of an application, it doesn't come without a price: GPOSs are generally not suited to optimally deal with heterogeneous architectures. In addition, considering the SPPs as merely co-processors is no longer sufficient, especially with modern HMC-SoCs containing several DSPs that can score more than one giga-floating-point operation-per-second (GFLOPS).

Two major problems concerning these solutions were highlighted. Firstly, none of them allows the development of complex, critical applications or their optimal execution on modern HMC-SoCs. Secondly, in both of these approaches the distribution of the application on the different processors has to be done manually. To cope with these limitations, other solutions based on the modeling of both the application and the operating system are being researched.

The automatic application generation and RTOS modeling solution starts from a functional model of the application written using a high-level representation and a specific model of computation. This model can then be transformed in a hierarchical task graph (HTG) and extended using real-time operating system elements such as a scheduler or an interrupt manager. This HTG can be simulated in order to get the configuration that best suits the application's requirements. Next, the application's code is generated from the HTG model and compiled using the programming interfaces of the available external libraries. Finally, the compiled code is linked to these libraries. This process is repeated for each processor of the target platform. The boot-up sequence and low-level debug of the application are equivalent to those of the BSP approach presented earlier.

This automated code generation solution allows the fast development of complex applications by automating the creation of the application's code and hiding software construction details. However, the algorithm of the application needs to be compatible with one of the supported models of computation. However, non-predictable behaviors such as distributed communications can hardly be simulated and consequently automatically generated. In those cases, manually written functions are still required.

The solution using component-based operating systems shares the same benefits as the GPOS-based solution. It is suited for the development of complex applications and it increases their portability by providing a stable API on all the processors and hardware architectures supported by the OS. In the same manner, it reduces the development times when multiple hardware platforms are targeted or when the application already exists and makes use of an API supported by the OS. The huge benefits of the component-based approach resides in its software architecture and its programming model.

The software architecture of this solution allows the software designer to use only the components required by the application and nothing more, dramatically reducing the final memory footprint of the application. It is also more flexible, since only the component's interface is accessible to the developer. The implementation of two components sharing the same interface can be different in every way, providing that its behavior is respected. The programming model of this solution allows the software designer to reuse existing application codes if the APIs used in the application are exported by existing components. It also speeds up the development cycle of an application since its dependences are automatically resolved. Last but not least, it guarantees the same programming interface on all the processors present on the targeted platform with the same level of flexibility.

Review Questions and Answers

[Q_1] **Why is the development of an application for a MC-SoC difficult?**

It is not difficult to program an application for a MC-SoC. In fact, it can be programmed as easily as any other computing platform: most of them are able to run a general-purpose operating system that offers convenient development environments. What is difficult is to efficiently take advantage of the hardware, and this for two main reasons: the platform embeds multiple computing cores and they might not all be the same.

An application can be designed without considerations for these specificities. However, only one core at a time and, if the cores are heterogeneous, only cores of the same family can be used for its execution, resulting in a considerable waste of processing power.

[Q_2] **Why do I need to parallelize my application?**
In order to take advantage of the multiple computation cores, you need to split your application into small pieces and register them as execution threads for the operating system to distribute them over the cores. This operation is not easy since each thread must be well balanced in terms of computing power and communication bandwidth.

[Q_3] **Why is my GPOS not suited to operate my HMC-SoC?**
A GPOS considers heterogeneous cores as hardware accelerators that can only be accessed through device drivers. Though this approach works well with real hardware accelerators, it is not suited to efficiently operate full-fledged, computation-oriented cores such as DSPs or ASIPs. Its main drawback is its latency. The GPOS needs to control each step of the interaction with the heterogeneous core (reset, send data, start, stop, fetch data), resulting in huge delays between each computation.

Bibliography

[1] State of embedded market survey. Technical report, Embedded Systems Design, 2006.

[2] Altera. Introduction to the NiosII Software Build Tool. http://www.altera.com/literature/hb/nios2/n2sw_nii52014.pdf.

[3] Atmel Corporation. Microsoft-certified Windows CE BSP for AT91SAM9261. http://www.atmel.com/dyn/products/view_detail.asp?ref=&FileName=Adeneo_5_22.html&Family_id=605.

[4] Felice Balarin, Yosinori Watanabe, Harry Hsieh, Luciano Lavagno, Claudio Passerone, and Alberto L. Sangiovanni-Vincentelli. Metropolis: An integrated electronic system design environment. *IEEE Computer*, 36(4):45–52, 2003.

[5] Prithviraj Banerjee, U. Nagaraj Shenoy, Alok N. Choudhary, Scott Hauck, C. Bachmann, Malay Haldar, Pramod G. Joisha, Alex K. Jones, Abhay Kanhere, Anshuman Nayak, S. Periyacheri, M. Walkden, and David Zaretsky. A MATLAB® compiler for distributed, heterogeneous, reconfigurable computing systems. In *Proc. of the IEEE Symp. on Field-Programmable Custom Computing Machines (FCCM)*, pages 39–48, 2000.

[6] Joseph Buck, Edward A. Lee, and David G. Messerschmitt. Ptolemy: A framework for simulating and prototyping heterogeneous systems, 1992.

[7] Lukai Cai, Daniel Gajski, and Mike Olivarez. Introduction of system level architecture exploration using the specc methodology. In *Proc. of IEEE Int. Symp. on Circuits and Systems (ISCAS) (5)*, pages 9–12, 2001.

[8] Roy H. Campbell and See mong Tan. Choices: an object-oriented multimedia operating system. In *In Fifth Workshop on Hot Topics in Operating Systems, Orcas Island*, pages 90–94. IEEE Computer Society, 1995.

[9] Delft University of Technology. The Artemis Project. http://ce.et.tudelft.nl/artemis, 2009.

[10] DSpace, Inc. dSpace. http://www.dspaceinc.com.

[11] Basant Kumar Dwivedi, Anshul Kumar, and M. Balakrishnan. Automatic synthesis of system on chip multiprocessor architectures for process networks. In *Proc. of the 2nd IEEE/ACM/IFIP Int. Conf. on Hardware/Software Codesign and System Synthesis (CODES+ISSS)*, pages 60–65, 2004.

[12] eCos Centric Limited. The eCos Operating System. http://ecos.sourceware.org/.

[13] Jean-Philippe Fassino, Jean-Bernard Stefani, Julia L. Lawall, and Gilles Muller. Think: A software framework for component-based operating system kernels. In *USENIX Annual Technical Conference, General Track*, pages 73–86, 2002.

[14] Bryan Ford, Godmar Back, Greg Benson, Jay Lepreau, Albert Lin, and Olin Shivers. The Flux OSKit: A substrate for kernel and language research. In *SOSP*, pages 38–51, 1997.

[15] Eran Gabber, Christopher Small, John L. Bruno, José Carlos Brustoloni, and Abraham Silberschatz. The Pebble component-based operating system. In *USENIX Annual Technical Conference*, pages 267–282, 1999.

[16] Xavier Guérin and Frédéric Pétrot. A system framework for the design of embedded software targeting heterogeneous multi-core SoCs. In *Proc. Int'l Conf. on Application-Specific Systems, Architectures, and Processors (ASAP)*, 2009.

[17] Sang-Il Han, Soo-Ik Chae, Lisane Brisolara, Luigi Carro, Katalin Popovici, Xavier Guerin, Ahmed A. Jerraya, Kai Huang, Lei Li, and Xiaolang Yan. Simulink®-based heterogeneous multiprocessor SoC design flow for mixed hardware/software refinement and simulation. *Integration, The VLSI Journal*, 2008.

[18] IEC. The 61508 Safety Standard. Technical report, 2005.

[19] IEEE Computer Society/Test Technology. Standard Test Access Port and Boundary Scan Architecture. Technical report, 2001.

[20] Paul Lieverse, Pieter van der Wolf, Kees A. Vissers, and Ed F. Deprettere. A methodology for architecture exploration of heterogeneous signal processing systems. *VLSI Signal Processing*, 29(3):197–207, 2001.

[21] Linux Kernel Organization, Inc. The Linux Kernel. http://www.kernel.org.

[22] Anthony Massa. *Embedded Software Development with eCos*. Prentice Hall Professional Technical Reference, 2002.

[23] Micrium. The microC-OS/II Operating System. http://www.micrium.com/products/rtos/kernel/rtos.html.

[24] Microsoft. The Windows CE Operating System. http://www.microsoft.com/windowsembedded.

[25] Claudio Passerone. Real time operating system modeling in a system level design environment. In *Proc. of IEEE Int. Symp. on Circuits and Systems (ISCAS)*, 2006.

[26] Frédéric Pétrot and Pascal Gomez. Lightweight implementation of the POSIX threads API for an on-chip MIPS multiprocessor with VCI interconnect. In *Proc. Int. ACM/IEEE Conf. Design, Automation and Test in Europe (DATE)*, pages 20051–20056, 2003.

[27] Andy D. Pimentel, Cagkan Erbas, and Simon Polstra. A systematic approach to exploring embedded system architectures at multiple abstraction levels. *IEEE Trans. Computers*, 55(2):99–112, 2006.

[28] QNX Software System. The QNX Operating System. http://www.qnx.com.

[29] Renesas. uCLinux SH7670 Board Support Package. http://www.renesas.com/fmwk.jsp?cnt=bsp_rskpsh7670.htm&fp=/products/tools/introductory_evaluation_tools/renesas_starter_kits/rsk_plus_sh7670/child_folder/&title=uCLinux%20SH7670%20Board%20Support%20Package.

[30] RTCA. DO-178B, Software Considerations in Airborne Systems and Equipment Certification. Technical report, 1992.

[31] Symbian, Ltd. The Symbian Operating System. http://www.dmoz.org/Computers/Mobile_Computing/Symbian/Symbian_OS/.

[32] Tensilica. Xtensa configurable processors. http://www.tensilica.com/.

[33] Texas Instrument. OMAP35x WinCE BSP. http://focus.ti.com/docs/toolsw/folders/print/s1sdkwce.html.

[34] The Fedora Project. SELinux. http://fedoraproject.org/wiki/SELinux.

[35] The MathWorks. Real-Time Workshop. http://www.mathworks.com/products/rtw/.

[36] Ric Vilbig. Jtag Debug: Everything You Need to Know. Technical report, Mentor Graphics, 2009.

[37] WindRiver. The VxWorks Operating System. http://www.windriver.com.

[38] Xenomai. The Xenomai Hard-RT Kernel Extension. http://www.xenomai.org.

[39] Xilinx. Generating Efficient Board Support Package. http://www.nuhorizons.com/FeaturedProducts/Volume3/articles/Xilinx_BoardSupport_Article.pdf.

10

Autonomous Power Management Techniques in Embedded Multi-Cores

Arindam Mukherjee, Arun Ravindran, Bharat Kumar Joshi, Kushal Datta and Yue Liu

Electrical and Computer Engineering Department
University of North Carolina
Charlotte, NC, USA
{amukherj, aravindr, bsjoshi, kdatta, yliu42}@uncc.edu

CONTENTS

10.1	Introduction	338
	10.1.1 Why Is Autonomous Power Management Necessary?	339
	10.1.1.1 Sporadic Processing Requirements	339
	10.1.1.2 Run-time Monitoring of System Parameters	340
	10.1.1.3 Temperature Monitoring	340
	10.1.1.4 Power/Ground Noise Monitoring	341
	10.1.1.5 Real-Time Constraints	341
10.2	Survey of Autonomous Power Management Techniques	342
	10.2.1 Clock Gating	342
	10.2.2 Power Gating	343
	10.2.3 Dynamic Voltage and Frequency Scaling	343
	10.2.4 Smart Caching	344
	10.2.5 Scheduling	345
	10.2.6 Commercial Power Management Tools	346
10.3	Power Management and RTOS	347
10.4	Power-Smart RTOS and Processor Simulators	349
	10.4.1 Chip Multi-Threading (CMT) Architecture Simulator	350
10.5	Autonomous Power Saving in Multi-Core Processors	351
	10.5.1 Opportunities to Save Power	353
	10.5.2 Strategies to Save Power	354
	10.5.3 Case Study: Power Saving in Intel Centrino	356
10.6	Power Saving Algorithms	358

        10.6.1   Local PMU Algorithm 358
        10.6.2   Global PMU Algorithm 358
10.7   Conclusions 360
Review Questions 362
Bibliography 363

10.1   Introduction

Portable embedded systems place ever-increasing demands on high-performance, low-power microprocessor design. Recent years have witnessed a dramatic transition in the expectations from, and the capabilities of, embedded systems. This in turn, has triggered a paradigm shift in the embedded processor industry, forcing manufacturers of embedded processors to continually alter their existing roadmap to incorporate multiple cores on the same chip. From a modest beginning of dual and quad cores that are currently available in the 45 and 32 nm technologies, multi-core processors are expected to include hundreds of cores in a single chip in the near future. In SuperComputing 2008, Dell announced that it will release a workstation containing an 80-core processor around 2010 [1], and Intel is planning a 256-core processor in the near future. While the industry focus is on putting higher numbers of cores on a single chip, the key challenge is to optimally architect these processors for low power operations while satisfying area and often stringent real-time constraints, especially in embedded platforms. This trend, together with unpredictable interrupt profiles found in modern embedded systems, motivates the need for smart power saving features in modern embedded processors.

Earlier embedded processor micro-architects had been designing energy-efficient processors to extend battery life. Features such as clock gating, banked caches with gateable regions, cache set prediction, code compression to save area, dynamic voltage and frequency scaling (DVFS), and static sleep (power-gated) modes are all matured concepts in embedded-processor systems. Unfortunately, the full promise of these techniques has been hindered by slow off-chip voltage regulators and circuit controllers that operate in the time scale of 10 mV/ms, and thus lack the ability to adjust to different voltages at small time scales. Recent availability of on-chip power saving circuits with response characteristics of the order of 10 mV/ns [41] has made it feasible for embedded processor designers to explore fast, per-core DVFS and power gating, and additional power saving by fine grain intra-core power gating and clock gating. The fundamental challenge that remains to be solved is realizing the vertical integration of embedded code development, scheduling and autonomous power-management hardware.

This chapter is organized into different sections and subsections as follows. In the next subsection we discuss the necessity of autonomous power management, followed by a background survey of different power saving strategies in Section 10.2. Commercial power management tools are discussed in Section 10.2.6 followed by an in-depth look into modern real time operating systems and their roles in power management in Section 10.3. The roles of processor and RTOS simulators which are critical for future research in this area are explained in Section 10.4, where we also present CASPER, an integrated embedded system and RTOS simulation platform that we are currently developing. Section 10.5 uses an embedded processor as an example to explain our proposed autonomous power saving schemes for multi-core processors, while Section 10.6 details some of the algorithms involved and outlines some of our on-going research and the necessities of further work in this area. Finally we draw conclusions in Section 10.7.

10.1.1 Why Is Autonomous Power Management Necessary?

A real-time system is one in which the correctness of the system depends not only on the logical results, but also on the time at which the results are produced. Many real-time systems are embedded systems, or components of a larger system. Such real-time systems are widely used for safety-critical applications where an incorrect operation of the system can lead to loss of life or other catastrophes. The safety-critical information processed by such applications has extremely high value for very short durations of time. Moreover, a large number of embedded applications have sporadic processing requirements in which tasks have widely varying release times, execution times, and resource requirements. The challenge of modern embedded computing is to satisfy such real-time constraints while managing power dissipation for extending battery life and keeping the system thermal-safe by dynamically managing hot-spots on chip, and noise-safe by reducing power-ground bounce for security and correctness. Since power management and dynamic task scheduling to meet real-time constraints are key components of embedded systems, we shall henceforth refer to both as techniques for autonomous power management in this chapter.

10.1.1.1 Sporadic Processing Requirements

Existing power management algorithms are based on deterministic workload and resource models, and work for deterministic timing constraint. They produce unacceptably low degrees of success for applications with sporadic processing requirements. Moreover, they cannot capture the system level behavior of the hardware platform and variable amounts of time and resources consumed by the system software and application interfaces because they are made for statically configured systems (i.e., systems in which applications are partitioned and processors and resources are statically allocated to the partitions). Consequently, when it comes to meeting real-time constraints, existing power management algorithms are overly pessimistic when applied to sporadic applications, especially in large, open run-time environments built on commodity computers, networks, and system software.

10.1.1.2 Run-time Monitoring of System Parameters

The future multi-core embedded processor will be a network of hundreds or thousands of heterogeneous cores, some of which will be general-purpose, some highly application-specific cores, and some reconfigurable logic to exploit fine-grained parallelism, all connected to hierarchies of distributed on-chip memories by high speed on-chip networks. These networks will not only interconnect shared memory processor (SMP) cores, but also on-chip clusters of SMP cores which will most likely communicate using message-passing-interface (MPI) like protocols. The different cores will satisfy different power-performance criteria in future many-core chips. The resulting system level power and performance uncertainties caused by unpredictable system level parameters like communication cost, memory latencies and misses, and kernel execution and idle times will be impossible to predict either statically or probabilistically, as now achieved by existing power management algorithms [19]. Hence, the need for run-time monitoring of system level parameters and dynamic power management in multi-core embedded processors.

10.1.1.3 Temperature Monitoring

Power management techniques like dynamic voltage and frequency scaling (DVFS) have been widely used for power and energy optimization in embedded system design. As thermal issues become increasingly prominent, however, run-time thermal optimization techniques for embedded systems will be required. The authors of [43] propose techniques for proactively optimizing DVFS with system thermal profile to prevent run-time thermal emergencies, minimize cooling costs, and optimize system performance. They formulate minimization of application peak temperature in the presence of real-time constraints as a nonlinear programming problem. This provides a powerful framework for system designers to determine a proper thermal solution and provide a lower bound on the minimum temperature achievable by DVFS. Furthermore, they examine the differences between optimal energy solutions and optimal peak temperature solutions. Experimental results indicate that temperature-unaware energy consumption can lead to overall high temperatures. Finally, a thermal-constrained energy optimization (i.e., power management) procedure is proposed to minimize system energy consumption under a constraint on peak temperature. However, the optimization is static and assumes a pre-deterministic knowledge of the task profile. Run-time thermal sensing [29] and power management techniques are being incorporated in emerging multi-core embedded processors executing real-life complex embedded applications.

10.1.1.4 Power/Ground Noise Monitoring

With the emergence of multi-core embedded processors and complex embedded applications, circuits with increasingly higher speed are being integrated at an increasingly higher density. Simultaneously, operating voltage is being reduced to lower power dissipation, which in turn leads to lowering circuit noise margins. This combined with high frequency switching and high circuit density causes large current demand and voltage fluctuations in the on-chip power distribution network due to IR-drop, L di/dt noise, and LC resonance. This is commonly referred to as power-ground noise [60]. Power-ground noise changes gate delays and can lead to errors in signal values. Therefore, power-ground integrity is a serious challenge in designing multi-core embedded processors. This problem is further compounded by the fact that switching currents and the consequent power-ground noise are dependent on particular embedded applications and the corresponding data which can be sporadic. Hence, any pre-deterministic modeling of such noise will be inaccurate, especially for future complex multi-core embedded platforms. Run-time measurement of power-ground noise [30] and a power management scheme which considers this data in any dynamic optimization are critical for accurate and safe embedded computing in the future.

10.1.1.5 Real-Time Constraints

There exists a strong correlation between scheduling in real-time operating systems (RTOS) and power saving features in embedded systems. Traditional RTOS schedulers use either static scheduling algorithms with static priorities or dynamic scheduling algorithms based on static priorities [21]. In the latter case, task priorities are determined *a priori* using loose inaccurate bounds of task periodicity like rate monotonic scheduling (RMS) [24] for example. In RMS, the task which arrives more frequently gets a higher priority for scheduling, irrespective of real-time constraints and the state of the processor system parameters (mentioned in Section 10.1.1.2 above). During run-time, the RTOS checks for interrupts and dynamically schedules tasks according to a static RMS prioritized list. All the existing techniques for dynamic priority-based dynamic scheduling algorithm, such as the earliest deadline first (EDF) algorithm [24], do not utilize state of system parameters. However, for the emerging complex embedded multi-core processors, it is critical to consider this information to estimate worst-case execution time of a task. Integration of power saving features which cause changes in the system parameters and scheduling algorithms which need to consider this information, are vital for achieving low power real-time operations in complex embedded systems.

10.2   Survey of Autonomous Power Management Techniques

10.2.1   Clock Gating

Gated clocking is a commonly applied technique used to reduce power by gating off clock signals to registers, latches, and clock regenerators. Gating may be done when there is no required activity to be performed by logic whose inputs are driven from a set of storage elements. Since new output values from the logic will be ignored, the storage elements feeding the logic can be blocked from updating to prevent irrelevant switching activity in the logic. Clock gating may be applied at the function unit level for controlling switching activity by inhibiting input updates to function units such as adders, multipliers and shifters whose outputs are not required for a given operation. Entire subsystems like cache banks or functional units may be gated off by applying clock gating in the distribution network. This provides further savings in addition to logic switching activity reduction since the clock signal loading within the subsystem does not toggle. Overhead associated with generation of the enable signal must be considered to ensure that power saving actually occurs, and this generally limits the granularity at which clock gating is applied. It may not be feasible to apply clock gating to single storage elements due to the overhead in generating the enable signal, although self-gating storage elements have been proposed that compare current and next state values to enable local clocking [59]. If the switching rate of input values is low relative to the clock, a net power saving may be obtained. The notion of disabling the clocks to unused units to reduce power dissipation in microprocessors has been discussed in [32] and [62]. In the CAD community, similar techniques have been demonstrated at the logic level of design. Guarded evaluation seeks to dynamically detect which parts of a logic circuit are being used and which are not [61]. Logic pre-computation seeks to derive a pre-computation circuit that under special conditions does the computation for the remainder of the circuit [20]. Both of these techniques are analogous to conditional clocking, which can be used at the architectural level to reduce power by disabling unused units. [31] showed that clock gating can significantly reduce power consumption by disabling certain functional units if instruction decode indicates that they will not be used. The optimization proposed in [27] watches for small operand values and exploits them to reduce the amount of power consumed by the integer unit. This is accomplished by an aggressive form of clock gating based on operand values. When the full width of a functional unit is not required, we can save power by disabling the upper bits. With this method the authors show that the amount of power consumed by the integer execution unit can be reduced for the SPECint95 suite with little additional hardware.

10.2.2 Power Gating

Historically, the primary source of power dissipation in CMOS transistor devices has been the dynamic switching due to charging/discharging load capacitances. Chip designers have relied on scaling down the transistor supply voltage in subsequent generations to reduce this dynamic power dissipation due to a much larger number of on-chip transistors, a consideration critical for designing low power embedded processors. However, lower supply voltages have to be coupled with lower transistor threshold voltages [51] to maintain high switching speeds required for complex embedded applications. The International Technology Roadmap for Semiconductors [13] predicts a steady scaling of supply voltage with a corresponding decrease in transistor threshold voltage to maintain a 30 percent improvement in performance every generation. The drawback of threshold scaling is an increase in leakage power dissipation due to an exponential increase in subthreshold leakage current even when the transistor is not switching. [25] estimates a factor of 7.5 increase in leakage current and a five fold increase in total leakage power dissipation in every chip generation; hence, the need for power gating to save leakage power.

Power gating is a circuit level technique which reduces leakage power by effectively turning off the supply voltage to the logic elements, when they have been idling for a certain long duration of time. Power gating may be implemented using NMOS or PMOS transistors, presenting a trade-off among area overhead, leakage reduction, and impact on performance. By curbing leakage, power gating enables high performance through aggressive threshold-voltage- scaling which has been considered problematic because of inordinate increase in leakage.

A novel power gating mechanism for instruction caches has been proposed in [68], which dynamically estimates and adapts to the required instruction cache size, and turns off the supply voltage to the unused SRAM cells of the cache. Similarly, power gating may be applied to any idling core, or cache bank, or functional units in a multi-core processor. However the increase and decrease of supply voltage as part of power gating is typically done over hundreds and thousands of clock cycles to avoid sudden increase or decrease of current when gates switch on or off respectively. These current spikes lead to L di/dt noise as mentioned in Section 10.1.1.4. Literature review shows that in the 90 nm technology, the maximum acceptable switching rate is 10 mV per 10 ns for off-chip control and 10 mV per ns for on-chip gating.

10.2.3 Dynamic Voltage and Frequency Scaling

Dynamic voltage and frequency scaling (DVFS) was introduced in the 1990s [45], offering great promise to dramatically reduce power consumption in large digital systems (including processor cores, memory banks, buses, etc.) by adapting both voltage and frequency of the system with respect to changing workloads [38, 55, 57, 66]. DVFS control algorithms can be implemented at different levels, such as in the processor microarchitecture [46], the operating system scheduler [39], or through compiler algorithms [37, 67].

Unfortunately, the full promise of DVFS has been hindered by slow off-chip voltage regulators that lack the ability to adjust to different voltages at small time scales. Modern implementations are limited to temporally coarse-grained adjustments governed by the operating system [10, 16].

In recent years, researchers have turned to multi-core processors as a way of maintaining performance scaling while staying within tight power constraints. This trend, coupled with diverse workloads found in modern systems, motivates the need for fast, per-core DVFS control. In recent years, there has been a surge of interest in on-chip switching voltage regulators [18, 35, 54, 64]. These regulators offer the potential to provide multiple on-chip power domains in future multi-core embedded processors. An on-chip regulator, operating with high switching frequencies, can obviate bulky filter inductors and capacitors, allow the filter capacitor to be integrated entirely on the chip, place smaller inductors on the package, and enable fast voltage transitions at nanosecond timescales. Moreover, an on-chip regulator can easily be divided into multiple parallel copies with little additional overhead to provide multiple on-chip power domains. However, the implementation of on-chip regulators presents many challenges including regulator efficiency and output voltage transient characteristics, which are significantly impacted by the system-level application of the regulator. In [41], the authors describe and model these costs, perform a comprehensive analysis of a CMP system with on-chip integrated regulators, and propose an off-line integer linear programming based DVFS algorithm using the multi-core processor simulator [17]. They conclude that on-chip regulators can significantly improve DVFS effectiveness and lead to overall system power savings in a CMP, but architects must carefully account for overheads and costs when designing next-generation DVFS systems and algorithms.

10.2.4 Smart Caching

Cache memories in embedded processors play significant role in determining the power- performance metric. In this section we will discuss two methods of saving power in embedded smart caches: cache set prediction and low power cache coherence protocols. However, since the focus of this chapter is autonomous power management, and since smart caching strategies are typically pre-determined and not run-time variable, we will not dwell on this topic beyond an introduction for the sake of completeness. These caching techniques can be used in conjunction with any autonomous power management technique that we discuss in this chapter.

In [50], the authors use two previously proposed techniques, way prediction [22, 28] and selective direct mapping [22], to reduce L1 dynamic cache power while maintaining high performance. Way prediction and selective direct mapping predict the matching way number and provide the prediction prior to the cache access, instead of waiting on the tag array to provide the way number as done by sequential access. Predicting the matching way enables the techniques not only to attain fast access times but also to achieve power reduction. The techniques reduce power because only the predicted way is accessed. While these techniques were originally proposed to improve set-associative cache access times, this is the first paper to apply them to reducing power.

Power efficient cache coherence is discussed in [52]. Snoop-based cache coherence implementations employ various forms of speculation to reduce cache miss latency and improve performance. This section examines the effects of reduced speculation on both performance and power consumption in a scalable snoop-based design. The authors demonstrate that significant potential exists for reducing power consumption by using serial snooping for load misses. They report only a 6.25 percent increase for average cache miss latency for SPLASH2 benchmark [65] while achieving substantial reductions in snoop-related activity and power dissipation.

10.2.5 Scheduling

Dynamic voltage supply (DVS) and dynamic voltage and frequency scaling (DVFS) techniques have led to drastic reductions in power consumption. However, supply voltage has a direct impact on processor speed, and hence, on the real-time performance of an embedded system. Thus classic task scheduling, frequency scaling and supply voltage selection have to be addressed together. Scheduling offers another level of possibilities for achieving energy and power efficient systems, especially when the system architecture is fixed or the system exhibits a very dynamic behavior. For such dynamic systems, various power management techniques exist and are reviewed for example in [23]. Yet, these mainly target soft real-time systems, where deadlines can be missed if the quality of service is kept. Several scheduling techniques for soft real-time tasks, running on DVS processors have already been described, for example in [49]. Power reductions can be achieved even in hard real-time systems, where no deadline can be missed, as shown in [34, 36]. Task level voltage scheduling decisions can further reduce the power consumption. Some of these intra-task scheduling methods uses several re-scheduling points inside a task, and are usually compiler assisted [42, 47, 56]. Alternatively, fixing the schedule before the task starts executing as in [34, 36] eliminates the internal scheduling overhead, but with possible loss of power efficiency. Statistics can be used to take full advantage of the dynamic behavior of the system, both at task level [47] and at task-set level [69]. [33] employs stochastic data to derive efficient voltage schedules without the overhead of intra-task re-scheduling for hard real-time scheduling techniques, where every deadline has to be met.

TABLE 10.1: Power Gating Status Register

Power Management Methods	Pros	Cons
Clock Gating	Simple additional gating logic	Leakage power dissipation Medium power/ ground noise
Power Gating	No leakage	Complex additional p/g switching logic High power/ground noise
DVFS	Good controllability between power and performance Low p/g noise	Complex additional on-chip p/g voltage regulators required
Smart Caching	Software controlled Some level of optimization possible between power and performance	Cache logic increases Verification of coherence protocols difficult
Scheduling	Global power optimization possibly unlike all other methods Good control over p/g noise	Kernel or user code has to be changed

10.2.6   Commercial Power Management Tools

Dynamic power can be controlled by the user application program, by the operating system, or by hardware (Table 10.1). Two of the most prominent and universally used commercial power management software suites used for embedded applications are discussed in this section. Processors such as the Transmeta Crusoe, Intel StrongARM and XScale processors, and IBM PowerPC 405LP allow dynamic voltage and frequency scaling of the processor core in support of dynamic power management strategies. Aside from the Transmeta system, all of the processors named above are highly integrated system-on-a-chip (SoC) processors designed for embedded applications. Dynamic power in these processors is controlled by the operating system. The IBM Low-Power Computing Research Center, IBM Linux Technology Center and MontaVistaTM Software [11] have developed a general and flexible dynamic power management scheme for embedded systems. This software attempts to standardize a dynamic power management and policy framework that will support different power management strategies, either under control of operating system components or user-level policy managers, which in turn will enable further research and commercial developments in this area. The framework is applicable to a broad class of operating systems and hardware platforms, including IBM PowerPC 405LP. MontaVista's primary interest is enabling dynamic power management capabilities for the Linux operating system.

Another prevalent real-time operating system (RTOS) with built-in power management features is VxWorks [15] from Wind River. VxWorks provides a complete, flexible, scalable, optimized embedded development, debugging and run-time platform that is built on open standards and industry-leading tools. It is the industry's most prevalent commercial RTOS, and tightly integrated run-time performance with power optimization. VxWorks 6.6 and 6.7 versions are built on a highly scalable, deterministic, and hard real-time kernel, and handles multi-core symmetric and asymmetric multiprocessing (SMP/AMP) for high performance, low costs, low power consumption, faster time-to-market. The VxLib software library that is part of the VxWorks integrated development environment (IDE) has API functions for user specified power management for multi-core embedded processors.

10.3   Power Management and RTOS

Since power dissipation and real-time performance are highly dependent on the particulars of the embedded platform and its application, a generic power management architecture needs to be flexible enough to support multiple platforms with differing requirements. Part of this flexibility comes from supporting pluggable power management strategies that allow system designers to easily tailor power management for each application. We believe that smart power management for emerging multi-core embedded processors and the complex systems they are increasingly incorporated in can only be achieved by a combination of several factors. These include autonomous dynamic power sensing and control logic in hardware, RTOS controlled task scheduling and power management, and by auto-tuner controlled active power management at the system level to meet real-time constraints. An RTOS for multi-core embedded processors should include the real-time kernel, the multi-core support, the file system, and the programming environment. The real-time kernel provides local task management, scheduling, timing primitives, memory management, local communication, interrupt handling, error handling, and an interface to hardware devices. The multi-core support includes inter-core communication and synchronization, remote interrupts, access to special purpose processors, and distributed task management. The file system provides access to secondary storage, such as disks and tapes, and to local-area-networks. The programming environment provides the tools for building applications; it includes the editor, compiler, loader, debugger, windowing environment, graphic interface, and command interpreter (also called a shell). The level of support provided for each part of the operating system (OS) varies greatly among RTOS. Similarly, the auto-tuner's job is to schedule tasks at the system level between different multi-core processors, co-ordinate the processors with sensors, actuators, memory banks and input/output devices at the system level, manage communication between these modules, observe system level parameters as mentioned in Section 10.1.1.2, and actively manage tasks and inter-process communications for optimal power-performance operation of the embedded system as a whole. With future embedded systems projected to have multiple operating systems for different processors and even for different cores in the same processor, virtualization will be an important component of future RTOSs and auto-tuners.

Although excellent results have been obtained with kernel-level approaches for DVFS, authors of [26] believe that the requirements for simplicity and flexibility are best served by leaving the workings of the DVFS system completely transparent to most tasks, and even to the core of the OS itself. These considerations led to the development of a software architecture for policy-guided dynamic power management called DPM. It is important to note at the outset that DPM is not a DVFS algorithm, nor a power-aware operating system such as described in [70], nor an all-encompassing power management control mechanism such as the advanced configuration power interface (ACPI) [6]. Instead, DPM is an independent module of the operating system concerned with active power management. DPM policy managers and applications interact with this module using a simple API, either from the application level or the OS kernel level. Although not as broad as ACPI, the DPM architecture does extend to devices and device drivers in a way that is appropriate for highly integrated SOC processors. A key difference with ACPI is the extensible nature of the number of power manageable states possible with DPM. While DPM is proposed as a generic feature for a general purpose operating system, so far the practical focus has been the implementation of DPM for Linux. DPM implementations are included in embedded Linux distributions for the IBM PowerPC 405LP and other processors.

Advanced sensor-based control applications, such as robotics, process control, and intelligent manufacturing systems have several different hierarchical levels of control, which typically fall into three broad categories: servo levels, supervisory levels, and planning levels. The servo levels involve reading data from sensors, analyzing the data, and controlling electromechanical devices, such as robots and machines. The timing of these levels is critical, and often involves periodic processes ranging from 1 Hz to 1000 Hz. The supervisory levels are higher level actions, such as specifying a task, issuing commands like turn on motor 3 or move to position B, and selecting different modes of control based on data received from sensors at the servo level. Time at these levels is a factor, but not as critical as for the servo levels. In the planning levels time is usually not a critical factor.

Examples of processes at this level include generating accounting or performance logs of the real-time system, simulating a task, and programming new tasks for the system to take on. In order to develop sensor-based control applications, a multitasking, multiprocessing, and flexible RTOS has been developed in [8]. The Chimera II RTOS has been designed as a local OS within a global/local OS framework to support advanced sensor-based control applications. The global OS provides the programming environment and file system, while the local OS provides the real-time kernel, multi-core support, and an interface to the global OS. For many applications the global OS may be non-real-time, such as UNIX or Mach. However, the use of a real-time global OS such as Alpha OS [40] and RT-Mach [63] can add real-time predictability to file accesses, networking, and graphical user interfaces.

Most commercial RTOS, including iRMX III [14], OS-9 [9], and pSOS+ [2], do not use the global/local OS framework, and hence they provide their own custom programming environment and file system. The environments, including the editors, compilers, file system, and graphics facilities are generally inferior to their counterparts in UNIX-based OS. In addition, since much development effort for these RTOS goes into the programming environment, they have inferior real-time kernels as compared to other RTOS. Some commercial RTOS, such as VRTX [3] and VxWorks [15], do use the global/local OS framework. However, as compared to Chimera, they provide very little multiprocessor support, and their communications interface to the global OS is limited to networking protocols, thus making the communication slow and inflexible. The commercial RTOSs only provide basic kernel features, such as static priority scheduling and very limited exception handling capabilities, and multiprocessor support is minimal or non-existent. Previous research efforts in developing an RTOS for sensor-based control systems include Condor [48], the Spring Kernel [58], Sage [53], and Harmony [44]. They have generally only concentrated on selected features for the real-time kernel, or were designed for a specific target application. Chimera differs from these systems in that it not only provides the basic necessities of an RTOS, but also provides the advanced features required for implementation of advanced sensor-based control systems, which may be both dynamically and statically reconfigurable. A comparison of the various RTOSs can be found at [4].

10.4 Power-Smart RTOS and Processor Simulators

To study the power-performance effects of different power management strategies in multi-core embedded processors, we have developed CASPER [7], a cycle accurate simulator for the embedded multi-core processors, which can simulate a wide range of multi-threading architectures as well.

10.4.1 Chip Multi-Threading (CMT) Architecture Simulator

We have implemented a CMT architecture simulator for performance, energy and area analysis (CASPER) [7], which targets the SPARCV9 instruction set. CASPER is a multi-threaded (and hence, fast) parameterized cycle-accurate architecture simulator, which captures, in every clock cycle, the states of (i) the functional blocks, sub-blocks and register files in all the cores, (ii) shared memories and (iii) interconnect network. Architectural parameters such as number of cores, number of hardware threads per core (virtual processors), register file size and organization, branch predictor buffer size and prediction algorithm, translation lookaside buffer (TLB) size, cache-size and coherence protocols, memory hierarchy and management, and instruction queue sizes, to name a few, are parameterized in CASPER. The processor architecture is a hierarchical design containing functional blocks, such as instruction fetch unit (IFU), decode and branch unit (DBU), execution unit (EXU) and load-store unit (LSU). These blocks contain functional sub-blocks, such as L1 instruction cache, load miss queue, translation lookaside buffer, and so on. A selected point in the architectural design space defines the structural and/or algorithmic specifications of each one of these functional blocks in CASPER, which can then be simulated and evaluated for power and performance. The shared memory subsystem can be configured to consist of either L2 cache or both L2 and L3 unified caches. The interconnection network is also parameterized.

CPI Calculations For a given set of architectural parameters, CASPER uses counters in each core to measure the number of completed instructions (I_{core}) every second. Separate counters are used in each hardware strand to count the completed instructions (I_{strand}) every second. Assuming that the processor clock frequency is 1.2GHz, the total number of clock cycles per second is 1.2G. CPI-per-core is calculated as ($1.2G/I_{core}$). CPI-per-strand is calculated as ($1.2G/I_{strand}$).

Current Calculations CASPER also collects the current profile information of different architectural components based on their switching characteristics, for a target application in every clock cycle. Each component can be in either one of three possible switching states: active (valid data and switching), static (valid data but not switching) and idle (power down), which contributes to the overall dynamic current characteristics of the processor. The current calculator in CASPER uses (i) the pre-characterized average and peak current profiles of different architectural components for different operating states, including switching and leakage currents, and (ii) the cycle-accurate switching states of the different components which are obtained during simulations, to calculate the dynamic average and peak currents drawn by the processor.

Power Calculations The average and peak power for every simulation cycle for an architecture can be calculated by multiplying the supply voltage with the average and peak current in that cycle. The peak power dissipation over an entire simulation is the maximum of the peak power dissipated in all cycles, and the average power dissipation is found by averaging the average power dissipated in all cycles. This data will be used to statistically model the dependence of power dissipation on architectural parameters.

Verification and Dissemination CASPER has been verified against the open-sourced commercial SPARCV9 function simulator (SAM T2). Currently, CASPER is able to simulate instructions from instruction trace files generated from SAM T2. To the best of our knowledge, CASPER is the only project where such a flexible parameterized cycle-accurate processor simulator has been developed and open-sourced for the entire research community through the OpenSPARC Innovation Contest [12], and has won the first prize for the submission that makes the most substantial contribution to the OpenSPARC community. CASPER can be requested for research use through our Web site [7]. We are currently in the process of (i) adding new functional routines to simulate autonomous hardware monitoring and power saving features in CASPER, (ii) generalizing CASPER to handle any multi-core processor, and (iii) adding a front-end RTOS macro-simulator which will allow RTOS designers to incorporate custom power-aware scheduling algorithms. Hence, CASPER will enable embedded processor and RTOS designers to study the impacts of different multi-core processor architectures and power management (including autonomous hardware power saving and RTOS scheduling) schemes on the performance of real-time embedded systems.

10.5 Autonomous Power Saving in Multi-Core Processors

Consider the pipelined microarchitecture of one hardware thread in a multi-core embedded variant of the UltraSPARC T1 processor shown in Figure 10.1. We plan to use this example for discussing where and how we can potentially save power.

Figure 10.2 shows the trap logic unit associated with every core in the processor. Traps achieve vectored transfer of control of software from lower to higher privilege modes, e.g., user mode to supervisor or hypervisor mode. In UltraSPARC T1, a trap may be caused by a Tcc instruction, an instruction-induced exception, a reset, an asynchronous error, or an interrupt request. Typically a trap causes the SPARC pipeline to be flushed. The processor state is saved in the trap register stack and the trap handler code is executed. The actual transfer of control occurs through a trap table that contains the first eight instructions of each trap handler. The virtual base address of the table for traps to be delivered in privileged mode is specified in the trap base address (TBA) register. The displacement within the table is determined by the trap type and the current trap level. The trap handler code finishes

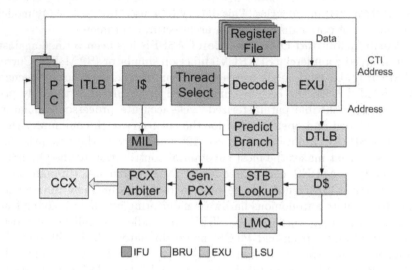

FIGURE 10.1: Pipelined micro-architecture of an embedded variant of Ultra-SPARC T1.

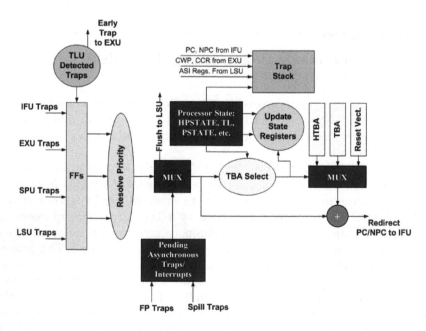

FIGURE 10.2: Trap logic unit.

execution when a DONE or RETRY instruction is encountered. Traps can either be synchronous or asynchronous with the SPARC core pipeline.

The figure illustrates the trap control and data flow in the TLU with respect to the other hardware blocks of the SPARC core. The priorities of the incoming traps from the IFU, EXU, LSU and TLU are resolved first. The type of the resolved trap is determined. According to the trap type and if no other interrupts or asynchronous traps with higher priorities are pending in the queue, a flush signal is issued to LSU to commit all previous unfinished instructions. The trap type also determines what processor state registers need to be stored into the trap register stack. The trap base address is then selected and is issued down the pipeline for further execution.

Figure 10.3 depicts the chip layout of a multi-core embedded processor with a variable number of cores, L2 cache banks, off-core floating point units (FPUs) and input-output logic, all interconnected by a network on chip. The CASPER simulation environment allows the designer to vary different architectural parameters.

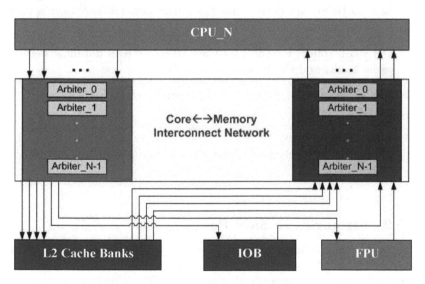

FIGURE 10.3: Chip block diagram.

10.5.1 Opportunities to Save Power

For the above multi-core embedded processor, we have identified the following power saving candidates (PSCs) at the core and chip levels:

1. Register files, which are thread-specific units. Each thread has one 160 double-word (64 bits) register file and achieves substantial savings in power when a task on a thread is blocked or idling.

2. Load miss queue (LMQ) which is used to queue data when there is a data cache miss; the LMQ is shared between threads and the power saving is small.

3. Branch predictor: branch history table can be thread-specific, leading to substantial power savings.

4. Entire core when all tasks in all threads in the core are blocked or idle, or when no task has been scheduled onto any thread in a core, producing major power savings.

5. Trap unit of a core for hardware and software interrupts. The percentage of trap instructions for typical network processing SPECJBB applications on the UltraSPARC T1 is less than 1% of all instructions according to our observation. This implies that the trap unit is a good PSC. Note that even though the rest of the trap logic can be in a power saving mode most of the time, the trap-receiving input receiving queues will have to be always active, but the queue power dissipation is comparatively negligible.

6. DMA controllers for the L2 caches which control dataflow between the cache banks and the input-output buffers.

7. The instruction and data queues between the cores and the L2 cache banks.

8. Cache miss path logic which is activated only when there is a cache miss in on-chip L2 caches when off-chip cache or main memory has to be accessed.

10.5.2   Strategies to Save Power

Now consider the following autonomous hardware power saving schemes for the above PSCs: (i) power gating (data is not retained), (ii) clock gating (data is retained on normal operation), and (iii) DVFS (simultaneous voltage and frequency scaling). DVFS is only used for an entire core or for a chip level component like a DMS controller, the interconnect network, a cache bank, an input-output buffer or an on-chip computation unit like the FPU in Figure 10.3. However, power and clock gating can be done both for components inside a core and for chip level components. Figure 10.4 shows a proposed hierarchical power saving architecture at both intra-core (local power management) and global chip levels. Above the dashed line, the local power management unit (LPMU) operates inside a core, observes the content of the power status registers (PSRs) which are associated with different PSCs, executes a power saving algorithm, and modifies the value in the corresponding power control registers (PCRs) to activate or deactivate power saving. The PCR contents are read by the on-chip analog voltage and clock regulators which use that data to control DVFS, power gating and clock gating on the PSCs. Note that the LPMU does not directly control core-wide power savings like DVFS. Instead, the LPMU signals the global power management unit (GPMU) through core control status registers (CSRs), which in turn, implement core level power saving through core control registers (CCRs). The PSRs inside the core are updated by the trap logic and the decoder which signal the impending activation of the PSC when certain interrupts have to be serviced or certain instructions are decoded. Similarly, the PSCs themselves can update their PSRs to signal the impending power saving due to prolonged inactivity (idle or blocked status) which is better observed locally inside a core.

Below the dashed line and outside the cores, is the chip level GPMU which reads the on-chip sensor data on thermal hot spots and power-ground noise which are globally observable phenomena, and makes intelligent power saving decisions about the cores and other chip level components. The GPMU interacts with the cores and other components through core status registers (CSRs) and core control registers (CCRs). Core-wide power gating, clock gating and DVFS are controlled by the GPMU. Figure 10.5 shows details of the GPMU's interactions (CR and SR denote control and status registers respectively), while Tables 10.2 through 10.4 show possible contents of the CSRs (64 bits wide). Note that for the sake of this discussion, we logically treat any chip level component as a core.

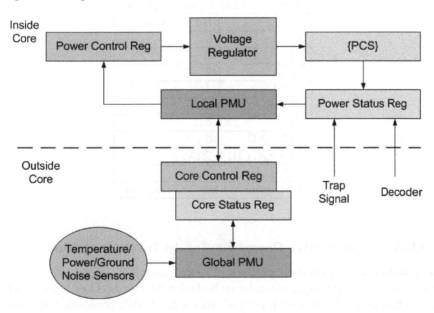

FIGURE 10.4: Architecture of autonomous hardware power saving logic.

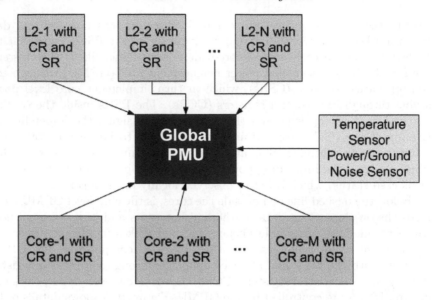

FIGURE 10.5: Global power management unit.

TABLE 10.2: Power Gating Status Register

Power Gating Status Register	
Field	Bit Position
MUL	0
DIV	1
MIL	2
LMQ	3
CORE	4
TLU	5
STRAND	6
STRAND_ID	7:15
CORE_ID	16:20
Remaining 43 bits are not used	

10.5.3   Case Study: Power Saving in Intel Centrino

A commercial embedded processor which partially implements the autonomous power management scheme is the Intel Centrino Core Duo [5], which is the first general-purpose chip multiprocessing (CMP) processor Intel has developed for the mobile market. The core was designed to achieve two main goals: (1) maximize the performance under the thermal limitation the platform allows, and (2) improve the battery life of the system relative to previous

TABLE 10.3: Clock Gating Status Register

Clock Gating Status Register	
Field	Bit Position
MUL	0
DIV	1
MIL	2
LMQ	3
CORE	4
TLU	5
STRAND	6
STRAND_ID	7:15
CORE_ID	16:20
Remaining 43 bits are not used	

TABLE 10.4: DVFS Status Register

DVFS Status Register	
Field	Bit Position
MUL	0:1
DIV	2:3
MIL	4:5
LMQ	6:7
CORE	8:9
TLU	10:11
STRAND	12:13
STRAND_ID	14:21
CORE_ID	22:29
Remaining 35 bits are not used	

generations of processors. Note that the OS views the Intel Core Duo processor as two independent execution units, and the platform views the whole processor as a single entity for all power management-related activities. Intel chose to separate the power management for a core from that of the full CPU and platform. This was achieved by making the power and thermal control unit part of the core logic and not part of the chipset as before. Migration of the power and thermal management flow into the processor allows the use of a hardware coordination mechanism in which each core can request any power saving state it wishes, thus allowing for individual core savings to be maximized. The CPU power saving state is determined and entered based on the lowest common denominator of both cores' requests, portraying a single CPU entity to the chipset power management hardware and flows. Thus, software can manage each core independently (based on the ACPI [6] protocol mentioned in Section 10.3), while the actual power management adheres to the platform and CPU shared resource restrictions. The ACPI power management protocol was not developed for complex multi-core processors with complex dependencies between cores, and their unpredictable effects on system level parameters (Section 10.1.1.2). Hence the need for developing new power management schemes which will better integrate hardware power saving logic with OS controlled scheduling in emerging multi-core embedded processors.

The Intel Core Duo processor is partitioned into three domains. The cores, their respective Level-1 caches, and the local thermal management logic operate individually as power management domains. The shared resources including the Level-2 cache, bus interface, and interrupt controllers, form yet another power management domain. All domains share a single power plane and a single-core PLL, thus operating at the same frequency and voltage levels. This is a fundamental restriction compared to our fine-granularity power saving scheme. However, each of the domains has an independent clock distribution (spine). The core spines can be gated independently, allowing the most basic per-core power savings. The shared-resource spine is gated only when both cores are idle and no shared operations (bus transactions, cache accesses) are taking place. If needed, the shared-resource clock can be kept active even when both cores' clocks are halted, thereby serving L2 snoops and interrupt controllers message analysis. The Intel Core Duo technology also introduces the enhanced power management features including dynamic L2 resizing; dynamically resizing/shutting down of the L2 cache is needed in preparation for DeepC4 state in order to achieve lower voltage idle state for saving power.

10.6 Power Saving Algorithms

10.6.1 Local PMU Algorithm

The pseudo code of a self-explanatory LPMU algorithm is proposed below (algorithm 1). The LPMU manages clock and power gating for intra-core components, and signals the GPMU for core-wide DVFS and power gating so that the GPMU can make globally optimal decisions. The given pseudo-codes are suggested templates for designers and they contain plug-and-play modular functions.

10.6.2 Global PMU Algorithm

Pseudo code of a proposed GPMU algorithm is outlined below (algorithm 2). Note that when thermal and power-ground noise sensor readings are greater than certain pre-determined thresholds, the GPMU will clock gate or DVFS certain cores.

Algorithm 1 Pseudo-Code for Local Power Management Unit

1: **while** *simulation* = *TRUE* **do**
2:   **for all** PSC_i **do**
3:     /* a set flag indicates that the PSC should be activated */
      $detect = \textbf{read\_trap\_decoder}(PSC_i)$;
4:     **if** $detect = TRUE$ **then**
5:       /* read the power control register of PSC_i */
        $\textbf{read\_reg}(PCR_i)$;
        /* if the PSC had power or clock gating active */
6:       **if** $\textbf{check\_PG\_CG}(PSC_i) = TRUE$ **then**
7:         /* initiate the deactivation of power-gating or clock-gating of PSC_i */
          $\textbf{wake\_up}(PSC_i)$;
8:       **end if**
9:     /* t_{access} is the average memory access time of entire Core – this is a locally observable and distinguishable phenomenon */
10:     **else if** $t_{access} <= T_{mem}$ **then**
11:       $\textbf{read\_reg}(PCR_i)$;
      /* when a core is in DVFS, all PSCs in the core will reflect it in their PCR contents */
12:       **if** $\textbf{check\_DVFS}(PSC_i) = TRUE$ **then**
13:         /* increases VF level of PSC_i; value of t_{access} determines the DVFS level*/
          $\textbf{speed\_up}(PSC_i, t_{access})$;
14:       **end if**
15:     /* detect = FALSE && $t_{access} > T_{mem}$ */
16:     **else**
17:       /* read the Power Status Register PSR_i */
        $\textbf{read\_reg}(PSR_i)$;
18:       **if** PSC_i has not been used in the last pre-determined T_{PG} clock cycles **then**
19:         /* this function starts the power gating process by writing appropriate codes into the power control register PCR_i */ $\textbf{start\_PG}(PSC_i)$;
20:       **else if** PSC_i has not been used in the last pre-determined T_{CG} clock cycles **then**
21:         /* This function starts the clock-gating process by writing appropriate codes into the power control register PCR_i */
          $\textbf{start\_CG}(PSC_i)$;
22:         /* Note that similar to power-gating, clock-gating cannot be done in 1 clock cycle in order to reduce p/g bounce. There should be a CG rate of x mA/s (switching current change per ns) */
23:         /* pre-determined time T_{mem} */
24:       **else if** average memory access time of entire Core ($t_{access} > T_{mem}$) **then**
25:         /* This function starts the DVFS process by signalling the GPMU which writes appropriate codes into the power control reg PCR_i; Value of t_{access} determines the DVFS level */
          $\textbf{start\_DVFS}(PSC_i, t_{access})$;
26:       **end if**
27:     **end if**
28:   **end for**
29:   $\textbf{advance\_simulation\_cycle}()$;
30: **end while**

The GPMU can also switch a core off (power gating) when the core is in idle status. We have introduced an element of fairness in the GPMU algorithms when there is a choice of cores to save power, the GPMU uses fairness by trying to prioritize power saving in cores which were not recently power saved. The fairness can also be dynamically determined in conjunction with the RTOS and its scheduler. Our algorithm also allows for an external power saving mode to extend battery life; the user can set a percentage reduction in battery power usage, and the GPMU can control DVFS in different cores accordingly.

A simple version of the estimate_sensitivity() function can be written to pick the variable which causes the smallest positive/negative change in power, and start/stop clock gating or DVFS for a single core in that simulation cycle. However, opportunities exist for better algorithms. Similarly, the reader is encouraged to write detailed pseudo codes for all the functions used in the pseudo codes above. Another on-going work we have is the integration of the above LPMU and GPMU algorithms with CASPER (as mentioned in Section 10.4.1). We are also in the process of generating pre-characterized power dissipation data (used by the function read_power_lib above) for different DVFS, CG and PG conditions in different cores for target applications. This data will be stored in memory, and read into the GPMU on boot-up. We are also investigating circuit designs for the sensors, voltage regulators and clock gate controls that are required for autonomous on-chip power saving. A key metric for evaluating any power management scheme will be to perform a cost-benefit analysis where the cost of optimization in extra area, power and delay of the power saving logic should be less than the benefits gained from power saving in the entire system. These trade-offs determine the granularity and methods of power savings embedded processors, the LPMU and the GPMU algorithms, and the integrated RTOS scheduling methods (as mentioned in Section 10.4).

10.7   Conclusions

Power and thermal management are becoming more challenging than ever before in all segments of computer-based systems. While in the server domain, the cost of electricity drives the need for low-power systems, in the embedded market, we are more concerned about battery life, thermal management and noise margins. The increasing demand for computational power in modern embedded applications has led to the trend of incorporating multi-core processors in emerging embedded platforms. These embedded applications require high frequency switching which leads to high power dissipation, thermal hot spots on chips, and power-ground noise resulting in data corruption and timing faults. On the other hand, cooling technology has failed to improve at the same rate at which power dissipation has been increasing, hence the need for aggressive on-chip power management schemes.

Algorithm 2 Pseudo-Code for Global Power Management Unit

Require: /* The library contains information about the power dissipations of all cores at all DVFS levels and core-level CG and PG conditions */
read_power_lib();

1: **while** $simulation = TRUE$ **do**
2:   **for all** $Core_j$ **do**
3:     **if** $status(Core_j) = IDLE$ **then**
4:       **start_PG**($Core_j$);
5:     **end if**
6:     **if** $status(Core_j) = READY$ **then**
7:       /* removes power-gating of $Core_j$ */
      wake_up($Core_j$);
8:     **end if**
9:   **end for**
10:   **if** **detect_external_power_saving**$() = TRUE$ **then**
11:     $PS =$ **calculate_power_to_save**();
12:     **if** $PS > small\_positive\_number$ **then**
13:       $Set\_of\_core\_DVFS =$ **select_cores_DVFS_levels**(PS);
14:       /* This function selects the minimum reduction in voltage and frequency levels for all cores or its subset, to satisfy the PS requirement. The goal is to have minimum speed impact in all cores by distributing the power saving over many cores. */
15:     **end if**
16:   **end if**
17:   **if** **new_sensor_reading_available**$()and($**read_sensor**$() > Threshold)$ **then**
18:     /* select the set of cores as power saving candidates in the vicinity of the sensor */
    $VCores =$ **select_PSC_cores**();
19:     /* This function has the following variables to change: (i) voltage and frequency levels in different cores for DVFS, (ii) clock gating different cores */
    $PS\_cores =$ **select_cores_CG_DVFS_levels**($VCores$);
20:     /* the goal is to estimate what change in a minimum set of variables (above) will cause the sensor reading to come down to just below but close to the threshold - this can be done using a function **estimate_sensitivity**() inside the function **select_cores_CG_DVFS_levels**($VCores$). */
21:   **else** **if** **new_sensor_reading_available**$()and($**read_sensor**$() < (Threshold - pre - determinedconstant))$ **then**
22:     /* select the set of cores as power saving candidates in the vicinity of the sensor */
    $VCores =$ **select_PSC_cores**();
23:     $PS\_cores = PS\_cores -$ **remove_cores_CG_DVFS_levels**($VCores$);
24:     /* This function changes – Voltage and Frequency levels for DVFS, and Clock-gating in different cores.The goal is to estimate what change in a minimum set of variables (above) will cause the sensor reading to go up to just below but close to the threshold - this can be done using the same function **estimate_sensitivity**() inside the function **remove_cores_CG_DVFS_levels**($VCores$). */
25:   **end if**
26:   **advance_simulation_cycle**();
27: **end while**

Moreover, modern embedded applications are characterized by sporadic processing requirements and unpredictable on-chip performance which make it extremely difficult to meet hard real-time constraints. These problems, coupled with complex interdependencies of multiple cores on-chip and their effects on system level parameters such as memory access delays, interconnect bandwidths, task context switch times and interrupt handling latencies, necessitate autonomous power management schemes. Future multi-core embedded processors will integrate on-chip hardware power saving schemes with on-chip sensing and hardware performance counters to be used by future RTOS. It is very likely that dynamic priority dynamic schedulers and auto-tuners will be integral components of future dynamic power management software. In this chapter we have presented the state-of-the-art in this area, described some on-going research that we are conducting, and suggested some future research directions. We have also described and provided links to CASPER, a top-down integrated simulation environment for future multi-core embedded systems.

Review Questions

[Q_1] Why are autonomous power management techniques necessary?

[Q_2] What is the advantage of run-time monitoring of system parameters?

[Q_3] What are the different techniques to save power in multi-core embedded platforms at the hardware level?

[Q_4] Discuss the different techniques to save power at the operating system level for embedded platforms.

[Q_5] What is smart caching?

[Q_6] How can scheduling affect the energy-delay product of an embedded multi-core processor?

[Q_7] Explain the power management features in existing power-managed RTOSs?

[Q_8] What is CASPER? What are the described power-saving features in CASPER?

[Q_9] What are the different power-saving techniques used in modern microprocessors?

[Q_10] What is the advantage of having a local power-management unit (LPMU) inside a core of a microprocessor ?

Bibliography

[1] http://www.theinquirer.net/inquirer/news/612/1049612/dell-talks-about-80-core-processor.

[2] http://en.wikipedia.org/wiki/PSOS.

[3] www.mentor.com/embedded.

[4] http://www.dedicated-systems.com/encyc/buyersguide/rtos/evaluations/docspreview.asp.

[5] www.intel.com/products/centrino.

[6] Advanced configuration and power interface specification. http://www.acpi.info.

[7] CASPER: CMT (chip multi-threading) architecture simulator for performance, energy and area analysis (SPARC V9 ISA). http://www.coe.uncc.edu/ kdatta/casper/casper.php.

[8] Chimera homepage:. http://www.cs.cmu.edu/ãml/chimera/chimera.html.

[9] Microware. www.microware.com.

[10] Mobile pentium iii processors - intel speedstep technology.

[11] Montavista embedded linux software. http://www.mvista.com.

[12] OpenSPARC community innovation awards contest . http://www.opensparc.net/community-innovation-awards-contest.html.

[13] Semiconductor industry association. the international technology roadmap for semiconductors (itrs). http://www.semichips.org.

[14] Tenasys. http://www.tenasys.com/products/irmx.php.

[15] Wind River VxWorks Platform 3.7. http://www.windriver.com/products/product-overviews/PO_VE_3_7_Platform_0109.pdf.

[16] Transmeta, 2002. Crusoe Processor Documentation.

[17] Sesc simulator, January 2005. http://sesc/sourceforege.net.

[18] S. Abedinpour, B. Bakkaloglu, and S. Kiaei. A multistage interleaved synchronous buck converter with integrated output filter in 0.18 um SiGe process. *IEEE Transactions on Power Electronics*, 22(6):2164–2175, Nov. 2007.

[19] L. Abeni and G. Buttazzo. Qos guarantee using probabilistic deadlines. *Proceedings of the 11th Euromicro Conference on Real-Time Systems, 1999*, pages 242–249, 1999.

[20] M. Alidina, J. Monteiro, S. Devadas, A. Ghosh, and M. Papaefthymiou. Precomputation-based sequential logic optimization for low power. *IEEE Transactions on Very Large Scale Integration (VLSI) Systems*, 2(4):426–436, Dec 1994.

[21] F. Balarin, L. Lavagno, P. Murthy, A. Sangiovanni-Vincentelli, C.D. Systems, and A. Sangiovanni. Scheduling for embedded real-time systems. *IEEE Design & Test of Computers*, 15(1):71–82, Jan-Mar 1998.

[22] Brannon Batson and T. N. Vijaykumar. Reactive-associative caches. In *PACT '01: Proceedings of the 2001 International Conference on Parallel Architectures and Compilation Techniques*, pages 49–60, Washington, DC, USA, 2001. IEEE Computer Society.

[23] Luca Benini and Giovanni de Micheli. System-level power optimization: techniques and tools. *ACM Trans. Des. Autom. Electron. Syst.*, 5(2):115–192, 2000.

[24] E. Bini and G.C. Buttazzo. Schedulability analysis of periodic fixed priority systems. *Computers, IEEE Transactions on*, 53(11):1462–1473, Nov. 2004.

[25] S. Borkar. Design challenges of technology scaling. *Micro, IEEE*, 19(4):23–29, Jul-Aug 1999.

[26] B. Brock and K. Rajamani. Dynamic power management for embedded systems [soc design]. In *Proceedings of IEEE International Systems-on-Chip (SoC) Conference, 2003*, pages 416–419, Sept. 2003.

[27] David Brooks and Margaret Martonosi. Value-based clock gating and operation packing: dynamic strategies for improving processor power and performance. *ACM Trans. Comput. Syst.*, 18(2):89–126, 2000.

[28] B. Calder, D. Grunwald, and J. Emer. Predictive sequential associative cache. In *Proceedings of 2nd International Symposium on High-Performance Computer Architecture, 1996*, pages 244–253, Feb 1996.

[29] Poki Chen, Chun-Chi Chen, Chin-Chung Tsai, and Wen-Fu Lu. A time-to-digital-converter-based CMOS smart temperature sensor. *IEEE Journal of Solid-State Circuits*, 40(8):1642–1648, Aug. 2005.

[30] F. de Jong, B. Kup, and R. Schuttert. Power pin testing: making the test coverage complete. *Proceedings of International Test Conference, 2000*, pages 575–584, 2000.

[31] R. Gonzalez and M. Horowitz. Energy dissipation in general purpose microprocessors. *Journal of Solid-State Circuits, IEEE*, 31(9):1277–1284, Sep 1996.

[32] M.K. Gowan, L.L. Biro, and D.B. Jackson. Power considerations in the design of the alpha 21264 microprocessor. *Proceedings of Design Automation Conference, 1998.*, pages 726–731, Jun 1998.

[33] Flavius Gruian. Hard real-time scheduling for low-energy using stochastic data and DVS processors. In *ISLPED '01: Proceedings of the 2001 International Symposium on Low Power Electronics and Design*, pages 46–51, New York, NY, USA, 2001. ACM.

[34] Flavius Gruian and Krzysztof Kuchcinski. Lenes: task scheduling for low-energy systems using variable supply voltage processors. In *Proceedings of the 2001 Conference on Asia South Pacific Design Automation (ASP-DAC '01)*, pages 449–455, New York, NY, USA, 2001. ACM.

[35] P. Hazucha, G. Schrom, Jaehong Hahn, B.A. Bloechel, P. Hack, G.E. Dermer, S. Narendra, D. Gardner, T. Karnik, V. De, and S. Borkar. A 233-mhz 80%-87% efficient four-phase dc-dc converter utilizing air-core inductors on package. *Journal of Solid-State Circuits, IEEE*, 40(4):838–845, April 2005.

[36] Inki Hong, Miodrag Potkonjak, and Mani B. Srivastava. On-line scheduling of hard real-time tasks on variable voltage processor. In *Proceedings of the 1998 IEEE/ACM International Conference on Computer-Aided Design*, pages 653–656, New York, NY, USA, 1998. ACM.

[37] Chung-Hsing Hsu and Ulrich Kremer. The design, implementation, and evaluation of a compiler algorithm for CPU energy reduction. In *Proceedings of the ACM SIGPLAN 2003 Conference on Programming Language Design and Implementation*, pages 38–48, New York, NY, USA, 2003. ACM.

[38] Canturk Isci, Alper Buyuktosunoglu, Chen-Yong Cher, Pradip Bose, and Margaret Martonosi. An analysis of efficient multi-core global power management policies: Maximizing performance for a given power budget. *39th Annual IEEE/ACM International Symposium on Microarchitecture*, pages 347–358, Dec. 2006.

[39] T. Ishihara and H. Yasuura. Voltage scheduling problem for dynamically variable voltage processors. *International Symposium on Low Power Electronics and Design*, pages 197–202, Aug 1998.

[40] E.D. Jensen and J.D. Northcutt. Alpha: a nonproprietary OS for large, complex, distributed real-time systems. In *Proceedings of IEEE Workshop on Experimental Distributed Systems, 1990*, pages 35–41, Oct 1990.

[41] Wonyoung Kim, M.S. Gupta, Gu-Yeon Wei, and D. Brooks. System level analysis of fast, per-core DVFS using on-chip switching regulators. In *IEEE 14th International Symposium on High Performance Computer Architecture, 2008*, pages 123–134, Feb. 2008.

[42] Seongsoo Lee and Takayasu Sakurai. Run-time voltage hopping for low-power real-time systems. In *Proceedings of the 37th Conference on Design Automation*, pages 806–809, New York, NY, USA, 2000. ACM.

[43] Yongpan Liu, Huazhong Yang, R.P. Dick, H. Wang, and Li Shang. Thermal vs energy optimization for dvfs-enabled processors in embedded systems. In *Proceedings of 8th International Symposium on Quality Electronic Design*, pages 204–209, March 2007.

[44] S. A. Mackay, W. M. Gentleman, D. A. Stewart, and M. Wein. Harmony as an object-oriented operating system. Technical report, SIGPLAN Notices, 1989.

[45] P. Macken, M. Degrauwe, M. Van Paemel, and H. Oguey. A voltage reduction technique for digital systems. In *Digest of Technical Papers IEEE International Solid-State Circuits Conference, 1990*, pages 238–239, Feb. 1990.

[46] D. Marculescu. On the use of microarchitecture-driven dynamic voltage scaling. In *Proceedings of Workshop on Complexity-Effective Design, in Conjunction with Intl. Symp. on Computer Architecture (ISCA)*, 2000.

[47] Daniel Mosse, Hakan Aydin, Bruce Childers, and Rami Melhem. Compiler-assisted dynamic power-aware scheduling for real-time applications. In *Proceedings of Workshop on Compilers and Operating Systems for Low Power (COLP)*, October 2000.

[48] S. Narasimhan, D.M. Siegel, and J.M. Hollerbach. Condor: a revised architecture for controlling the utah-mit hand. In *IEEE International Conference on Robotics and Automation, 1988*, pages 446–449, vol. 1, Apr 1988.

[49] T. Pering, T. Burd, and R. Brodersen. The simulation and evaluation of dynamic voltage scaling algorithms. In *Proceedings of International Symposium on Low Power Electronics and Design, 1998*, pages 76–81, Aug 1998.

[50] Michael D. Powell, Amit Agarwal, T. N. Vijaykumar, Babak Falsafi, and Kaushik Roy. Reducing set-associative cache energy via way-prediction and selective direct-mapping. In *Proceedings of the 34th Annual ACM/IEEE International Symposium on Microarchitecture*, pages 54–65, Washington, DC, USA, 2001. IEEE Computer Society.

[51] Jan M. Rabaey, Anantha Chandrakasan, and Borivoje Nikolic. *Digital Integrated Circuits.* 2nd ed., Saddle River, N.J., Prentice Hall, January 2003.

[52] C. Saldanha and M. Lipasti. *Power Efficient Cache Coherence (High Performance Memory Systems).* Madison, USA, Springer-Verlag, 2003.

[53] L. Salkind. The SAGE operating system. In *Proceedings of IEEE International Conf. on Robotics and Automation, 1989,* pages 860–865, vol. 2, May 1989.

[54] G. Schrom, P. Hazucha, J. Hahn, D.S. Gardner, B.A. Bloechel, G. Dermer, S.G. Narendra, T. Karnik, and V. De. A 480-mhz, multi-phase interleaved buck dc-dc converter with hysteretic control. In *IEEE 35th Annual Power Electronics Specialists Conference,* pages 4702–4707, Vol. 6, Jun 2004.

[55] G. Semeraro, G. Magklis, R. Balasubramonian, D.H. Albonesi, S. Dwarkadas, and M.L. Scott. Energy-efficient processor design using multiple clock domains with dynamic voltage and frequency scaling. In *Proceedings of 8th International Symposium on High-Performance Computer Architecture, 2002,* pages 29–40, Feb. 2002.

[56] Dongkun Shin, Jihong Kim, and Seongsoo Lee. Intra-task voltage scheduling for low-energy hard real-time applications. *IEEE Design & Test of Computers,* 18(2):20–30, Mar/Apr 2001.

[57] T. Simunic, L. Benini, A. Acquaviva, P. Glynn, and G. de Micheli. Dynamic voltage scaling and power management for portable systems. In *Proceedings of Design Automation Conference, 2001,* pages 524–529, 2001.

[58] J. A. Stankovic and K. Ramamritham. The design of the spring kernel. In *Tutorial: hard real-time systems,* pages 371–382, Los Alamitos, CA, USA, 1989. IEEE Computer Society Press.

[59] A.G.M. Strollo, E. Napoli, and D. De Caro. New clock-gating techniques for low-power flip-flops. In *Proceedings of the 2000 International Symposium on Low Power Electronics and Design,* pages 114–119, 2000.

[60] K.T. Tang and E.G. Friedman. Simultaneous switching noise in on-chip CMOS power distribution networks. *Very Large Scale Integration (VLSI) Systems, IEEE Transactions,* 10(4):487–493, Aug 2002.

[61] V. Tiwari, S. Malik, and P. Ashar. Guarded evaluation: pushing power management to logic synthesis/design. *IEEE Transactions on Computer-Aided Design of Integrated Circuits and Systems,* 17(10):1051–1060, Oct 1998.

[62] V. Tiwari, D. Singh, S. Rajgopal, G. Mehta, R. Patel, and F. Baez. Reducing power in high-performance microprocessors. In *Proceedings of Design Automation Conference, 1998*, pages 732–737, June 1998.

[63] Hideyuki Tokuda, Tatsuo Nakajima, and Prithvi Rao. Real-time Mach: towards a predictable real-time system. In *Proceedings of USENIX Mach Workshop*, pages 73–82, 1990.

[64] J. Wibben and R. Harjani. A high efficiency dc-dc converter using 2nh on-chip inductors. In *Proceedings of IEEE Symposium on VLSI Circuits, 2007.*, pages 22–23, June 2007.

[65] Steven Cameron Woo, Moriyoshi Ohara, Evan Torrie, Jaswinder Pal Singh, and Anoop Gupta. The SPLASH-2 programs: characterization and methodological considerations. In *Proceedings of the 22nd Annual International Symposium on Computer Architecture*, pages 24–36, New York, NY, USA, 1995. ACM.

[66] Qiang Wu, P. Juang, M. Martonosi, and D.W. Clark. Voltage and frequency control with adaptive reaction time in multiple-clock-domain processors. In *Proceedings of 11th International Symposium on High-Performance Computer Architecture*, pages 178–189, Feb. 2005.

[67] Fen Xie, Margaret Martonosi, and Sharad Malik. Compile-time dynamic voltage scaling settings: opportunities and limits. In *Proceedings of the ACM SIGPLAN 2003 Conference on Programming Language Design and Implementation*, pages 49–62, New York, NY, USA, 2003. ACM.

[68] S. Yang, M.D. Powell, B. Falsafi, K. Roy, and T.N. Vijaykumar. An integrated circuit/architecture approach to reducing leakage in deep-submicron high-performance i-caches. In *The 7th International Symposium on High-Performance Computer Architecture*, pages 147–157, 2001.

[69] F. Yao, A. Demers, and S. Shenker. A scheduling model for reduced cpu energy. In *Proceedings of 36th Annual Symposium on Foundations of Computer Science*, pages 374–382, Oct. 1995.

[70] Heng Zeng, Carla S. Ellis, Alvin R. Lebeck, and Amin Vahdat. Ecosystem: managing energy as a first class operating system resource. *SIGPLAN Not.*, 37(10):123–132, 2002.

11

Multi-Core System-on-Chip in Real World Products

Gajinder Panesar, Andrew Duller, Alan H. Gray and Daniel Towner

picoChip Designs Limited
Bath, UK
{gajinder.panesar, andy.duller, alan.gray, daniel.towner}@picochip.com

CONTENTS

11.1	Introduction	370
11.2	Overview of picoArray Architecture	371
	11.2.1 Basic Processor Architecture	371
	11.2.2 Communications Interconnect	373
	11.2.3 Peripherals and Hardware Functional Accelerators	373
	11.2.3.1 Host Interface	373
	11.2.3.2 Memory Interface	374
	11.2.3.3 Asynchronous Data/Inter picoArray Interfaces	374
	11.2.3.4 Hardware Functional Accelerators	374
11.3	Tool Flow	375
	11.3.1 picoVhdl Parser (Analyzer, Elaborator, Assembler)	376
	11.3.2 C Compiler	376
	11.3.3 Design Simulation	378
	11.3.3.1 Behavioral Simulation Instance	379
	11.3.4 Design Partitioning for Multiple Devices	381
	11.3.5 Place and Switch	381
	11.3.6 Debugging	381
11.4	picoArray Debug and Analysis	381
	11.4.1 Language Features	382
	11.4.2 Static Analysis	383
	11.4.3 Design Browser	383
	11.4.4 Scripting	385
	11.4.5 Probes	387

     11.4.6   FileIO 387

11.5   Hardening Process in Practice 388

     11.5.1   Viterbi Decoder Hardening 389

11.6   Design Example 392

11.7   Conclusions 396

Review Questions 396

Bibliography 397

11.1   Introduction

In a field where no single standard exists, wireless communications systems are typically designed using a mixture of DSPs, FPGAs and custom ASICs, resulting in systems that are awkwardly parallel in nature. Due to the fluid nature of standards, it is very costly to enter the market with a custom ASIC solution. What is required is a scalable, programmable solution which can be used in most, if not all, areas. To this end picoChip created the picoArray™ and a rich toolset.

picoChip has produced several generations of devices based around the picoArray. These range from devices which may be connected to form systems containing many thousands of processors, for use in macrocell wireless base stations, to system-on-chip devices deployed in femtocells.

When devices are deployed in consumer equipment they come under increasing cost pressure; the final BoM (bill of materials) of a system can determine success or failure in a new market.

picoChip has addressed this by exploiting its architecture; functions which, in one generation, are implemented in software using a number of processors, are hardened in a subsequent generation but maintain the same programming paradigm. Three generations of device have been produced:

- First generation PC101: 430 programmable processors.

- Second generation PC102 [5]: 308 programmable processors, 14 accelerators for wireless specific operations such as correlation. Independent evaluation of this device as used in an OFDM system can be found in [1].

- Third generation PC20x [6][7][8]: A family of 3 devices with 273 programmable processors, an optional embedded host processor and 9 accelerators for a variety of DSP and wireless operations such as FFT, encryption and turbo decoding.

This chapter starts by describing the picoArray architecture which underlies all of these devices. Subsequently the development tool flow that has been created to support multi-device systems is explained together with the tools and methods that are needed to debug and analyse such systems. The specific example of a Viterbi decoder block is used to demonstrate the process that has been used to move from a fully programmable device (PC101) to a hybrid programmable/hardened device (PC20x). Finally, the use of the PC20x device, in a femtocell wireless access point, is used as a design example of how picoArray devices have been used to realize real-world products.

11.2   Overview of picoArray Architecture

The heart of all picoChip's devices is the picoArray. The picoArray is a tiled processor architecture in which many hundreds of heterogeneous processors are connected together using a deterministic interconnect. The interconnect consists of bus switches joined by picoBus™ connections. Each processor is connected directly to the picoBus above and below it. An enlarged view of part of the interconnect is shown in Figure 11.1. To simplify the diagram only two of the four vertical bus connections are shown.

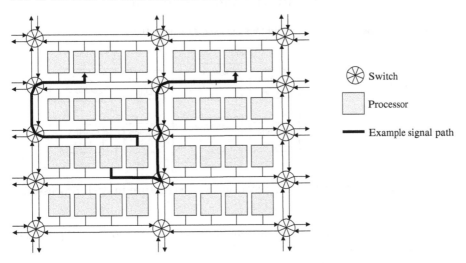

FIGURE 11.1: picoBus interconnect structure.

In fact, the picoBus is used to connect a variety of entities together and these can be processors, peripherals and accelerator blocks, all of which are referred to as array elements (AEs).

11.2.1   Basic Processor Architecture

There are three RISC processor variants, which share a common core instruction set, but have varying amounts of memory and differing additional

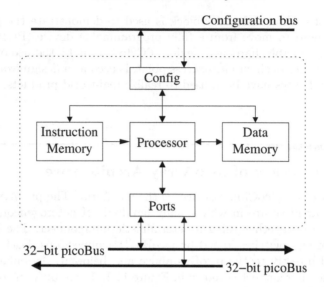

FIGURE 11.2: Processor structure.

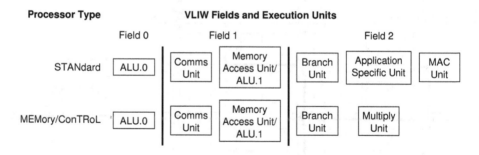

FIGURE 11.3: VLIW and execution unit structure in each processor.

instructions to implement specific wireless baseband control and digital signal processing functions.

Each of the processors in the picoArray is 16-bit, and uses 3-way VLIW scheduling. The basic structure of the processor is shown in Figure 11.2. Each processor has its own small memory, which is organized as separate data and instruction banks (i.e., Harvard architecture). The processor contains a number of communication ports, which allow access to the interconnect buses through which it can communicate with other processors. Each processor is programmed and initialized using a special configuration bus. The processors have a very short pipeline which helps programming, particularly at the assembly language level. The architecture of the three processor variants (STAN, MEM and CTRL) is shown in Figure 11.3.

11.2.2 Communications Interconnect

Within the picoArray, processors are organized in a two-dimensional grid, and communicate over a network of 32-bit unidirectional buses (the picoBus) and programmable bus switches. The physical interconnect structure is shown in Figure 11.1. The processors are connected to the picoBus by ports which contain internal buffering for data. These act as nodes on the picoBus, and provide a simple processor interface to the bus based on *put* and *get* instructions. The processors are essentially independent of the ports unless they specifically use a *put* or a *get* instruction.

The inter-processor communication protocol implemented by the picoBus is based on a time division multiplexing (TDM) scheme. There is no run-time bus arbitration, so communication bandwidth is guaranteed. Data transfers between processor ports occur during specific time slots, scheduled in software, and controlled using the bus switches. Figure 11.1 shows an example in which the switches have been set to form two different signals between processors. Signals may be point-to-point or point-to-multi-point. Data transfer will not take place until all the processor ports involved in the transfer are ready.

Communication time slots throughout the picoBus architecture are allocated according to the bandwidth required. Faster signals are allocated time-slots more frequently than slower signals. The user specifies the required bandwidth for a signal by giving a rate at which the signal must communicate data. For example, a transfer rate might be described as @4, which means that every fourth time-slot has been allocated to that transfer.

The default signal transfer mode is synchronous; data is not transferred until both the sender and receiver ports are ready for the transfer. If either is ready before the other then the transfer will be retried during the next available time slot. If, during a *put* instruction no buffer space is available then the processor will sleep (hence reducing power consumption) until space becomes available. In the same way, if during a *get* instruction there is no data available in the buffers then the processor will also sleep. Using this protocol ensures that no data can be lost.

11.2.3 Peripherals and Hardware Functional Accelerators

In addition to the general purpose processors, there are a number of other AEs that are connected to the picoBus. The following set of peripherals and hardware functional accelerators can serve as parts of a picoArray:

11.2.3.1 Host Interface

The Host or microprocessor interface is used to configure the picoArray device and to transfer data to and from the picoArray device using either a register transfer method or a DMA mechanism. The DMA memory-mapped interface has a number of ports mapped into the external microprocessor memory area. Two ports are connected to the configuration bus within the picoArray and the others are connected to the internal picoBus. These enable the external microprocessor to communicate with the internal AEs using signals on the picoBus.

11.2.3.2 Memory Interface

Each processor in the picoArray has local memory for data and instruction storage. However, an external memory interface is provided to supplement the on-chip memory. This interface allows processors within the core of the picoArray to access external memory across the internal picoBus.

11.2.3.3 Asynchronous Data/Inter picoArray Interfaces

These interfaces can be configured in one of two modes: either the inter picoArray interface (IPI) mode or the asynchronous data interface (ADI) mode. The choice of interface mode is made for each interface separately during device configuration.

- Inter picoArray interface

  The IPI interfaces are bidirectional and designed to allow each picoArray to exchange data with other picoArrays through their IPIs. Using this feature, multiple picoArray devices can be connected together to implement highly complex and computationally intensive signal processing systems. The IPI interface operates in full duplex, sending and receiving 32-bit words. The 32-bit words on the on-chip picoBus are multiplexed as two 16-bit data on the interface itself.

- Asynchronous data interface

  The asynchronous data interface (ADI) allows data to be exchanged between the internal picoBus and external asynchronous data streams such as those input and output by data converters or control signals between the base band processor and the RF section of a wireless base station.

11.2.3.4 Hardware Functional Accelerators

The first generation device employed no functional accelerators and all the AEs were programmable. This flexibility had enormous advantages when systems were developed for wireless standards which were in a state of flux, and the main goal was to provide the required functionality in the shortest time.

In subsequent generations of device however, considerations of cost and power consumption increased in importance relative to flexibility. Therefore, the decision was taken to provide optimized hardware for some important functions, whose definition was sufficiently stable and where the performance gain was substantial. For example, in the second generation device, the PC102, this policy led to the provision of multiple instances of a single accelerator type, called a functional accelerator unit (FAU), which was designed to support a variety of correlation and error correction algorithms.

For the third generation device, the PC20x, a wider range of functions were hardened, but fewer instances of each accelerator were provided, as this device family focused on a narrower range of applications and hence the requirements were more precisely known. Examples of functions which have been hardened into array elements are fast Fourier transforms, Reed-Solomon decoders, and Viterbi decoders.

11.3 Tool Flow

The picoArray is programmed using picoVhdl, which is a mixture of VHDL [10], ANSI/ISO C and a picoChip-specific assembly language. The VHDL is used to describe the structure of the overall system, including the relationship between processes and the signals which connect them together. Each individual process is programmed in conventional C or in assembly language. A simple example is given below.

```
entity Producer is              -- Declare a producer process
   port (channel:out integer32@8);   -- 32-bit output signal
end entity Producer;            --    with @8 rate

5  architecture ASM of Producer is    -- Define the 'Producer' in ASM
   begin MEM                    -- use a 'MEM' processor type
      CODE                      -- Start code block
         COPY.0 0,R0 \ COPY.1 1,R1   -- Note use of  VLIW
      loopStart:
10       PUT R[0,1],channel \ ADD.0 R0,1,R0  -- Note communication
         BRA loopStart
      ENDCODE;
   end;                         -- End Producer definition.

15 entity Consumer is           -- Declare a consumer
      port (channel:in integer32@8);  -- 32-bit input signal
   end;

   architecture C of Consumer is    -- Define the 'Consumer' in C
20 begin STAN                   -- Use a 'STAN' processor
      CODE
      long array[10];           -- Normal C code

      int main() {              -- 'main' function - provides
25       int i = 0;             --    entry point

         while (1) {
```

```
          array[i] = getchannel();        -- Note use of communication.
          i = (i + 1) % 10;
30    }

      return 0;
    }
    ENDCODE;
35  end Consumer;                          -- End Consumer definition

    use work.all;                          -- Use previous declarations
    entity Example is                      -- Declare overall system
    end;
40
    architecture STRUCTURAL of Example is  -- Structural definition
      signal valueChannel: integer32@8;    -- One 32-bit signal...
    begin
        producerObject: entity Producer    -- ...connects Producer
45         port map (channel=>valueChannel);
        consumerObject: entity Consumer    -- ...to Consumer
           port map (channel=>valueChannel);
    end;
```

The toolchain converts the input picoVhdl into a form suitable for execution on one or more picoArray devices. It comprises a compiler, an assembler, a VHDL parser, a design partitioning tool, a place-and-switch tool, a cycle-accurate simulator and a debugger. The relationship of these components is shown in Figure 11.4. The following sections briefly examine each of these tools in turn.

11.3.1   picoVhdl Parser (Analyzer, Elaborator, Assembler)

The VHDL parser is the main entry point for the user's source code. A complete VHDL design is given to the parser, which coordinates the compilation and assembly of the code for each of the individual processes. An internal representation of the machine code for each processor and its signals is created. Static source code analysers may also be run at this to detect common coding issues.

11.3.2   C Compiler

The C compiler is an official port of the GNU compiler collection (GCC) [15]. The compiler is invoked by the elaborator whenever a block of C code is encountered in the VHDL source code. It is not simply a question of invoking the compiler on the block of code contained between the VHDL's CODE/ENDCODE since there are several ways in which the C code must be coupled to the VHDL environment in which it operates:

VHDL Types: VHDL allows types to be named and created in the source itself. The elaborator is responsible for making these types available in

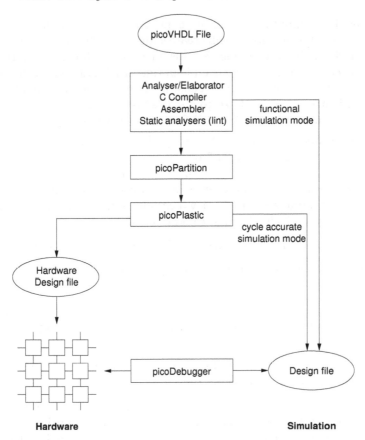

FIGURE 11.4: Tool flow.

the C code. This is achieved by creating equivalent C type definitions for each VHDL type and passing these to the compiler, along with the source code itself.

VHDL Constants and Generics: VHDL source code allows constants and generics to be defined for each entity (a generic is a constant whose value is defined on a per-entity basis). As with types, the elaborator generates appropriate C definitions for each constant and generic and passes these to the compiler, along with the source code itself. In addition, since generics are often used as a way of parameterising the entity's source code, C pre-processor #define statements can also be generated to allow conditional compilation within the C source code.

Ports: To enable C code to communicate over the signals associated with a VHDL entity, the elaborator creates a set of special port functions. Consider the example source code given at the start of this section. On line 16 an input signal called `channel` is created. To allow this signal to be used, the elaborator will create a function called `getchannel`, a call to which can be seen on line 28. This function is defined to be `inline` and it will call a special compiler built-in (intrinsic) mapping to the underlying assembly instruction `GET`. Thus, port communication functions can be efficiently compiled down to a single instruction, and these instructions can even be VLIW scheduled for further efficiency gains.

VHDL allows arrays of ports to be defined, as well as individual ports. In assembly language, `GET` or `PUT` instructions must be hard-coded to specific port numbers, and the user cannot dynamically select which port to use. In C however, when arrays of ports are used, a special support library is provided which allows the code to efficiently index ports dynamically, leading to more flexibility.

In addition to the above features, the C compiler also provides:

- Built-in (intrinsic) functions for accessing the various special instructions that the processors provide

- A wide range of hand-written assembly libraries for efficiently performing memory operations (string operations, memory copies and moves, etc.)

- Support for division, and higher bit-width arithmetic, which is not available through the 16-bit instruction set

The GCC compiler on which the picoChip compiler is based is designed primarily for 32-bit general purpose processors capable of using large amounts of memory. Supporting 16-bit embedded processors with only a few kilobytes of memory is a challenge. However, the compiler is able to do a surprisingly good job even under these constraints.

11.3.3 Design Simulation

The simulator is capable of simulating any design file, and can operate in two major modes:

Functional: In this mode the user's design is simulated without reference to the physical placement of processes on processors, or signals on buses. The communication across the signals is assumed to be achievable in a single clock cycle and there is no limit to the number of AEs that can comprise a system. In addition, each AE is capable of using the maximum amount of instruction memory (64 KB since they are 16-bit processors). These attributes mean that such simulations need not be executable on picoArray hardware. The importance of this mode is twofold. Firstly, to allow exploration of algorithms prior to decomposing the design to make it amenable for use on a picoArray. Secondly, to allow the "hardening" process to be explored (see Section 11.5).

Back-annotated: This mode allows the modeling of a design once it has been mapped to a real hardware system. This can consist of a number of picoArray devices connected via IPI connections. In this case, the simulation of the design will have knowledge of the actual propagation delays across the picoBus and will also model the delays inherent in the IPI connections between devices.

The simulator core is cycle-based, with simulated time advancing in units of the 160 MHz clock. For the programmable AEs in the system, the simulation accurately models the processing of the instructions, including the cycle-by-cycle behavior of the processor's pipeline, and communications over the picoBus through the hardware ports. A crucial aspect of the simulation system is the provision of "behavioral simulation instances" (BSIs) which allow arbitrary blocks which connect to the picoBus to be simulated together with the standard programmable AEs. Their operation is detailed in the next section. The simulator is controlled through the debugger interface, which is described in Section 11.3.6.

Importantly, the same simulation system was used to provide a "golden reference" during the design and verification of all picoArray devices.

11.3.3.1   Behavioral Simulation Instance

A behavioral simulation instance (BSI) has a number of purposes. It allows an arbitrary behavioral model to be encapsulated within a framework which allows connection directly to the picoBus. Its uses are

- To model hardware peripheral blocks (i.e., non programmable AE instances such as an external memory interface)

- To model hardware functional accelerators (HFAs, see Section 11.2.3.4)

- To allow arbitrary input/output to be used as part of a simulation (e.g., FileIO, see Section 11.4.6)

- To allow users to build custom blocks within the simulation

A BSI is an instance of a C++ class which provides a model of an abstract function in a form which can be used as part of a cycle-based simulation. In its most basic form a BSI comprises a core C++ function called from an interface layer which models its communication with the picoBus via hardware ports, as shown in Figure 11.5. It is created from a picoVhdl entity containing C++ code sections which describes the construction and initialization of the instance, and its behavior when the simulation is clocked. The C++ has access to the data from the hardware port models via communication functions similar to those provided by the C compiler. A program generator combines these individual code sections with "boilerplate" code to form the complete C++ class.

The following example is about the most trivial useful BSI it is possible to produce. Its function is to accept null-terminated character strings on an

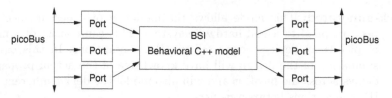

FIGURE 11.5: Behavioral simulation instance

input port and send them to the standard output of the simulation, each string being stamped with the simulation time at which its first bytes were received and with a name which identifies the instance receiving it.

```
   entity Console is
     generic (name:string:="CONSOLE"; -- Identifier for the messages
              slotRate:integer:=2);   -- rate of the input signal
     port (data:in integer32@slotRate);
 5 end entity Console;

   architecture BEHAVIOURAL of Console is
   begin NONE
     SIM_DATA CODE
10   std::string output;
     uint64_t latchCycles; // Remembers start time of message
     ENDCODE;

     SIM_START CODE
15   output = "";
     latchCycles = 0;
     ENDCODE;

     SIM_CASE data CODE
20   if (output.empty())
       latchCycles = getSimTime();
     integer32 data = getdata();
     for (int i=0; i<4; ++i)
     {
25     char c = (data >> (i * 8)) & 0xff;
       if (c == 0)
       {
         std::cout << latchCycles << ": " << name << ": " << output << std::endl;
         output.clear();
30       latchCycles = getSimTime();
       }
       else
         output.push_back (c);
     }
35   ENDCODE;

   end Console;
```

The C++ code at lines 10-11 of the example defines the member data which each instance will have, and the code at lines 15-16 initializes this data at the start of simulation. The code at lines 20-34 is called every time data is available in the buffers of the input hardware port. The call to the communication function 'getdata' at line 22 reads an item from the port.

11.3.4 Design Partitioning for Multiple Devices

If a design requires more processors than are available in a single picoArray, the design must be partitioned across multiple devices. This process is mostly manual, with the user specifying which AEs map to which device. This means that some signals will have to be present on multiple devices and an IPI will have to be used to handle the off-chip communication. However, the partitioning tool automatically performs this task and uses an appropriate pair of IPIs (one from each of the devices involved) to support the communication.

11.3.5 Place and Switch

Once a design has been partitioned between chips an automatic process akin to place-and-route in an ASIC design has to be performed for each device. This assigns a specific processor to each AE instance in the design and routes all of the signals which link AE instances together. The routing must use the given bandwidth requirements of signals specified in the original design. The routing algorithm should also address the power requirements of a design by reducing the number of bus segments that signals have to traverse, enabling unused bus segments to be switched off. This process is performed using the picoPlastic (PLace And Switch To IC) tool.

When a successful place-and-switch has been achieved, a hardware design file is produced, which contains the necessary information required to load the design onto real hardware.

11.3.6 Debugging

The debugger allows an entire design to be easily debugged, either as a simulation or using real hardware. The debugger supports common symbolic debugging operations such as source code stepping, variable views, stack traces, conditional breakpoints and much more. The debugger also supports a range of more specialized tools for debugging and analyzing large-scale multi-core systems, and these are discussed in greater detail in Section 11.4.

11.4 picoArray Debug and Analysis

The debug and analysis of parallel systems containing perhaps thousands of processors requires specialized tool support. This support can be broadly split into two classes: static and dynamic. Static support includes language features designed to prevent bugs from being introduced in the first place, or analysis modes in the assembler, compiler and debugger to visualize how data flows through the system. Dynamic support includes features such as ways of viewing the activity in the system, or being able to 'probe' communications signals to see when data is transferred between processes. We now discuss the main features provided by the tools to aid debug and analysis.

11.4.1 Language Features

The language features aid debug through three main features: strong type checking, fixed process creation, and bandwidth allocation. These features are designed to prevent bugs from being introduced into the source code.

Strong type checking is used to ensure that whenever data is communicated from one process to another, the data will be interpreted by both producer and consumer in the same way. Types are selected from a library of built-in types, or by the users defining their own types. At the structural level, processes will be defined with ports of specific types, and they will be connected with signals which must match the port types. Within a process, any data which is "put" or "get" from a port must be of the correct type. For processes written in C, this is achieved by synthesising the available types using C encoding rules (e.g., using typedef's, union's, and struct's), and hence tying into the C compiler's type system. Thus, end-to-end communication of data can only occur when all processes and signals agree on the type format. This makes integration of independently developed components easy since any discrepancies in type formats will be detected at compile time, when they are easily fixed.

The structural VHDL used to define a system requires the number of processes, and their interconnections to be fixed at compile time. During compilation, the tools will allocate each process to its own processor and schedule the signals connecting the processes onto the picoBus interconnection fabric. Because of this compile-time scheduling, non-deterministic run-time effects, such as process scheduling and bus contention, have been eliminated. This makes it easier to integrate systems. If problems are found, it also makes the reproduction of the problems, their debugging, and the verification of their fixes easier.

In addition to specifying fixed signals connecting processes, the signals are also allocated bandwidth. This is achieved using a language notation which allows the frequency of communication over the signal to be specified. Processes requiring high signal bandwidths will use high frequencies (e.g., every 4 cycles), while processes requiring low bandwidth will use low frequencies (e.g., every 1024 cycles).

11.4.2   Static Analysis

The picoTools provide two types of static analysis features: automatic static code analysis (commonly known as 'linting'), and visualization of static information to aid program comprehension.

Static source code analyzers are used to analyze the original source code to try to find bugs or problems without actually executing the code. The analysis of C source code in picoTools is performed using the commercially available FlexeLint tool [14], which is invoked during the compilation stage. The analysis of assembly source code is performed by the assembler itself. In addition to the normal checks that an assembler would perform, the assembler also builds extra data structures, commonly found in compilers, such as control-flow graphs, def-use chains, and data-dependence graphs [12]. From these it can perform a set of checks for common problems, such as overwriting unused values or detecting when conditional branches are never taken. No commercial static analyzer is available for assembly source code, so this tool is specific to picoChip.

Visualization of static program information can help the user understand why the program behaves as it does [16]. The debugger provides several GUIs which aid the programmer in this way. For example, suppose a register in an assembly program contains an unexpected value. The debugger allows the user to see all places where that value may have been written by using the 'where-defined' analysis illustrated in Figure 11.6. This feature is implemented using data-flow analysis techniques, allowing it to be very accurate. The alternative way of finding where a data value might be defined would be by textual search, but this would show all occurrences of the particular register, rather than those which actually contribute to the value, and such a search would also be fooled if the register has an alias. The where-defined analysis is complemented by other similar analyses.

11.4.3   Design Browser

The design browser is a tool which allows the user's logical design to be viewed graphically and can be used both during simulation and when executing a design on hardware. The following different graphical views are possible:

- Hierarchical

- Flat with a given scope

- As strongly connected components (SCCs)

The hierarchical view mirrors the structural hierarchy that was created by the user and allows each level of this hierarchy to be explored.

There are times when the user wishes to see more of a design than is permitted by the hierarchy view, and the "flat" display provides this. If displayed from the root of the design, the entire design is shown at once. Alternatively, by displaying from a scope other than the root, sub-trees of the design can be viewed. The major difference between this and the hierarchy display is that, from a given scope, all of the leaf AE instances are displayed.

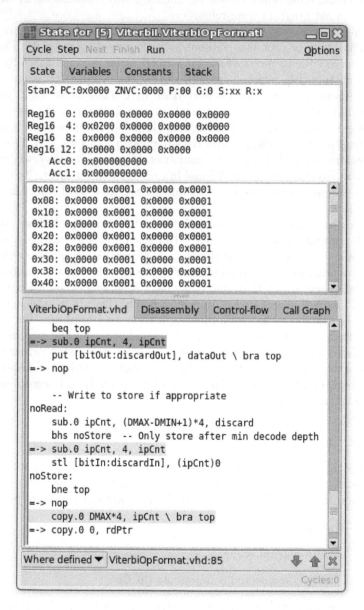

FIGURE 11.6: Example of where-defined program analysis.

The final view comes from thinking of a design as a directed graph and then showing a single level of hierarchy by producing the strongly connected components (SCCs). Each of the components can be viewed on its own. The importance of the SCC view is that from the root level the graph becomes acyclic (a directed acyclic graph, or DAG) and therefore this gives advantages when trying to debug a system which has deadlock, live-lock or data through-put problems. It separates out the parts of the design that contain feedback from those that are simply pipeline processing. An example of this is shown in Figure 11.7.

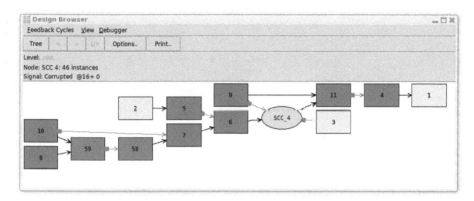

FIGURE 11.7: Design browser display.

In addition to these static features the design browser can provide dynamic information about each AE instance in a design, such as whether it is processing or waiting for a communications operation. This allows the user to quickly visualize which parts of the system are stalled or caught in deadlock.

11.4.4   Scripting

While debugging large parallel systems, operations such as viewing the source code or variable values of individual processes become too low level; this is analogous to debugging a compiled process by inspecting its raw disassembly and register values. For large parallel systems it is more convenient to be able to abstract the debugger to provide a higher, system-level interface. Such an interface allows the details of individual processes to be hidden, and replaced by system-specific displays instead. Clearly, it is impossible for picoChip to provide interfaces for every possible system, so instead the debugger can be programmed using Tcl/Tk [13]. This allows users to create their own system-specific interfaces, built on top of the picoChip debugger. Figure 11.8 shows an example of a WiMax system interface.

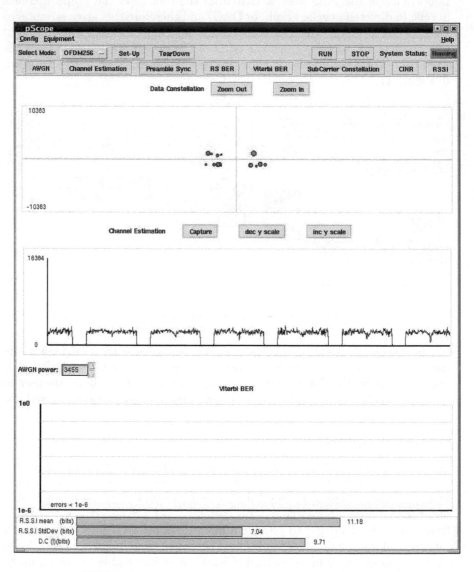

FIGURE 11.8: Diagnostics output from 802.16 PHY.

11.4.5   Probes

Probes are special-purpose processes which the debugger inserts into the user's design to monitor existing signals. They allow otherwise unused processors to be utilized for adding extra debug, analysis, or assertion-checking. Probe processes cannot view the internal details of other processors, but by monitoring the data traveling between processes over signals, they can still perform useful debug and analysis work.

Probes work by taking advantage of the hardware's ability to have one-to-many signals. Suppose a one-to-one signal must be monitored. The tools would insert the extra probe process on to a nearby unused processor, and extend the probed signal to include the probe processor, thus making the probed signal a one-to-many signal. The probe itself can be configured so that it does not actively participate in the communications traveling over the signal, but it may still see the data communicated. The original processes are unaffected by this change (both in terms of latency and bandwidth), but the probe is now able to monitor all communication over that signal.

Probes are implemented as processes, and so can run at full hardware speed. This enables probes to be used to debug systems in real-time. One use for probes is to allow real-time signal traces to be performed. Other uses include signal assertions, and on-chip analysis.

Signal assertion probes can be used to check that the data passing over a signal conforms to some compile-time specified property. For example, all signals in picoArray devices have pre-allocated bandwidth. A signal assertion probe could be attached to a signal to record the bandwidth actually used, thus allowing signals which have been allocated too much bandwidth for their actual requirements to be detected.

Probes can be used to perform on-chip analysis of signal data, rather than having to transport the data off-chip for later analysis. For example, during the development of a picoChip base station, a probe was created which performed bit-error rate (BER) computation on signals. These BER probes could be used to monitor the performance of the base station's Viterbi decoders in real-time, under different system loads.

11.4.6   FileIO

When testing and debugging, it is useful to be able to inject data into the system from a data file, or to record data generated by the system in a file. picoTools provide source and sink FileIO AEs which allow connections to a file to be made easily. A source FileIO has an output signal, over which successive samples from the associated file are sent. A sink FileIO has an input signal from which samples are read, and written to the associated file.

The FileIO AEs have two implementations which operate in either simulation mode, or in hardware mode. The simulation implementation of FileIO AEs can read and write files directly. On hardware, a programmable AE is used as a buffer to manage the transfer. For example, to handle a source FileIO, data from a file is written to the AE's memory, and a program on the AE used to send that data out over a signal. Once the memory has been emptied of data samples, it is refilled with the next block of data from the file. A sink FileIO operates in an equivalent manner, filling the AE's memory, and then writing the data to a file once the AE is full. Although the implementation in simulation and hardware is different, the interface is identical, and designs can be moved between simulation and hardware modes without requiring any changes.

In addition to simple files, providing a continuous stream of data into or out of a system, Timed FileIO may be used to mimic more complex IO operations which occur at specific times. For example, a timed source FileIO reads from a file in which each data sample is tagged with the time (absolute, or relative to a preceeding sample) at which that sample should be injected into the system. Similarly, output Timed FileIO may be used to generate a time-stamped file which records the time at which data samples where received by the FileIO AE.

Probes and Timed FileIO may be used very effectively together. A real hardware system may be monitored using time-stamped probes, which record what data passes over a signal and when. This allows the data used by a real system under realistic conditions to be captured. This information can then be converted into timed FileIO, which effectively replays real-time data into a test system, or into a simulation, allowing off-line debug and analysis.

11.5   Hardening Process in Practice

As OEM products evolve and move into the consumer sector they come under increasing price pressure. Second and subsequent generation products tend to require a lower BoM (bill of materials) as well as more features. To this end, picoChip's third generation device family, the PC20x, integrates an ARM926 subsystem which is used to run a variety of control software. This would typically have been done in a processor, external to the baseband device. Examples of control software include:

- An IEEE 802.16 MAC for a WiMAX base station. The WiMAX MAC software layer provides an interface between the higher transport layers and the physical layer (the SDR in the picoArray). The MAC layer takes packets from the upper layers and organizes them into MAC protocol data units (MPDUs) for transmission over the air. For received transmissions, the MAC layer does the reverse.

- Collapsed RNC stack in a WCDMA femtocell. This allows any standard 2G/3G mobile phone to communicate with the radio network.

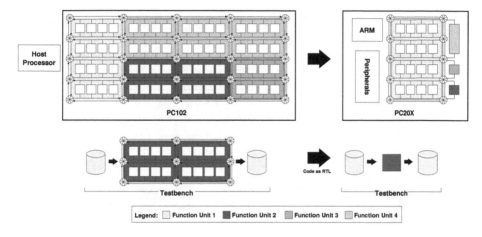

FIGURE 11.9: Hardening approach.

As mentioned in Section 11.2.3.4 another area addressed by the PC20x was the hardening of certain key functions previously implemented in software. Figure 11.9 shows the method used to harden the software functions. It shows the software functions implemented in a PC102 system which were migrated to fixed function hardware blocks wrapped with ports in the PC20x (in essence, a hardware realization of the BSI concept shown in Figure 11.5).

As can be seen in Figure 11.9, the software functions were verified in a testbench using the simulator. These functions can be viewed as executable models for the RTL. The hardware blocks are implemented using RTL but crucially the same testbench is used for the silicon verification. This method exploits the techniques described in Sections 11.4.6 and 11.3.3. By doing this, the hardware implementation will match against the golden reference, which in turn was used in a commercially deployed system for at least one generation.

The hardened blocks include:

- A Turbo decoder which is UMTS and IEEE 802.16 compliant

- A Viterbi decoder which is IEEE 802.16, UMTS, and GSM compliant

- A Reed-Solomon decoder which is IEEE 802.16 compliant

- An FFT/IFFT acceleration block

- A cryptography engine, which supports DES/3DES and AES in various modes

11.5.1   Viterbi Decoder Hardening

This section illustrates the hardening process using the Viterbi decoder as an example. The example design combines a Viterbi decoder with a testbench

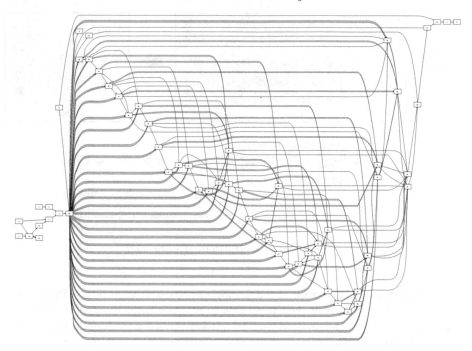

FIGURE 11.10: Software implementation of Viterbi decoder and testbench.

which can drive it at about 10 Mbps, comprising a random data generator, noise generator and output checking, all themselves implemented in software using other AEs. Control parameters for the test and result status indication are communicated to the user via the host processor. This testbench uses 11 AEs (4 MEM, 7 STAN) in addition to the host processor.

On PC101, the Viterbi decoder itself was also implemented entirely in software, and requires 48 AEs (1 MEM, 47 STAN). Note that this is for a Viterbi decoder which is only capable of 10 Mbps, to match its environment. Figure 11.10 shows a flat schematic of this design, produced by the design browser described in Section 11.4.3. For this and the other schematics referred to in this section, only the overall complexity of the design is of interest. The labeling of the individual AEs is uninformative. Signal flow is predominantly from left to right here, and also in Figures 11.11 and 11.12. The complexity and picoBus bandwidth requirements of this software implemented design are considerable.

On PC102, the FAU hardware accelerator was used to implement the core trellis decode function. The modified version of the 10 Mbps Viterbi decoder and testbench is shown in Figure 11.11. The decoder itself now requires 4 instances of the hardware accelerator and only 8 other AEs (1 MEM, 7 STAN), a saving of almost 40 AEs.

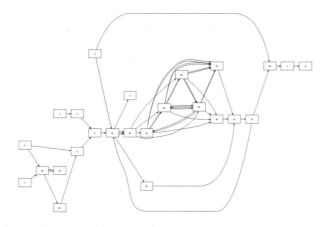

FIGURE 11.11: Partially hardened implementation of Viterbi decoder and testbench.

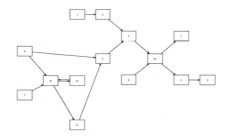

FIGURE 11.12: Fully hardened implementation of Viterbi decoder and testbench.

Finally, on PC20x a hardware accelerator is provided which implements the complete Viterbi decoder function. This is shown in Figure 11.12. Here the whole Viterbi decoder is reduced to a single AE instance. Moreover, this accelerator is actually capable of operating at over 40 Mbps, and is able to support multiple standards, including IEEE 802.16-2004 [11] multi-user mode, largely autonomously, which means that, in a more demanding application than this example, its use represents an even greater saving of resources. Table 11.1 contains figures for a PC102 implementation of such a "full featured" Viterbi decoder, but its schematic is too large to be included.

The quantitative benefits of the hardening process are illustrated by Table 11.1, which shows estimates of the transistor counts for Viterbi decoders of two different capabilities on each generation of chip. Area and power estimates are not compared, since different fabrication processes were used for

TABLE 11.1: Viterbi Decoder Transistor Estimates (in millions of transistors)

	MEMs @1.0M trans.	STANs @250k trans.	FAUs @1.0M trans.	Viterbi AEs @4.0M trans.	Total
10 Mbps Viterbi					
PC101	1	47			11.75
PC102	1	8	4		6.75
PC20x				1	4
40 Mbps Viterbi					
PC101					N/A
PC102	39	147	18		93.75
PC20x				1	4

the different generations of picoArray device, rendering any such comparison meaningless. The most meaningful comparison is for the 40 Mbps case, where similar functionality is being compared, and the reduction in transistor count is a factor of 23.

11.6 Design Example

There are several markets addressed by the PC20x and the latest one that has attracted the most interest is femtocells. Femtocells are low-power wireless access points that operate in licensed spectrum to connect standard mobile devices to a mobile operator's network using residential DSL or cable broadband connections [3] [2].

Femtocells are home base stations and to be cost effective they must be plug-and-play in the same way that consumers can install and use WiFi access points. This means using an existing network connection such as the public Internet. The radio access networks in use today comprise hundreds of base stations connected to a single radio network or base station controller (RNC/BSC). The interface is the Iub (interface between 3G base station and a radio network controller) running the ATM protocol over dedicated leased line [2].

A typical femtocell system can be seen in Figure 11.13. This can be implemented on a single PC202 picoArray device with an integrated ARM926 core giving an extremely cost effective, fully programmable solution which allows rapid customization, optimization and upgrades.

As can be seen the main interface to the backhaul (i.e., the mobile operator's network) is ethernet, and there is a requirement for synchronization (to ensure the femtocell is in synchronism with the network) and a radio interface. Such a system will be housed in a low cost box similar to the one shown in Figure 11.14. In the spirit of keeping cost to a minimum the PCB within the

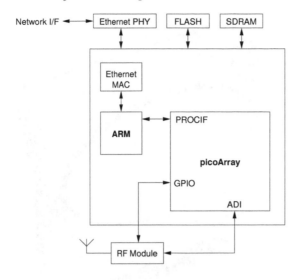

FIGURE 11.13: Femtocell system.

box needs to be relatively cheap to manufacture and populate. Figure 11.15 shows an example reference design from picoChip. In addition to the PC202, the PCB itself has:

Oscillator to clock the PC202 and the radio

Ethernet Phy interface to the backhaul

SDRAM DDR2 used to store data for both the ARM and the picoArray

FLASH used to store the program image and environment variables

Radio I/Q digital data interface subsystem

Miscellaneous components including DC power supply, antenna LED, connectors plus some switches

The rest of the system is realized by software executing within the PC20x. The ARM would typically execute the network interface software and the picoArray implements WCDMA software defined radio (SDR) [4].

The main features of the SDR are:

- Full compliance with 3GPP Release 6 TS25 series standards [9] including support for HSDPA (high speed downlink packet access) and HSUPA (high speed uplink packet access). This means supporting high speed data rates of up to 7.2 Mbps in the downlink and up to 1.46 Mbps in the uplink.

FIGURE 11.14: Femtocell.

- Support for up to four users in a 200 m Femtocell, each with AMR (adaptive multi-rate) voice data along with an HSDPA and an HSUPA session. This means that those four users will be able to share the total high speed packet switched data bandwidth in the uplink and the downlink whilst being able to handle a number of voice channels (up to 4) at lower rates. The combinations of the latter are varied but a share of the 384 kbps can be apportioned to the users.

- Convolutional and turbo FEC encoding/decoding. These can be used for different users and indeed different channels.

- Support of all downlink and uplink transport and physical channel types required by 3GPP FDD standards channels including HSPDA/HSUPA.

- Demodulation and despreading of uplink physical channels with up to four rake fingers per DCH (dedicated channel) per HSDPA user with integrated searcher.

- In addition to algorithms for channel estimation, MAC-hs and MAC-e scheduling and low-latency power control, the SDR provides MAC-b support for control of broadcast messages over the BCH (broadcast channel). This comprises a sophisticated scheduling sub-layer which is executed on the ARM, to dynamically configure the modem resources in response to uplink and downlink traffic.

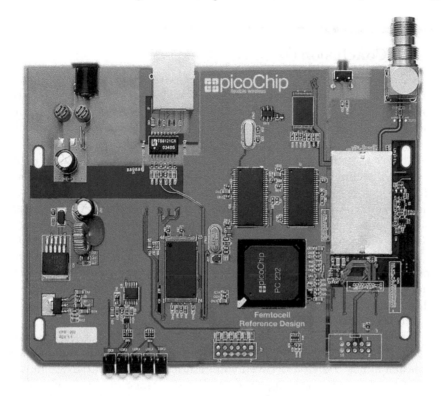

FIGURE 11.15: Femtocell reference board.

- Support of all baseband measurements required by 3GPP Release 6 standards and OAM-defined hardware alarm gathering and reporting via host-processor interface. As the modem is implemented in software these measurements are extracted by the scheduler executing on the ARM whilst maintaining the control and data channels.

- Filter and gain control features to allow a simple interface to DAC/ADC and reduce cost in RF subsystem.

In order to ensure the SDR and the PC20x would operate as a femtocell, the team at picoChip employed all the tools described in Section 11.3. The whole system was decomposed into manageable units and the tools used to write elements of the SDR. These were verified on the golden reference: the simulator. Testbenches were used to verify the modules within the SDR. BSIs of the HFAs were created and again the modules utilizing these were simulated using the simulator. The testbenches used were transferred into the RTL verification to ensure the silicon implementation of HFAs matched with the functional behavior of the BSIs.

11.7 Conclusions

In order to address its target markets in wireless communications, picoChip has created a family of picoArray devices which provide the computational power required by these applications and allow designers to trade off flexibility and cost. This family of devices is now in production and has been deployed in a wide range of wireless applications by a number of companies.

This chapter has explained the challenges which face designers applying multi-core techniques to wireless applications, and described the various facilities which the picoChip tools provide to allow these challenges to be met and overcome. Amongst the most important of these facilities are design visualization and data-flow analysis tools, non-intrusive data monitoring with "probes" and simulated replay of realistic system data via timed FileIO.

The process of behavioral modeling has also been developed to aid in the decomposition of designs and to allow the exploration of future architectures and take informed decisions regarding the "hardening" of functional blocks. This approach to cost and power reduction leverages the standard picoBus interface, allowing the programming paradigm to remain the same.

The hardening process was illustrated using a Viterbi decoder, showing that there are many ways that the hardening can be done, thus allowing a variation in the trade-off between transistor count and flexibility. The behavioral model-based hardening process allows a range of these options to be explored before devices are fabricated. This architectural hardening process has been used to produce a progression of commercially deployed devices.

picoChip is a member of the Femto Forum [3] and has worked with its customers to produce a WCDMA femtocell. The scope of the work was aimed at producing a system which could be used in the consumer premises equipment (CPE) market. As part of this work, picoChip has developed the first part of this system solution: a WCDMA PHY. This in conjunction with the PC20x device provides an ideal implementation of a first generation femtocell. As femtocells become deployed in larger and larger numbers the pressure of cost and added features will make the case for the shift to a second generation of femtocells. In order to address this, picoChip has already used much of the work described in this chapter to develop its fourth generation device: the PC302. This will cater for higher data rates as well more users which in turn has required more hardening of critical parts of the WCDMA PHY. The PC302 is now available.

Review Questions

[Q.1] The telecommunications industry has multiple standards, and these are fluid in nature. What are the implications of entering such markets with

ASIC solutions, and why might programmable solutions be used instead?

[Q.2] The picoBus is time-division multiplexed, with signals allocated at compile time. Describe the pros and cons of static allocation compared to buses with run-time arbitration.

[Q.3] Early picoArray devices used software implementations of functions such as Viterbi decoders and FFTs. Later devices have hardened these into silicon blocks. Explain the implications of hardening to cost and flexibility.

[Q.4] The VHDL and C source languages used to program picoArrays are *strongly typed*. Explain what strong typing means, and how it can aid development.

[Q.5] The picoArray toolchain software can be scripted using Tcl. What are the benefits of using scripting in large-scale multi-core systems?

[Q.6] During development, integration and testing of systems using the picoArray toolchain, the same test files may be used on hardware and software. Describe why this is important.

[Q.7] 'Probes' are special-purpose processes which utilize unused AEs. Probes are commonly used for recording the data that passes over a signal. What else could probes be used for?

[Q.8] picoArray array elements (processors) use a Harvard architecture. Discuss the benefits and costs of such an approach.

[Q.9] The use of behavioral simulation instances (BSIs) is an important part of the picoChip approach to "hardening" blocks of software into hardware functional accelerators (HFAs). Describe the advantages and drawbacks of such an approach.

[Q.10] picoChip produces not just silicon devices. Discuss what these are, what the challenges are and why it is important for real product deployment.

Bibliography

[1] BDTI communications benchmark (OFDM)^{TM}
http://www.bdti.com/products/services_comm_benchmark.html.

[2] The case for home basestations. http://www.picochip.com.

[3] Femto forum. http://www.femtoforum.org/femto.

[4] PC8209 HSDPA/HSUPA UMTS Femtocell PHY product brief. http://www.picochip.com.

[5] picoArray PC102 product brief. http://www.picochip.com.

[6] picoArray PC202 product brief. http://www.picochip.com.

[7] picoArray PC203 product brief. http://www.picochip.com.

[8] picoArray PC205 product brief. http://www.picochip.com.

[9] 3GPP. *3GPP Specification TS25 Series*.

[10] Peter Ashenden. *The Designer's Guide to VHDL*. Morgan Kaufmann, San Francisco, 1996.

[11] IEEE. *802.16 IEEE Standard for Local and Metropolitan Area Networks*.

[12] Steven S. Muchnick. *Advanced Compiler Design and Implementation*. Morgan Kaufmann Publishers, 1997.

[13] John K. Ousterhout. *Tcl and the Tk Toolkit*. Addison-Wesley Professional, May 1994.

[14] Gimpel Software. FlexeLint for C/C++. http://www.gimpel.com/.

[15] Richard Stallman. Using and porting the GNU compiler collection. ISBN 059510035X, http://gcc.gnu.org/onlinedocs/gcc/, 2000.

[16] Andreas Zeller. *Why Programs Fail: A Guide to Systematic Debugging*. Morgan Kaufmann, October 2005.

12

Embedded Multi-Core Processing for Networking

Theofanis Orphanoudakis

University of Peloponnese
Tripoli, Greece
fanis@uop.gr

Styllanos Perissakis

Intracom Telecom
Athens, Greece
sper@intracom.gr

CONTENTS

12.1	Introduction	400
12.2	Overview of Proposed NPU Architectures	403
	12.2.1 Multi-Core Embedded Systems for Multi-Service Broadband Access and Multimedia Home Networks	403
	12.2.2 SoC Integration of Network Components and Examples of Commercial Access NPUs	405
	12.2.3 NPU Architectures for Core Network Nodes and High-Speed Networking and Switching	407
12.3	Programmable Packet Processing Engines	412
	12.3.1 Parallelism	413
	12.3.2 Multi-Threading Support	418
	12.3.3 Specialized Instruction Set Architectures	421
12.4	Address Lookup and Packet Classification Engines	422
	12.4.1 Classification Techniques	424
	12.4.1.1 Trie-based Algorithms	425
	12.4.1.2 Hierarchical Intelligent Cuttings (HiCuts)	425
	12.4.2 Case Studies	426
12.5	Packet Buffering and Queue Management Engines	431

12.5.1   Performance Issues   433

         12.5.1.1   External DRAM Memory Bottlenecks   . . .   433

         12.5.1.2   Evaluation of Queue Management Functions: INTEL IXP1200 Case   434

12.5.2   Design of Specialized Core for Implementation of Queue Management in Hardware   435

         12.5.2.1   Optimization Techniques   439

         12.5.2.2   Performance Evaluation of Hardware Queue Management Engine   440

12.6   Scheduling Engines   442

         12.6.1   Data Structures in Scheduling Architectures   443

         12.6.2   Task Scheduling   444

                12.6.2.1   Load Balancing   445

         12.6.3   Traffic Scheduling   450

12.7   Conclusions   453

Review Questions and Answers   455

Bibliography   459

12.1   Introduction

While advances in wire-line and wireless transmission systems have provided ample bandwidth surpassing customer demand at least for the near future, the bottleneck for high-speed networking and enhanced service provisioning has moved to processing. Network system vendors try to push processing at the network edges employing various techniques. Nevertheless, the networking functionality is always proliferating as more and more intelligence (such as multimedia content delivery, security applications and quality of service (QoS)- aware networks) is demanded. The future Internet is expected to provide a data-centric networking platform providing services beyond today's expectations for shared workspaces, distributed data storage, cloud and grid-computing, broadcasting and multi-party real-time media-rich communications and many types of e-services such as sophisticated machine-machine interaction between robots, e-health, and interactive e-learning. Thus, the model of routing/switching devices has been augmented to enable the introduction of value added services involving complex network processing over multiple protocol stacks and the raw data forwarding functionality has been left only as the major task of large core switches that are exclusively assigned with this task. To cope with this demand, system designers have leaned on micro-electronic technology to embed network processing functions in either fixed or programmable hardware as much as possible. This led to a new generation of multi-core embedded systems specifically designed to tackle network processing application requirements.

In the past the power required for the processing of protocol functions at wire speed was usually obtained either by generic microprocessors (also referred to as central processing units, CPUs) designed with the flexibility to perform a variety of functions, but at a slower speed, or application specific integrated circuits (ASICs) designed to meet a specific functional requirement with high efficiency. Notwithstanding the requirement for high capacity and high quality of the offered services, the development cost of such systems (affected by the system component cost, application development cost, time-to-market as well as time-in-market) remains a critical factor in the development of such platforms. Hybrid programmable system-on-chip (SoC) devices integrating either generalized or task-specific processing cores called in general network processing units (NPUs) have recently deposed generic CPU-based products from many networking applications, extending the scalability (i.e., time-in-market) and performance of these products, therefore reducing cost and maximizing profits. In general NPUs can be defined as programmable embedded multi-core semiconductor systems optimized for performing wire speed operations on packet data units (PDUs). The development of such complex devices with embedded CPUs and diverse IP blocks has introduced a new paradigm in micro-electronics design as well in exploitation, programming and application porting on such devices.

The requirements of applications built on NPU-based devices are expanding dramatically. They must accommodate the highest bit rate on the one hand while coping with protocol processing of increased complexity on the other. In general the functionality of these protocols that spans the area of network processing can be classified as shown in Figure 12.1.

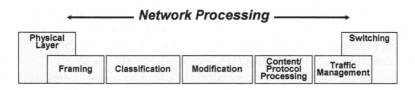

FIGURE 12.1: Taxonomy of network processing functions.

Physical layer processing and traffic switching are mostly related to the physical characteristics of the communications channel and the technology of the switching element used to interconnect multiple network nodes. The physical layer processing broadly includes all functions related to the conversion from transport media signals to bits. These functions include reception of electronic/photonic/RF signals and are generally classified in the different sub-layers such as the physical medium dependent (PMD), physical medium attachment (PMA) and physical coding sub-layer (PCS) resulting in appropriate signal reception, demodulation, amplification and noise compression, clock recovery, phase alignment, bit/byte synchronization and line coding. Switching includes the transport of PDUs from ingress to egress ports based on classification/routing criteria. For low rate applications switching is usually implemented through shared memory architectures, whereas for high rate applications through crossbar, bus, ring or broadcast and select architectures (the latter especially applied in the case of optical fabrics). The most demanding line-rates that motivated the wider introduction of NPUs in networking systems range in the order of 2.5 to 40 Gbps (OC-48, OC-192 and OC-768 data rates of the synchronous optical networking standard: SONET).

Framing and deframing includes the conversion from bits to PDUs, grouping bits into logical units. The protocols used are classified as Layer 2 protocols (data link layer of the OSI reference architecture) and the logical units are defined as frames, cells or packets. PDU conversion may also be required in the form of segmentation and reassembly. Most usually some form of verification of the PDU contents also needs to be applied to check for bit and field errors requiring the generation/calculation of checksums. In the more general case the same functionality is extended to all layers of the protocol stack since all telecommunication protocols employ some packetization and encapsulation techniques that require the implementation of programmable field extraction and modification, error correction coding and segmentation and reassembly (including buffering and memory management) schemes.

Classification includes the identification of the PDUs based on pattern matching to perform field lookups or policy criteria, also called rules. Many protocols require the differentiation of packets based on priorities, indicated in bits, header fields or multi-field (layers 2 up to 7 information fields) conditions. Based on the parsing (extraction of bits/fields) of the PDUs, pattern matching in large databases (including information about addresses, ports, flow tables etc.) is performed. Modification facilitates actions on PDUs based on classification results. These functions perform marking/editing of PDU bits to implement network address translation, ToS (type of service), CoS (class of service) marking, encapsulation, recalculation of checksums etc.

Content/protocol processing (i.e., processing of the entire PDU payload) may be required in case of compression (in order to reduce bandwidth load through the elimination of data redundancy) and encryption (in order to protect the PDU through scrambling, using public/private keys etc.) as well as deep packet inspection (DPI) for application aware filtering, content based routing and other similar applications. Associated functions required in most cases of protocol processing include the implementation of memory management techniques for the maintenance of packet queues, management of timers and implementation of finite state machines (FSMs).

Traffic engineering facilitates differentiated handling for flows of PDUs characterized by the same ToS or CoS, in order to meet a contracted level of QoS. This requires the implementation of multiple queues per port, loaded based on classification results (overflow conditions requiring additional intelligent policing and packet discard algorithms) and served based on specific scheduling algorithms.

In the packet network world, the CPU traditionally assumes the role of packet processor. Many protocols for data networks have been developed with CPU-centered architectures in mind. As a result, there are protocols with variable length headers, checksums in arbitrary locations and fields using arbitrary alignments. Two major factors drive the need for NPUs: i) increasing network bit rates, and ii) more sophisticated protocols for implementing multi-service packet-switched networks. NPUs have to address the above communications system performance issues coping with three major performance-related resources in a typical data communication system:

1. Processing cores

2. System bus(es)

3. Memory

These requirements drive the need for multi-core embedded systems specifically designed to alleviate the above bottlenecks by assigning hardware resources to efficiently perform specific network processing tasks. NPUs mainly aim to reduce CPU involvement in the above packet processing steps, which represent more or less independent functional blocks and generally result in the high-level specification of an NPU as a multi-core system.

12.2 Overview of Proposed NPU Architectures

12.2.1 Multi-Core Embedded Systems for Multi-Service Broadband Access and Multimedia Home Networks

The low-cost, limited-performance, feature-rich range of multi-core NPUs can be found in market applications that are motivated by the trend for multi-service broadband access and multimedia home networks. The networking devices that are designed to deliver such kind of applications to the end users over a large mixture of networking technologies and a multitude of interfaces face stringent requirements for size, power and cost reduction. These requirements can only be met by a higher degree of integration without sacrificing though performance and the flexibility to develop new applications on the same hardware platform over time. Broadband access networks use a variety of access technologies, which offer sufficient network capacity to support high-speed networking and a wide range of services. Increased link capacities have created new requirements for processing capabilities at both the network and the user premises.

The complex broadband access environment requires inter-working devices connecting network domains to provide the bridge/gateway functionality and to efficiently route traffic between networks of diverse requirements and operational conditions. These gateways constitute the enabling technology for multimedia content to reach the end users, advanced services to be feasible, and broadband access networking to become technically feasible and economically viable. A large market share of these devices includes the field of home networks, including specialized products to interconnect different home appliances, such as PCs, printers, DVD players, TV, over a home network structure, letting them share broadband connections, while performing protocol translation (e.g., IP over ATM) and routing, enforcing security policies etc. The need for such functionality has created the need for a new device, the residential gateway (RG).

The RG allows consumers to network their PCs, so they can share Internet access, printers, and other peripherals. Furthermore, the gateway allows people to play multiplayer games or distribute movies and music throughout the home or outdoors, using the broadband connection. The RG also enables interconnection and interworking of different telephone systems and services, wired, wireless, analog and IP-based, and supports telemetry and control applications, including lighting control, security and alarm, and in-home communication between appliances [44].

A set of the residential gateway functions includes carrying and routing data and voice securely between wide area network (WAN) and local area network (LAN), routing data between LANs, ensuring only the correct data is allowed in and out of the premises, converting protocols and data, selecting channels for bandwidth-limited LANs, etc. [20]. RGs with minimal functionality can be transparent to multimedia applications (with the exception of the requirement for QoS support for different multimedia traffic classes). However, sophisticated RGs will be required to perform media adaptations (i.e., POTS to voice over IP VoIP) or stream processing (i.e., MPEG-4) as well as control functions to support advanced services like for example stateful inspection firewalls and media gateways. All of the above networking applications are based on a protocol stack implementation involving processing of several layers and protocols. The partitioning of these functions into system components is determined by the expected performance of the overall system architecture. Recent trends employ hardware peripherals as a means to achieve acceleration of critical time-consuming functions. In any case the implementation of interworking functions is mainly performed in software. It is evident though that software implementations fail to provide real-time response, a feature especially crucial for voice services and multimedia applications.

To better understand the system level limitations of a gateway supporting these kinds of applications for a large number of flows, we show in Figure 12.2 the available system/processor clock cycles per packet, for different clock frequencies, for three different link rates. Even in the best case where one processor instruction could be executed in each cycle (which is far from true due

to pipeline dependencies, cache misses etc.), the number of instructions that
can be executed per packet is extremely low compared to the required processing capacity of complex applications. Taking into account also the memory
bottlenecks of legacy architectures, it is evident that the overall system level
architecture must be optimized with respect to network processing in order
to cope with demanding services and multimedia applications.

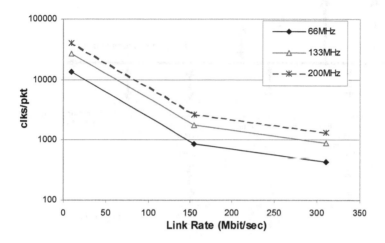

FIGURE 12.2: Available clock cycles for processing each packet as a function
of clock frequency and link rate in average case (mean packet size of 256 bytes
is assumed).

12.2.2 SoC Integration of Network Components and Examples of Commercial Access NPUs

Currently, the major trend in network processing architectures centers on
their implementation by integrating multiple cores resulting in a system-on-
chip (SoC) technology. SoC technology provides high integration of processors,
memory blocks and algorithm-specific modules. It enables low cost implementation and can accommodate a wide range of computation speeds. Moreover,
it offers a supporting environment for high-speed processor interconnection,
while input/output (I/O) speed and the remaining off-SoC system speed can
be low. The resulting architecture can be used for efficient mapping of a variety of protocols and/or applications. Special attention is focused on the edge,
access, and enterprise markets, due to the scales of the economy in these markets. In order to complete broadband access deployment, major efforts are
required to transfer the acquired technological know-how from the high-speed
switching systems to the edge and access domain by either developing chips
geared for the core of telecom networks that are able to morph themselves

into access players, or by developing new SoC architectures tailored for the access and residential system market.

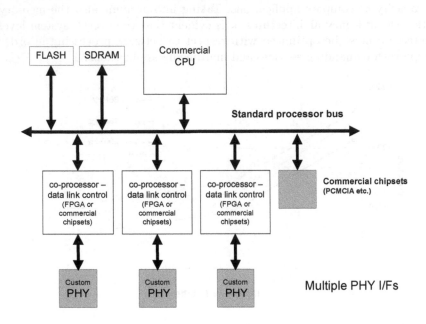

FIGURE 12.3: Typical architecture of integrated access devices (IADs) based on discrete components.

A common trend for developing gateway platforms to support multi-protocol and multi-service functionality in edge devices was until recently to use as main processing resources those of a commercial processor (Figure 12.3). Network interfaces were implemented as specialized H/W peripherals. Protocol processing was achieved by software implementations developed on some type of standard operating system and development platform. The main bottleneck in this architecture is apparently on one hand the memory bandwidth (due to the limited throughput of the main system memory) and on the other hand the limited speed of processing in S/W.

Driven by the conflicting requirements of higher processing power versus cost reduction, SoC architectures with embedded processor cores and increased functionality/complexity have appeared, replacing discrete component integrated access devices (IADs). Recent efforts to leverage NPUs in access systems aim to reduce the bottleneck of the central (CPU) memory. Furthermore, the single on-chip bus that interconnects all major components in typical architectures is another potential bottleneck. In an NPU-based architecture the bandwidth demands on this bus are reduced, because this bus can become arbitrarily wide (Figure 12.4) or alternatively the processor and peripheral buses can be separated. Therefore, such architectures are expected to scale better, being able to support network devices with higher throughput and more complex protocol processing than current gateways.

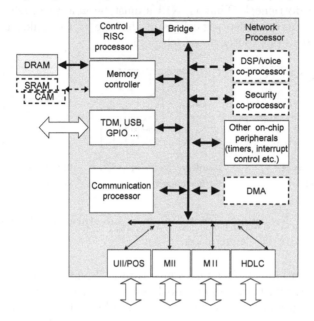

FIGURE 12.4: Typical architecture of SoC integrated network processor for access devices and residential gateways.

12.2.3 NPU Architectures for Core Network Nodes and High-Speed Networking and Switching

Beyond broadband access, the requirements for specialized multi-core embedded systems to perform network processing, as mentioned in the introduction of this section, have initially been considered in the context of replacing the high-performance but with limited programmability ASICs traditionally been developed to implement high-speed networking and switching in core network nodes. Core network nodes include IP routers, layer 3 fast, gigabit and 10 gigabit Ethernet switches, ATM switches and VoIP gateways. Next-generation embedded systems require a silicon solution that can handle the ever-increasing speed, bandwidth, and processing requirements. State-of-the-art systems need to process information implementing complex protocols and priorities at wire-speed and handle the constantly changing traffic capacity of the network. NPUs have emerged as the promising solution to deliver high capacity switching nodes with the required functionality to support the emerging service and application requirements. NPUs are usually placed on the data path between the physical layer and backplane within layer 3 switches or routers implementing the core functionality of the router and perform all the network traffic processing. NPUs must be able to support large bandwidth connections, multiple protocols, and advanced features without becoming a performance bottleneck. That is, NPUs must be able to provide wire-speed, non-blocking performance regardless of the size of the links, protocols and features enabled per router or switch port.

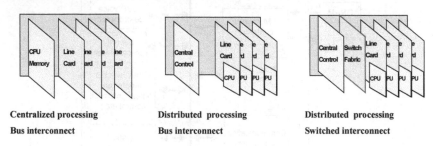

FIGURE 12.5: Evolution of switch node architectures: (a) 1st generation (b) 2nd generation (c) 3rd generation.

In the evolution of switching architectures, 1st and 2nd generation switches relied on centralized processing and bus interconnection-based architectures limiting local per port processing merely on physical layer adaptation. From the single CPU multiple line cards with single electrical backplane of 1st generation switches, technology advanced to distributed processing in its 2nd generation with one CPU per line card and a central controller for routing protocols and system control and management. A major breakthrough was the introduction of the switch fabric for inter-connection in the 3rd generation switches, to overcome the interconnection bandwidth problem, whereas the processing bottleneck was still treated with the same distributed architecture (Figure 12.5).

The PDU flow is shown in more detail in Figure 12.6. For the 1st generation switches shown in Figure 12.5 above, the network interface card (NIC) passes all data to CPU, which does all the processing, resulting in inexpensive NICs and overloaded interconnects (buses) and CPUs. The 2nd generation switches relieve the CPU overload by distributed processing, placing dedicated CPUs in each NIC; the interconnect bottleneck though remained. Finally 3rd generation switches introduced the switching fabric for efficient board-to-board communication over electronic backplanes.

NPUs mainly aim to reduce CPU involvement, used either in a centralized or distributed fashion and have been introducing the modifications to the architecture of Figure 12.6 as shown in Figure 12.7 below. In the centralized architecture (Figure 12.7a), the NIC passes all data to a high bandwidth NPU, which does all packet processing assuming the same protocol stack for all ports. Performance degrades with increased protocol complexity and increased numbers of ports. In a distributed architecture (Figure 12.7b) the CPU config-

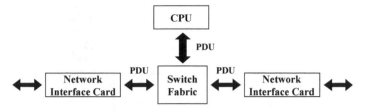

FIGURE 12.6: PDU flow in a distributed switching node architecture.

ures NPU execution and NPUs do all packet processing, possibly assisted by specialized traffic managers (TMs) for performing complex scheduling/shaping and buffer management algorithms. Each port can execute independent protocols and policies through a programmable NIC architecture.

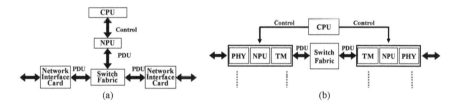

FIGURE 12.7: Centralized (a) and distributed (b) NPU-based switch architectures.

NPUs present a close coupling of link-layer interfaces with the processing engine, minimizing the overhead typically introduced in generic microprocessor-based architectures by device drivers. NPUs use multiple execution engines, each of which can be a processor core usually exploiting multi-threading and/or pipelining to hide DRAM latency and increase the overall computing power. NPUs may also contain hardware support for hashing, CRC calculation, etc., not found in typical microprocessors. Figure 12.8 shows a generic NPU architecture, which can be mapped to many of the NPUs discussed in the literature and throughout this chapter. Additional storage is also present in the form of SRAM (synchronous random access memory) and DRAM (dynamic random access memory) to store program data and network traffic. In general, processing engines are intended to carry out data-plane functions. Control-plane functions could be implemented in a co-processor, or a host processor.

An NPU's operation can be explained in terms of a representative application like IP forwarding, which could be tentatively executed through the following steps:

1. A thread on one of the processing engines handles new packets that arrive in the receive buffer of one of the input ports.

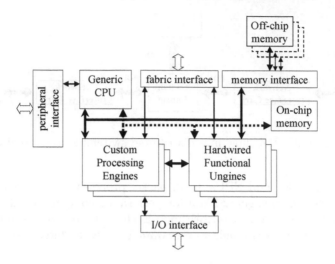

FIGURE 12.8: Generic NPU architecture.

2. The (same or alternative) thread reads the packet's header into its registers.

3. Based on the header fields, the thread looks up a forwarding table to determine to which output queue the packet must go. Forwarding tables are organized carefully for fast lookup and are typically stored in the high-speed SRAM.

4. The thread moves the rest of the packet from the input interface to packet buffer. It also writes a modified packet header in the buffer.

5. A descriptor to the packet is placed in the target output queue, which is another data structure stored in SRAM.

6. One or more threads monitor the output ports and examine the output queues. When a packet is scheduled to be sent out, a thread transfers it from the packet buffer to the port's transmit buffer.

The majority of the commercial NPUs fall mainly into two categories: The ones that use a large number of simple RISC (reduced instruction set computer) CPUs and those with a number (variable depending on their custom architecture) of high-end, special-purpose processors that are optimized for the processing of network streams. All network processors are system-on-chip (SoC) designs that combine processors, memory, specialized logic, and I/O on a single chip. The processing engines in these network processors are typically RISC cores, which are sometimes augmented by specialized instructions, multi-threading, or zero-overhead context switching mechanisms. The on-chip memory of these processors is in the range of 100KB to 1MB.

Within the first category we find:

- Intel IXP1200 [28] with six processing engines, one control processor, 200 MHz clock rate, 0.8-GB/s DRAM bandwidth, 2.6-Gb/s supported line speed, four threads per processor

- Intel IXP2400 and Intel IXP2800 [19] with 8 or 16 micro- engines, one control processor and 600 MHz or 1.6GHz clock rates, while also supporting 8 threads per processor

- Freescale (formerly Motorola) C-5 [6] with 16 processing units, one control processor, 200 MHz clock rate 1.6-GB/s DRAM bandwidth, 5-Gb/s supported line speed and four threads per processor

- CISCOs Toaster family [7] with 16 simple microcontrollers

All these designs generally adopt the parallel RISC NPU architecture employing multiple RISCs augmented in many cases with datapath co-processors (Figure 12.9(a)). Additionally they employ shared engines capable of delivering (N × port BW) throughput interconnected over an internal shared bus of 4 × total aggregate bandwidth capacity (to allow for at least two read/write operations per packet) as well as auxiliary external buses for implementing insert/extract interfaces to external controllers and control plane engines.

Although the above designs can sustain network processing from 2.5 to 10 Gbps, the actual processing speed depends heavily on the kind of application and for complex applications it degrades rapidly. Further, they represent a brute-force approach, in the sense that they use a large number of processing cores, in order to achieve the desired performance.

The second category includes NPUs like:

- EZChips NP1 [9] with a 240 MHz system clock that employs multiple specific-purpose (i.e., lookup) processors as shared resources without being tied to a physical port

- HiFns (formerly IBMs) PowerNP [17] with 16 processing units (pico-processors), one control processor, 133 MHz clock rate, 1.6-GB/s DRAM bandwidth, eight-Gb/s line speed and two threads per processor, as well as specialized engines for look-up, scheduling and queue management

These designs may follow different approaches most usually found as either pipelined RISC architectures including specialized datapath RISC engines for executing traffic management and switching functions (Figure 12.9(a)), or generally programmable state machines which directly implement the required functions (Figure 12.9(b)). Both these approaches have the feature that the internal data path bus is required to offer only 1 × total aggregate bandwidth.

Although, the aforementioned NPUs are capable of providing a higher processing power for complicated network protocols, they lack the parallelism of the first category. Therefore, their performance, in terms of bandwidth

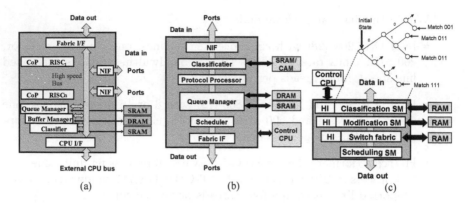

FIGURE 12.9: (a) Parallel RISC NPU architecture (b) pipelined RISC NPU architecture (c) state-machine NPU architecture.

serviced, is lower than the one of the first category whenever there is a large number of independent flows that should be processed.

Several of these architectures are examined in the next section, while the micro-architectures of several of the most commonly found co-processors and hardwired engines are discussed throughout this chapter.

12.3   Programmable Packet Processing Engines

NPUs are typical domain-specific architectures: in contrast to general purpose computing, their applications fall in a relatively narrow domain, with certain common characteristics that drive several architectural choices. A typical network processing application consists of a well-defined pipeline of sequential tasks, such as: decapsulation, classification, queueing, modification, etc. Each task may be of small to modest complexity, but has to be performed with a very high throughput, or repetition rate, over a series of data (packets), that most often are independent from each other. This independence arises from the fact that in most settings the packets entering a router, switch, or other network equipment, belong to several different flows. In terms of architectural choices, these characteristics suggest that emphasis must be placed on throughput, rather than latency. This means that rather than architecting a single processing core with very high performance, it is often more efficient to utilize several simpler cores, each one with moderate performance, but with a high overall throughput. The latency of each individual task, executed for each individual packet, is not that critical, since there are usually many independent data streams processed in parallel. If and when one task stalls, most of the time there will be another one ready to utilize the processing cycles made available. In other words, network processing applications are usually latency tolerant.

The above considerations give rise to two architectural trends that are common among network processor architectures: *multi-core parallelism*, and *multi-threading*.

12.3.1 Parallelism

The classic trade-off in computer architecture, that of performance versus cost (silicon area) manifests itself here as single processing engine (PE) performance versus the number of PEs that can fit on-chip. In application domains where there is not much inherent parallelism and more than a single PE cannot be well utilized, high single-PE performance is the only option. But where parallelism is available, as is the case with network processing, the trade-off usually works out in favor of many simple PEs. An added benefit of the simple processing core approach is that typically higher clock rates can be achieved. For these reasons, virtually all high-end network processor architectures rely on multiple PEs of low to moderate complexity to achieve the high throughput requirements common in the OC-48 and OC-192 design points. As one might expect, there is no obvious "sweet spot" in the trade-off between PE complexity and parallelism, so a range of architectures have been used in the industry.

Typical of one end of the spectrum are Freescale's C-port and Intel's IXP families of network processors (Figure 12.10). The Intel IXP 2800 [2][30] is based on 16 microengines, each of which implements a basic RISC instruction set with a few special instructions, contains a large number of registers, and runs at a clock rate of 1.4 GHz. The Freescale C-5e [30] contains 16 RISC engines that implement a subset of the MIPS ISA in addition to 32 custom VLIW processing cores (Serial Data Processors, or SDPs) optimized for bit and byte processing. Each RISC engine is associated with one SDP for the ingress path, that performs mainly packet decapsulation and header parsing, and one SDP for the egress path, that performs the opposite functions — those of packet composition and encapsulation.

Further reduction in PE complexity, with commensurate increase in PE count, is seen in the architecture of the iFlow Packet Processor (iPP) [30] by Silicon Access Networks. The iPP is based on an array of 32 simple processing elements called *Atoms*. Each Atom is a reduced RISC processor, with an instruction set of only 47 instructions. It is interesting to note, however, that many of these are custom instructions for network processing applications.

As a more radical case, we can consider the PRO3 processor [37]: its main processing engine, the reprogrammable pipeline module (RPM) [45] consists of a series of three programmable components: a field extraction engine (FEX), the packet processing engine proper (PPE), and a field modification engine (FMO), as shown in Figure 12.11. The allocation of tasks is quite straightforward: packet verification and header parsing are performed by FEX, general

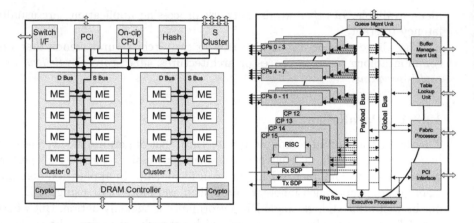

FIGURE 12.10: (a) Intel IXP 2800 NPU, (b) Freescale C-5e NPU.

processing on the PPE, and modification of header fields or composition of new packet headers is executed on the FMO. The PPE is based on a Hyperstone RISC CPU, with certain modifications to allow fast register and memory access (to be discussed in detail later). The FEX and FMO engines are barebones RISC-like processors, with only 13 and 22 instructions (FEX and FMO, respectively).

In another approach, a number of NPU architectures attempt to take advantage of parallelism at a smaller scale within each individual PE. *Instruction-level parallelism* is usually exploited by superscalar or Very-Long-Instruction-Word (VLIW) architectures. Noteworthy is EZchip's architecture [9][30], based on superscalar processing cores, that EZchip claims are up to 10 times faster on network processing tasks than common RISC processors. SiByte also promoted the use of multiple on-chip four-way superscalar processors, in an architecture complete with two-level cache hierarchy. Such architectures of course are quite expensive in terms of silicon area, and therefore only a relatively small number of PEs can be integrated on-chip. Compared to superscalar technology, VLIW is a lot more area-efficient, since it moves a lot of the instruction scheduling complexity from the hardware to the compiler. Characteristic of this approach are Motorola's SDP processors, mentioned earlier, 32 of which can be accommodated on-chip, along with all the other functional units.

Another distinguishing feature between architectures based on parallel PEs is the *degree of homogeneity*: whether all available PEs are identical, or whether they are specialized for specific tasks. To a greater or lesser degree, all architectures include special-purpose units for some functions, either fixed logic or programmable. The topic of subsequent sections of this chapter is to analyze the architectures of the more commonly encountered special-purpose units. At this point, it is sufficient to note that some of the known archi-

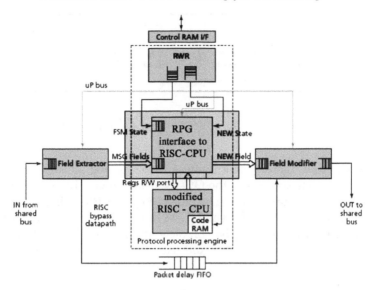

FIGURE 12.11: Architecture of PRO3 reprogrammable pipeline module (RPM).

tectures place emphasis on many identical programmable PEs, while others employ PEs with different variants of the instruction set and combinations of functional units tailored to different parts of the expected packet processing flow.

Typical of the specialization approach is the EZchip architecture: it employs four different kinds of PEs, or Task-OPtimized cores (TOPs):

- *TOPparse*, for identification and extraction of header fields and other keywords across all 7 layers of packet headers

- *TOPsearch*, for table lookup and searching operations, typically encountered in classification, routing, policy enforcement, and similar functions

- *TOPresolve*, for packet forwarding based on the lookup results, as well as updating tables, statistics, and other state for functions such as accounting, billing, etc.

- *TOPmodify*, for packet modification

While the architectures of these PEs all revolve around EZchip's superscalar processor architecture, each kind has special features that make it more appropriate for the particular task at hand.

Significant architectures along these lines are the fast pattern processor (FPP) and routing switch processor (RSP), initially of Agere Systems and currently marketed by LSI Logic. Originally, these were separate chips, that

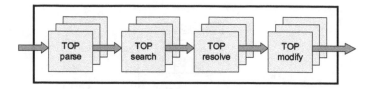

FIGURE 12.12: The concept of the EZchip architecture.

together with the Agere system inteface (ASI) formed a complete chipset for routers and similar systems at the OC-48c design point. Later they were integrated into more compact products, such as the APP550 single-chip solution (depicted in Figure 12.13) for the OC-48 domain and the APP750 two-chip set for the OC-192 domain. The complete architecture is based on a variety of specialized programmable PEs and fixed-function units. The PEs come in several variations:

- The packet processing engine (PPE), responsible for pattern matching operations such as classification and routing. This was the processing core of the original FPP processor.

- The traffic management compute engine, responsible for packet discard algorithms such as RED, WRED, etc.

- The traffic shaper compute engine, for CoS/QoS algorithms.

- The stream editor compute engine, for packet modification.

At the other end of the spectrum we have architectures such as Intel's IXP and IBM's PowerNP, that rely on multiple identical processing engines, that are interchangeable with each other. The PowerNP architecture [3][30] is based on the *dyadic packet processing unit* (DPPU), each of which contains two *picoprocessors*, or *core language processors* (CLPs), supported by a number of custom functional units for common functions such as table lookup. Each CLP is basically a 32-bit RISC processor. For example, the NP4GS3 processor, an instance of the PowerNP architecture, consists of 8 DPPUs (16 picoprocessors total) each of which may be assigned any of the processing steps of the application at hand. The same holds for the IXP and iFlow architectures, that, as mentioned earlier, consist of arrays of identical processing elements. The feature that differentiates this class of architectures from the previous is that for every task that needs to be performed on a packet, the "next available" PE is chosen, without constraints. This is not the case for the EZchip and Agere architectures, where processing tasks are tied to specific PEs.

Finally, we may distinguish a class of architectures that fall in the middle ground, and that includes the C-port and PRO3 processors, among others.

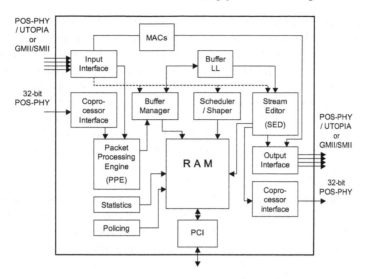

FIGURE 12.13: Block diagram of the Agere (LSI) APP550.

The basis of these architectures is an array of *identical* processing units, each of which consists of a number of *heterogeneous* PEs. Recall the combination of reduced MIPS RISC with the two custom VLIW processors that form the Channel Processor (CP) of the C-port architecture, or the Field Extractor, Packet Processing Engine, and Field Modifier, that together form the reprogrammable pipeline module (RPM) of PRO3. A CP or RPM can be repeated as many times as silicon area allows, for a near-linear increase in performance.

With all heterogeneous architectures, the issue of *load balancing* arises. What is the correct mix of the different kinds of processing elements, and/or, what is the required performance of each kind? Indeed, there is no simple answer that will satisfy all application needs. NPU architects have to resort to extensive profiling of their target applications, based on realistic traffic traces, to determine a design point that will be optimal for a narrow class of applications, provided of course that their assumptions on traffic parameters and processing requirements hold. The broader the target market is for a specific processor, the more difficult it is to attain a single mix of PEs that will satisfy all applications. On the contrary, with homogeneous architectures PEs can be assigned freely to different tasks according to application needs. This may even be performed dynamically, following changing traffic patterns and the mix of traffic flows with different requirements. Of course, for such flexibility one has to sacrifice a certain amount of performance that could be achieved by specialization.

In terms of communication between the processing elements, most NPU architectures avoid fancy and costly on-chip interconnection networks. To justify such a choice, one must consider how packets are processed within an

NPU. Processing of packets that belong to different flows is usually independent. On the other hand, the processing stages for a single packet most often form a pipeline, where the latency between stages is not that critical. Therefore, for most packet processing needs, some kind of shared memory will be sufficient. Note however that usage of an external memory for this purpose would cause a severe bottleneck at the chip I/Os, so on-chip RAM is the norm. For example, in the FPP architecture, a *block buffer* is used to hold 64-byte packet segments (or blocks) until they are processed by the Pattern Processing Engine, the Queue Engine, and other units. In the more recent APP550 incarnation of the architecture, all blocks share access to 3 MB of embedded on-chip DRAM. Similarly, in the PowerNP architecture, packets are stored in global on- and off- chip data stores and from there packet headers are forwarded to the next available PE for processing. No direct communication between PEs is necessary in the usual flow of processing.

There are of course more elaborate communication schemes than the above, with most noteworthy probably the IXP case. In this architecture, PEs are divided in two 8-PE clusters. The PEs of each cluster communicate with each other and with other system components over two buses (labeled D and S). No direct communication between the two clusters is possible. Each PE (Figure 12.14) has a number of registers, called *transfer registers*, dedicated to inter-PE communication. By writing to an output transfer register, a PE can directly modify the corresponding input transfer register of another PE. Furthermore, another set of registers is dedicated to nearest neighbor communication. With this scheme, each PE has direct access to the appropriate register of its neighbor. In this way, a very efficient ring is formed.

In the C-port family, a hierarchy of buses is also used. Three different buses, with bandwidths ranging from 4.2 to 34.1 Gbits/sec (on the C-5e), are used to interconnect all channel processors and other units with each other.

12.3.2   Multi-Threading Support

Turning now to the microarchitecture of the individual PEs, a prevailing trend in NPU architectures is multi-threading. The reason that most NPU vendors have converged to this technique is that it offers a good method to overcome the unavoidably long latency of certain operations. Table lookup is a characteristic one. It is often handled by specialized coprocessors and can take a large number of clock cycles to complete. But as with all complex SoCs, even plain accesses to external memories, such as the packet buffer, incur a significant latency. Multi-threading allows a processing element to switch to a new thread of execution, typically processing a different packet, every time a long-latency operation starts. It is important to note here that the nature of most network processing applications allows multi-threading to be very effective, since there will almost always be some packet waiting to be processed, and each packet can be associated with a thread. So, ready-to-run threads will almost always be available and most of the time long latency operations of one or more threads will overlap with processing of another thread. In this way, processing cycles will almost never get wasted waiting for long-running operations to complete.

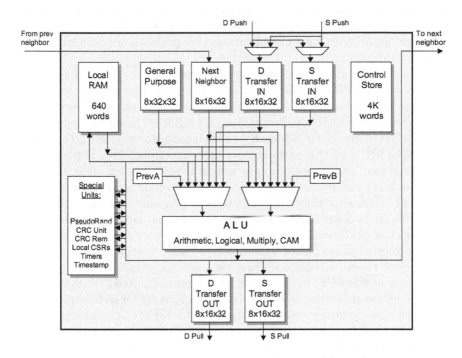

FIGURE 12.14: The PE (microengine) of the Intel IXP2800.

For multi-threading to be effective, switching between threads must be possible with very little or no overhead. Indeed, many network processor vendors claim zero-overhead thread switching. To make this possible, the structure of the PE is augmented with multiple copies of all execution state. By the term *state* we define the content of registers and memory, as well as the program counter, flags and other state bits, depending on the particular architecture. So, multi-threaded PEs typically have register files partitioned into multiple banks, one per supported thread, while local memory may also be partitioned per thread. Events that trigger thread switching can be a request to a coprocessor or an external memory access. On such an event, the current thread becomes inactive, a new thread is selected among those ready for execution, and the appropriate partition of the register file and related state is activated. When the long-running operation completes, the stalled thread will become ready again and get queued for execution.

A critical design choice is the number of supported threads per PE. If the PE does not directly support enough threads in hardware, the situation will often arise that all supported threads are waiting for an external access, in which case the processing cycles remain unused. Processors with shorter cycle times and more complex coprocessors (requiring longer to complete) or a slower external memory system will require more threads. On the other hand, the cost of supporting many threads can have a significant impact on both die area and cycle time. Therefore, this parameter must be chosen very judiciously, based on profiling of target applications and performance simulations of the planned architecture.

Most industrial designs offer good examples of multi-threading: Each picoprocessor in IBM's NP4GS3 supported two threads, a number that was apparently found insufficient and later raised to four in the more recent 5NP4G (marketed by HiFn). Threads also share a 4 KB local memory available within each DPPU of the NP4GS3, each one having exclusive access to a 1 KB segment. The iPP and IXP architectures are very similar with respect to multi-threading; each architecture supports eight threads per PE, each with its register file partition and other state. Thread switching is performed with zero overhead, when long-running instructions are encountered along the thread's execution path. Such instructions may be external memory accesses or complex functions executed on a coprocessor. The programmer also has the possibility to relinquish control by executing special instructions that will cause the current thread to sleep, waiting for a specific event. Finally, noteworthy is the case of the FPP, whose single PE supports up to 64 threads!

The PRO3 processor follows a different approach for overlapping processing with slow memory accesses. The FEX-PPE-FMO pipeline is organized in such a way that these processing engines almost always work out of local memory. The PPE's register file has two banks. One of them can be accessed directly by either FEX or FMO, at the same time that the PPE is executing, using the other bank. In addition, the PPE's local memory has two ports, one of which can be accessed by an external controller. When a packet arrives at the RPM, the FEX extracts all necessary fields from its headers, under program control. It then writes the values of these fields into one bank of the PPE register file. To retrieve per-flow state from off-chip memory, a flow identifier (FlowId) is constructed from the packet header, that is used as index to memory. State retrieved thus is written into the PPE's local memory over its external port. These actions can take place while the PPE is still processing the previous packet. When it finishes, the PPE does not need to output the results explicitly, since the FMO can pull the results directly out of the PPE's register file. A data I/O controller external to the PPE will also extract data from the PPE's local memory to update flow state in the off-chip RAM. All that the PPE needs to do is to switch the two partitions of the register file and local RAM and restart executing. The relevant header fields and flow state will already be present in its newly activated partitions of the register file and local RAM respectively. In this way, data I/O instructions are eliminated from the PPE code and computation largely overlaps with I/O (output of the previous packet's results and input of the next packet's data). With the PPE working on local memory (almost) all the time, there is very little motivation for multi-threading support. So, PRO3 PPEs do not need to support more than one thread.

12.3.3 Specialized Instruction Set Architectures

Finally, the instruction set architecture (ISA) is another area where vendors tend to innovate and differentiate from each other. While some vendors rely on more-or-less standard RISC instruction sets, it is recognized by many that this is not an efficient approach; instead, an instruction set designed from scratch and optimized for the special mix of operations common in packet processing can give a significant performance edge over a simple RISC ISA. This is easy to comprehend if one considers that RISC instruction sets have resulted from years of profiling and analyzing *general-purpose* computing applications; it is only natural to expect that a similar analysis on *networking* applications should be the right way to define an instruction set for an NPU.

Based on the above rationale, many NPU vendors claim great break-throughs in performance, solely due to such an optimized instruction set. AMCC has dubbed its ISA NISC (network instruction set computing) in analogy to RISC. EZchip promotes its Task Optimized Processing Core technology, with customized instruction set and datapath for each packet processing stage. Interestingly, both vendors claim a speedup over RISC-based architectures in the order of 10 times. Finally, Silicon Access, with its iFlow architecture, also based on a custom instruction set, claimed double the performance of its nearest competitor.

One can distinguish two categories of special instructions encountered in NPU ISAs: those that have to do with the coordination of multiple PEs and multiple processing threads working in parallel, and those that perform packet processing-oriented data manipulations. In the first category one can find instructions for functions such as thread synchronization, mutual exclusion, inter-process (or -thread) communication, etc. We can mention for example support in the IXP ISA for atomic read-modify-write (useful for mutual exclusion) and events, used for signalling between threads. Instructions that fall in this first category are also encountered in parallel architectures outside of the network processing domain. In the following we will focus on the data manipulation operations.

Arguably the most common kinds of operations in packet processing have to do with header parsing and modification: extraction of bit fields of arbitrary length from arbitrary positions in the header for the parsing stage, on packet ingress, or similar insertions for the modification stage, on packet egress. Many NPU architectures cater to accelerate such operations with custom instructions. For example, the IXP combines shifting with logical operations in one cycle, to speed-up the multiple shift-and-mask operations needed to parse a header. Also, the iFlow architecture supports single-cycle insertion and extraction of arbitrary bit fields. The same is true for the Field Extractor and Field Modifier in the PRO3 architecture.

Multi-way branches are also common when parsing fields such as packet type, or encoded protocol identifiers. With standard RISC instruction sets, a wide switch statement is translated into many sequential compare-and-branch statements. Custom ISAs accelerate this kind of code by special support for conditional branches. Silicon Access claimed to be able to speed up such cases by up to 100 times, with a technology dubbed *massively parallel branch acceleration* that allows such a wide switch to be executed in only two clock cycles. As another example, the IXP microengine includes a small CAM that can be used to accelerate multi-way branches, by allowing up to 16 comparisons to be performed in parallel, providing at the same time a branch target.

Predicated execution is another branch optimization technique, that is actually borrowed from the DSP world. It allows execution of certain instructions to be enabled or disabled based on the value of a flag. In this way, many conditional branch operations are avoided, something that can speed up significantly tight loops with many short if-then-else constructs. The CLP processor of the PowerNP architecture is an example of such an instruction set.

Finally, many architectures provide instructions for tasks such as CRC calculation and checksumming (1's complement addition), evaluation of hash functions, pseudorandom number generation, etc. Another noteworthy addition is support in the IXP architecture for efficient linked list and circular buffer operations (insert, delete, etc). Given that the use of such structures in networking applications is very common, such hardware support has a significant potential for overall code speedup.

12.4   Address Lookup and Packet Classification Engines

The problem of packet classification is usually the first that has to be tackled when packets enter a router, firewall, or other piece of network equipment. Before classification the system has no information regarding how to handle incoming packets. To maintain wire speed operation, it has to decide very quickly what to do with each new packet received: queue it for processing, and if so, to which queue? Discard it? Any other possibility? The classifier is the functional unit that will inspect the packet and provide the necessary information for such decisions.

In general, a classifier receives an unstructured stream of packets and by applying a configurable set of rules it splits this stream into parallel flows of packets, with all packets that belong to the same flow having something in common. The definition of this common feature is arbitrary. Historically it has been the destination port number (where classification served solely the purpose of forwarding). But more recently it may represent other notions, such as same QoS requirements, or type of security processing, or other. Whatever this common characteristic is, it implies that all packets of a flow will be processed by the router in the same manner, at least for the next stage (or stages) of processing. The decision as to how to classify each incoming packet depends on one (rarely) or multiple (more commonly) fields of the packet header(s) at various layers of the protocol hierarchy.

Classification is not an easy problem, especially given that it has to be performed at wire speed. Even in the case of simple route lookup based on the packet's destination IP address (probably the simplest special case of the problem) it is not trivial. Consider that an IPv4 address is 32 bits wide, with normally up to 24 bits used for routing. A naïve table implementation would contain 2^{24} entries, something prohibitive. However, such a table would be quite sparse, motivating implementations based on various kinds of data structures. The size of such a table would be a function of the active (valid) entries only. Unfortunately, this is still a large number. Up-to-date statistics maintained by [1] show that as of this writing, the number of entries in the Internet's core routers (known as the BGP table, from the Border Gateway Protocol) has exceeded 280,000 and is still rising. Searching such a table at wire speed at 10 Gbps is certainly a challenge; assuming a flow of minimum-size IP packets, only 32 nsec are available per search. Consider now that this is only a one-dimensional lookup. In more demanding situations classification has to be based on multiple header fields. Typical is the quintuple of source and destination IP addresses, source and destination port numbers, and layer 4 protocol identifier, often used to define a flow. Finally, such tables have to be updated dynamically in large metropolitan and wide area networks, more than 1000 times per second.

Classification also appears further down the processing pipeline, depending on the application. Classification based on the aforementioned quintuple is applicable to tasks such as traffic management, QoS assurance, accounting, billing, security processing, and firewalls, just to name a few. Classification can even be performed on packet payload, for example on URLs appearing in an HTTP message, for applications such as URL filtering and URL-based switching.

Formally, the problem of classification can be stated as follows: For any given packet, a search key or lookup key is defined as an arbitrary selection of N header fields (an *N-tuple*). A rule is a tuple of values, possibly containing wildcards, against which the key has to be matched. A *rule database* is a prioritized list of such rules. The task of classification is to find the highest priority rule that matches the search key. In most cases, the index of the matching rule is used as the *flow identifier* (flowID) associated with all packets that match the same rule. So, each rule defines a *flow*.

Wildcards usually take one of two forms: (i) prefixes, usually applicable to IP addresses. For example, the set of addresses 192.168.*.* is a 16-bit prefix. This is an effect of the way Classless Interdomain Routing (CIDR) [11] works and gives rise to a variety of longest-prefix matching (LPM) algorithms and (ii) ranges, most commonly used with port numbers, such as 100-150.

12.4.1　Classification Techniques

The simplest and fastest way to search a rule database is by use of a content-addressable memory (CAM). Indeed, CAMs are used often in commercial classification engines, even though they have certain disadvantages. In contrast to a normal memory, that receives an address and provides the data stored in that address, a CAM receives a data value and returns the address where this value is found. The entire array is searched in parallel, usually in a single clock cycle, the matching locations are identified by their address, and a priority encoder resolves potential multiple matches. One or more match addresses may be returned.

The growing importance of LPM matching has given rise to *Ternary CAMs*, or TCAMs, that support wildcarding. For every bit position in a TCAM, two actual bits are used: a care/don't care bit and the data bit. All the care/don't care bits of a memory address form a mask. In this way prefixes can be easily specified. For example the IP address prefix 192.168.*.* can be specified with data value 0xC0A80000 and mask 0xFFFF0000.

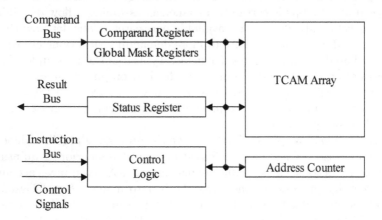

FIGURE 12.15: TCAM organization [Source: Netlogic].

Searching with a CAM becomes trivial. One needs only concatenate the relevant header fields, provide those to the CAM, and wait for the match address to be returned. The main disadvantage of CAMs (and even more so of TCAMs) is the silicon area required, which is several times larger than that of simple memory. This gives rise to high cost, limited overall capacity, and impact on overall system dimensions. Furthermore, the parallel search of the memory array causes a high power dissipation. In spite of these problems, TCAMs are not uncommon in commercial systems. They are certainly more appropriate in highest throughput systems (such as OC-48 and OC-192 core routers), which are also the least cost-sensitive.

For the cases where a TCAM is not deemed cost-efficient, a variety of algorithmic approaches have been proposed and applied in many practical systems.

Most of these approaches store the rule database in SRAM or DRAM in some kind of pointer-based data structure. A search engine then traverses this data structure to find the best-matching rule. In practical systems, this search engine may be fixed logic, although programmable units are also common, for reasons of flexibility.

In the following we briefly review two representative techniques. A good survey of algorithms can be found in [15]. When examining such algorithms, one needs to keep in mind that in addition to lookup speed, such algorithms must be evaluated for the speed and ease of incremental updates, and memory size and cost (e.g., whether they require SRAM or DRAM).

12.4.1.1 Trie-based Algorithms

Many of the most common implementations of the classifier database are based on the *trie* data structure [23]. A trie is a special kind of tree, used for creating dictionaries for languages with arbitrary alphabets, that is quite effective when words can be prefixes of other words (as is the case of IP address prefixes). When the alphabet is the set of binary digits, a trie can be used to represent a set of IP addresses and address prefixes. Searching for a prefix in a single dimension, as in the case of route lookup, is simple: just traverse the tree based on the digits of the search key, until either a match is found or the key characters are exhausted. Obviously, nodes lower in the tree take precedence, since they correspond to longer matches. The problem gets more interesting when multidimensional searches are required.

A *hierarchical* or *multilevel* trie can be thought of as a three-dimensional trie, where the third dimension corresponds to the different fields of an N-dimensional key. Lookup involves traversing all dimensions in sequence, so the lookup performance of the basic hierarchical trie search is O(Wd), where W is the key width and d the number of dimensions. The storage requirements are O(NdW), with N the number of rules. Finally, incremental updates are possible with complexity $O(d^2W)$. Details on the construction and lookup of hierarchical tries can be found in references such as [15].

Many variations of the basic algorithm have also been proposed. For example, for two-dimensional classifiers, the *grid-of-tries* algorithm [41] enhances the data structure with some additional pointers between nodes in the second dimension tries, so that no backtracking is needed and the search time is reduced to O(W). However, this comes at the expense of difficult incremental updates, so rebuilding the database from scratch is recommended. So, this algorithm is appropriate for relatively static classifiers only.

12.4.1.2 Hierarchical Intelligent Cuttings (HiCuts)

This is representative of a class of algorithms based on the geometric interpretation of classifiers. A two-dimensional classifier can be visualized as a set of rectangles contained in a box that is defined by the overall ranges of the two dimensions. For example, Figure 12.16 defines a classifier:

Rule	X	Y
R1	0*	*
R2	1*	*
R3	0*	1*
R4	10*	00*
R5	*	010
R6	000	00*
R7	001	00*

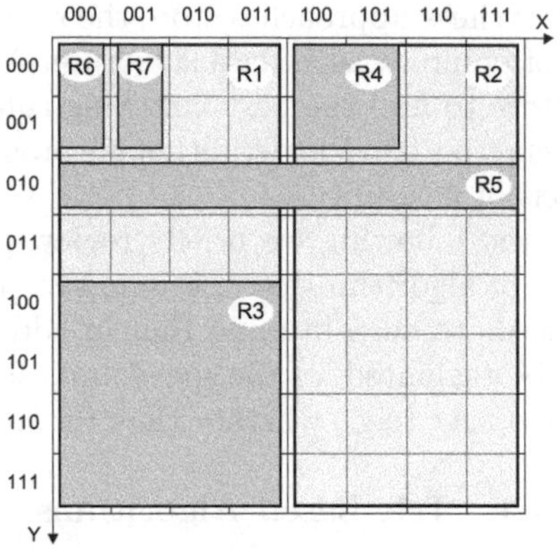

FIGURE 12.16: Mapping of rules to a two-dimensional classifier.

While we use here a two-dimensional example for the purpose of illustration, the algorithm generalizes to any number of dimensions. HiCuts [14] constructs a decision tree based on heuristics that aim to exploit the structure of the rules. Each node of the tree represents a subset of the space. A cut, determined by appropriate heuristics, is associated with each node. A cut partitions the space along one dimension into N equal parts, creating N children, each of which represents one N^{th} of the original box. Each node is also associated with all rules that overlap fully or partially with the box it represents. Cutting proceeds until all leaf nodes contain at most B rules, where B is a tunable parameter trading storage space for lookup performance. To match a given search key, the algorithm traverses the decision tree guided by the bits of the key, until it hits a leaf node. Then, the B or fewer rules that leaf contains are searched sequentially to determine the best match.

12.4.2 Case Studies

Finally we review some of the most representative classification/table lookup engines in the industry.

PowerNP. The Dyadic Packet Processing Unit (DPPU) of the PowerNP architecture [3] contains two RISC cores, along with two *Tree Search Engines* (TSEs), together with other coprocessors. The TSE is a programmable unit that supports table lookup in three modes: full match (for looking up structures like MAC address tables), longest-prefix match (for example for layer 3 forwarding) and software-managed trees, the most general kind of search. This last mode supports all the advanced search features, such as general N-tuple matching, and support for arbitrary ranges in any dimension (not just prefixes).

Operation of the TSE starts with a RISC core constructing the search key from the appropriate header fields. Then, it issues a request to one of the two TSEs of the DPPU to execute the search. The TSE first consults the *LuDefTable* (Lookup Definition Table), an on-chip memory that contains information about the available tables (where they are stored, the kind of search to do, key sizes, tree formats, etc). The TSE also has access to the system's *control store* (control memory) where tables are stored, among other data. The control store is a combination of on-chip memory with off-chip DDR SDRAM and ZBT SRAM (in the newer NP4GX, Fast Cycle RAM (FCRAM) is also used).

Typical performance numbers for the TSE of the NP4GS3 are from 8 to 12 million searches per second, depending on the type of search, a rate sufficient to support basic processing at an OC-48 rate (2.5 Gbps), with minimum size IP packets and one lookup per packet. In case higher performance is needed, the NP4GS3 also supports external CAM.

Agere. The primary role of Agere's (currently LSI Logic's) Fast Pattern Processor [30] is packet header parsing and classification. The Packet Processing Engine (PPE), the main programmable unit of the FPP, is programmed in Agere's own *Functional Programming Language* (FPL). As its name implies, FPL is a functional language, which is very appropriate for specifying patterns to be matched. Supposedly, it also generates very compact machine code, at least for the kinds of tasks encountered in packet classification. The FPP also uses a proprietary, patented search technique, that Agere has dubbed Pattern Matching Optimization. This technique places emphasis on fast lookups, which are executed in time bounded by the length of the key (pattern) and not by the size of the database.

The FPP processes data in 64-byte blocks. Complete processing of a packet involves two steps, or passes. When packets enter the FPP, they are first segmented to blocks and stored in the external packet buffer. At the same time, the first block of each packet is loaded into a context, an on-chip storage area that maintains short-term state for a running thread. With 64 threads supported in hardware, there are 64 contexts to choose from. Once basic first-pass processing is done, the packet is assigned to a *replay queue*, getting in line for the second pass. When a context is available it is loaded and the second pass starts. Once the second pass is over, the packet is sent downstream to the RSP, followed by the classification results that the PPE retrieved.

While the original FPP relied on SRAM for classification database storage, newer incarnations of the architecture, such as the 10 Gbps APP750NP, replace this with FCRAM, reducing the cost and at the same time achieving better performance than would be possible with regular DRAM.

Silicon Access. Silicon Access introduced the iFlow product family [30], consisting of several chips: packet processor, traffic manager, accountant (for billing etc) and not one but two search engines: the Address Processor (iAP) and the Classifier (iCL). Even though the company did not survive the slow-down of core network rollouts in the early 2000s, the architecture has several interesting features that makes it worth examining.

The two search engines are designed for different requirements. The Address Processor can perform pipelined, full or longest prefix matching operations on on-chip tree-based tables with keys ranging from 48 to 144 bits wide. On the other hand, the Classifier is TCAM-based and performs general range matching with keys up to 432 bits long. So, the iAP is more appropriate for operations like address lookup, while the more general classification problem is the task of the iCL.

The innovation of Silicon Access in the design of the iFlow chipset is undoubtedly the use of wide embedded DRAM, an architectural choice that in many applications eliminates the need for external CAMs and SRAMs, and even reduces the pressure on external DRAM. The two search engines rely entirely on on-chip memory. The iAP has a total of 52 Mbits of memory, holding 256K prefixes up to 48 bits long, 96 bits of associated data memory per entry, and a smaller 8K by 256 per-next-hop associated data memory. The iCL's 9.5 Mbits of total memory are organized as 36K entries by 144 bits of TCAM plus 128 bits associated data per entry. Of course, in both systems multiple table entries can be combined to cover each device's maximum key width. The much smaller amount of total memory in the iCL is unavoidable, given that most of it is organized as a TCAM, with much lower density than plain RAM. It is also noteworthy that all embedded memory in these devices is ECC protected, which makes them effective for high reliability applications. In terms of performance, the devices are rated at 100 Msps (iCL) and 65 Msps (iAP), allowing up to three or two, respectively, searches per minimum-size IP packet on a 10 Gbps link. The embedded memory-based architecture of iAP and iCL is of course both a curse and a blessing: on one hand it reduces the chip count, cost, and power dissipation of the system; on the other, it places a hard limit on the table sizes that can be implemented. Fortunately, multiple devices can be combined to form larger tables, although this is unlikely to be cost-effective.

In the following we give a few details about the organization and operation of the iAP, the more interesting of the two devices from an algorithmic standpoint [34]. The device has a ZBT SRAM interface, over which it is attached to a network processor, such as the iPP. The network processor performs regular memory writes to issue requests and memory reads to retrieve the results. iAP's operation is pipelined and with fixed latency, independent of the number of entries, prefix length or key length: it can start a new lookup every 2 clock cycles (at 133MHz), with a latency of 26 cycles.

The search algorithm, which is hardwired in fixed logic, takes advantage of the very wide on-chip RAMs that allow many entries to be read out in parallel and an equal number of comparisons to be made simultaneously. Prefixes are stored in RAM in ascending order, with shorter prefixes treated as larger than longer ones; for example, 11011* is larger than 1101101*. A three-level B-tree

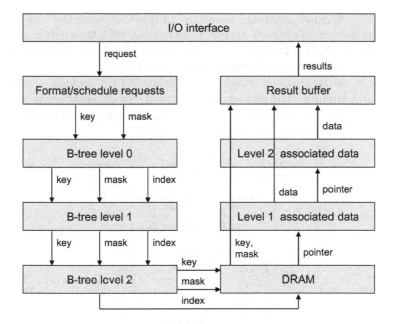

FIGURE 12.17: iAP organization.

in on-chip SRAM provides pointers to the complete list of prefixes, stored in on-chip DRAM. Traversing the three levels of the B-tree results in a pointer to a small subset of the prefix list. A small number of parallel comparisons there allows the correct prefix to be located.

We should also note that the architecture of the iAP allows table maintenance (insertions and deletions of prefixes) to be performed in parallel with the searches. The iAP can support up to 1 million updates per second, consuming only about 20 percent of the search bandwidth.

EZchip. EZchip stresses the implementation of classifier tables in DRAM, which helps reduce system cost, chip count and power dissipation. A second feature emphasized is the support for long lookups (for arbitrary length strings, such as URLs).

In EZchip's heterogeneous architecture, the processing engine (Figure 12.18) dubbed TOPsearch is the one responsible for table lookup operations [10]. This is primarily a fixed-logic engine, with a minimal instruction set designed to support chained lookups, where the result of one lookup, possibly combined with additional header fields, is used as the key for a new lookup. TOPsearch supports three types of lookups: direct, hash, and tree-based. In the latter case, the optimization employed to make operation at high link rates possible is to store the internal nodes of the tree on-chip in embedded DRAM, and the leaf nodes in external DRAM. The embedded memory organization, with a 256-bit wide interface, allows up to three levels of the tree to be traversed with a single memory access. Put together with the shorter access time of on-chip DRAM, this architecture provides a significant speedup compared with the more common external memory organization.

Finally, parallelism is employed: EZchip NPUs include a number of TOPsearch engines, that can work concurrently on different packets. With all the above optimizations, tables with over one million entries can be searched at up to a 30 Gbps link rate with the NP-3 NPU.

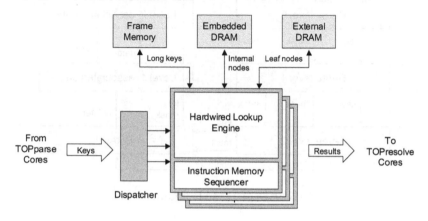

FIGURE 12.18: EZchip table lookup architecture.

Third-party search engines. A limited number of vendors specialize in search engines, without a full network processing chipset in their portfolio. The standardization of a coprocessor interface for search engines by the Network Processing Forum[1], dubbed LA-1(b) [13], helps in the integration of such third-party devices into systems built around NPU families of most major vendors.

Typical is the case of Netlogic Microsystems, which has been among the leading suppliers of CAM and TCAM devices. The obvious path toward on-chip integration has been to incorporate table lookup and maintenance logic into the TCAM device, thus transforming what was only table storage into a self-contained search engine. A number of variations on the theme are provided, ranging from plain address lookup engines for IP forwarding, to layer four classification engines supporting N-tuple lookup, all the way to layer seven processors for "deep packet inspection", for applications such as URL matching and filtering, virus signature recognition, stateful packet inspection, etc.

With this kind of specialized search engine, key matching is becoming increasingly sophisticated. For example, the above mentioned engines sup-

---
[1]Later merged into the Optical Internetworking Forum

port regular expression matching, an additional step up in complexity and sophistication from the longest-prefix matching and range lookup that we have discussed so far. In fact, a long search key may span the payload of more than one packet. The capability of on-the-fly inspection of packets all the way up to layer seven, combined with such sophisticated matching, leads to new applications, such as intrusion detection, general malware detection, application-based switching, etc. Standardization of interfaces, such as the LA-1, certainly fosters innovation in this field, since it allows more players to enter the market with alternative architectures.

12.5 Packet Buffering and Queue Management Engines

Most modern networking technologies (like IP, ATM, MPLS etc.) share the notion of connections or flows (we adopt the term *flow* hereafter) that represent data transactions in specific time spans and between specific end-points in the network for the implementation of networking applications. Furthermore scheduling among multiple per port, QoS and CoS queues requires the discrimination of packet data and the handling of multiple data flows with differentiated service requirements. Depending on the applications and algorithms used, the network processor typically has to manage thousands of flows, implemented as packet queues in the processor packet buffer [27]. Therefore, effective queue management is key to high-performance network processing as well as to reducing development complexity. In this section we focus on the review of potential implementations within a NPU architecture and performance evaluation of queue management, which is performed extensively in network processing applications and show how HW cores can be used to offload completely this task from other processing elements.

The requirements with regard to memory management implementations in networking applications stem from the fact that data packets need to be stored in an appropriate queue structure either before or after processing and be selectively forwarded. These queues of packets need to at least serve the first-in-first-out (FIFO) service discipline, while in many applications flexible access to their data is required (in order to modify, move, delete packets or part of a packet, which resides in a specific position in the queue, e.g., head or tail of the queue etc.). In order to efficiently cope with these requirements several solutions based on dedicated hardware have been proposed initially targeting high-speed ATM switching where the fixed ATM cell size favored very efficient queue management [39][33] [46] and later extended to management of queues of variable-size packets [18]. The basic advantage of these implementations in hardware is of course the higher throughput with modest implementation cost. On the other hand the functions they can provide (e.g., single versus double linked lists, operations in the head/tail of the queue, copy operations etc.)

needs to be selected carefully at the beginning of the design. Several trade-offs between dedicated hardware and implementations in software have been exposed in [48], in which specific implementations of such queue management schemes in ATM switching applications are examined.

As in many other communication subsystems, memory access bandwidth to the external DRAM-based packet data repository is the scarcest resource in NPUs. For this reason, the NPU architecture must be designed very carefully to avoid unnecessary data transfer across this memory interface. In an NPU architecture, each packet byte may traverse the memory interface up to four times, e.g., when encryption/decryption or deep packet parsing functions are performed. This is also the case for short packets such as TCP/IP acknowl-edgments, where the packet header is the entire packet, in order to perform the following operations: (a) write packet to buffer on ingress, (b) read head-er/packet into processing engines, (c) write back to memory, and (d) read for egress transmission.

This means that for small packets, which typically represent 40 percent of all Internet packets, the required memory interface capacities amount to 10, 40, or 120 Gb/s for OC-48, OC-192, or OC-768, respectively. Even the lowest of these values, 10 Gb/s, exceeds the access rate of todays commercial DRAMs. Complex memory-interleaving techniques that pipeline memory access and distribute individual packets over multiple parallel DRAM (dynamic RAM) chips can be applied for 10 Gb/s and possibly 40 Gb/s memory subsystems. At 120 Gb/s, todays 166 MHz DDR (double-data-rate) SDRAMs would require well over 360-bit-wide memory interfaces, or typically some 25 DDR SDRAM chips.

Several commercial NPUs follow a hybrid approach targeting the acceler-ation of memory management implementations by utilizing specialized hard-ware units that assist specific memory access operations, without providing a complete queue management implementation. The first generation of the Intel NPU family, the IXP1200, initially provided an enhanced SDRAM unit, which supported single byte, word, and long-word write capabilities using a read-modify-write technique and may reorder SDRAM accesses for best per-formance (the benefits of this will also be explored in the following section). The SRAM Unit of the IXP1200 also includes an 8-entry push/pop register list for fast queue operations. Although these hardware enhancements im-prove the performance of typical queue management implementations they cannot keep in pace with the requirements of high-speed networks. Therefore the next generation IXP-2400 provides high-performance queue management hardware that automates adding data to and removing data from queues [40]. Following the same approach the PowerNP NP4GS3 incorporates dedicated hardware acceleration for cell enqueue/dequeue operations in order to man-age packet queues [3]. The C-Port/Motorola C-5 NPU also provided mem-ory management acceleration hardware [6], still not adequate though to cope with demanding applications that require frequent access to packet queues. The next-generation Q-5 Traffic Management Coprocessor provided dedicated hardware designed to support traffic management for up to 128K queues at a rate of 2.5 Gbps [18]. In the rest of this section we review the most important performance requirements evaluating a set of alternative implementations that dictate the basic design choices when assigning specific tasks to embedded engines in a multi-core NPU implementation.

12.5.1   Performance Issues

12.5.1.1   External DRAM Memory Bottlenecks

A crucial design decision at such high rates is the choice of the buffer memory technology. Static random access memory (SRAM) provides high throughput but limited capacity, while DRAM offers comparable throughput and significantly higher capacity per unit cost; thus, DRAM is the prevalent choice among all NPUs for implementing large packet buffering structures. Furthermore, among DRAM technologies, DDR SDRAM is becoming very popular because of its high performance and affordable price. DDR technology can provide 12.8 Gbps peak throughput by using a 64-bit data bus at 100 MHz with double clocking (i.e., 200 Mbps/pin). A DIMM module provides up to 2 GB total capacity and it is organized into four or eight banks to provide interleaving (i.e., to allow multiple parallel accesses). However, due to bank-pre-charging periods (during which the bank is characterized as busy) successive accesses must respect specific timing requirements. Thus, a new read/write access to 64-byte data blocks to the same bank can be inserted every four clock-cycles, i.e., every 160 ns (with an access cycle of 40 ns). When a memory transaction tries to access a currently busy bank, we say that a bank conflict has occurred. This conflict causes the new transaction to be delayed until the bank becomes available, thus reducing memory utilization. In addition, interleaved read and write accesses also cause loss to memory utilization because they create different access delays. Thus, while the write access delay can be as low as 40 ns and the read access delay 60 ns, when write accesses occur after read accesses, the write access must be delayed by one access cycle.

It is worth demonstrating the impact of the above implications in DDR-DRAM performance in the overall aggregate throughput that can be provided under usual access patterns following the methodology presented in [36]. The authors in [36] simulated a behavioral model of a DDR-SDRAM memory under a random access pattern and estimated the impacts of bank conflicts and read-write interleaving on memory utilization. The results of this simulation for a range of available memory banks (1 to 16) are presented in the two left columns of Table 12.1.

The access requests assume aggregate accesses from two write and two read ports (a write and a read port from/to the network, a write and a read port from/to the internal processing element (PE) array). By serializing the accesses from the four ports in a simple/round-robin order (i.e., without optimization) the throughput loss presented in Table 12.1 is achieved. However, by

TABLE 12.1: DDR-DRAM Throughput Loss Using 1 to 16 Banks

Banks	No Optimization Throughput Loss		Optimization Throughput Loss	
	Bank conflicts	Bank conflicts + write-read interleaving	Bank conflicts	Bank conflicts + write-read interleaving
1	0.750	0.75	0.750	0.750
2	0.647	0.66	0.552	0.660
3	0.577	0.598	0.390	0.432
4	0.522	0.5	0.260	0.331
5	0.478	0.48	0.170	0.290
6	0.442	0.46	0.100	0.243
7	0.410	0.42	0.080	0.220
8	0.384	0.39	0.046	0.199
9	0.360	0.376	0.032	0.185
10	0.338	0.367	0.022	0.172
11	0.321	0.353	0.018	0.165
12	0.305	0.347	0.012	0.159
13	0.289	0.335	0.010	0.153
14	0.275	0.33	0.007	0.148
15	0.264	0.32	0.004	0.143
16	0.253	0.317	0.003	0.139

scheduling the accesses of these four ports in a more efficient manner, a lower throughput loss is achieved since a reduction in bank conflicts is possible. A simple way to do this is to effectively reorder the accesses of the four ports to minimize bank conflicts. The information for bank availability in order to appropriately schedule accesses is achieved by keeping the memory access history (i.e., storing the last three accesses). In case that more than one accesses are eligible (belong to a non-busy bank), the scheduler selects one of the eligible accesses in round-robin order. If no pending access is eligible, then the scheduler sends a no-operation to the memory, losing an access cycle. The results of this optimization are presented in the right side of Table 12.1. Assuming organization of eight banks, the optimized scheme reduces throughput loss by 50 percent with respect to the un-optimized scheme. Thus, it is evident that only a percentage of the nominal 12.8 Gbps peak throughput of a 64-bit/100 MHz DDR-DRAM can be utilized and the design of the memory controller must be an integral part of the memory management solution.

12.5.1.2 Evaluation of Queue Management Functions: INTEL IXP1200 Case

As described above, the most straightforward implementation of memory management in NPUs is based on software executed by one or more on-chip microprocessors. Apart from the memory bandwidth that was examined in isolation in the previous section, a significant factor that affects the overall performance of a queue management implementation is the combination of the processing and communication latency (communication with the peripheral memories and memory controllers) of the queue handling engine (either generic processor or fixed/configurable hardware) and the memory response latency. Therefore the overall actual performance can only be evaluated at system level. Using Intel's IXP1200 as an example representing a typical NPU architecture, the authors in [36] have also presented results regarding the maximum throughput that can be achieved when implementing memory management in IXP1200 software.

The IXP1200 consists of six simple RISC processing microengines [28] running at 200 MHz. According to [36], when porting the queue management software to the IXP RISC-engines, special care should be given so as to take advantage of the local cache memory (called scratch memory) as much as possible. This is because any accesses to the external memories use a very large number of clock cycles. One can argue that using the multi-threading capability of the IXP can hide this memory latency. However, as it was proved in [48], the overhead for the context switch, in the case of multi-threading, exceeds the memory latency and thus this IXP feature cannot increase the performance of the memory management system when external memories should be accessed. Even by using a very small number of queues (i.e., fewer than 16), so as to keep every piece of information in the local cache and in the IXPs registers, each microengine cannot service more than 1 million packets per second (Mpps). In other words, the whole of the IXP cannot process more than 6 Mpps. Moreover, if 128 queues are needed, and thus external memory accesses are necessary, each microengine can process at most 400 Kpps. Finally, for 1K queues the peak bandwidth that can be serviced by all six IXP microengines is about 300 Kpps [40]. The above throughput results are summarized in Table 12.2.

TABLE 12.2: Maximum Rate Serviced When Queue Management Runs on IXP 1200

No. of Queues	1 Microengine	6 Microengines
16	956 Kpps	5.6 Mpps
128	390 Kpps	2.3 Mpps
1024	60 Kpps	0.3 Mpps

12.5.2 Design of Specialized Core for Implementation of Queue Management in Hardware

Due to the performance limitations identified above, the only choice to achieve very high capacity (mainly in NPUs targeting core network systems and high-speed networking applications) is to implement dedicated embedded cores to offload the other PEs from queue management tasks. Such cores are implemented either as fixed hardware engines, designed specifically to accelerate the task of packet buffering and per-flow queuing, or as programmable HW cores with limited programmability extending to a range of operations indexed by means of an OPCODE that can be executed. In the remainder of this section we present the micro-architecture and performance details of such a specifically designed engine (originally presented in [36]) designed as a task-specific embedded core for NPUs supporting most of the requirements for queue and buffer management applications. The maintenance of queues of packets per flow in the design presented in [36] is undertaken by a dedicated data memory management controller (called DMM) designed to efficiently support per flow queuing, providing tens of gigabits per second throughput to an external buffer based on DDR-DRAM technology and many complex operations on these packet queues. The classification of packets into flows is considered part of the protocol processing accomplished prior to packet buffering by a specific processing module denoted as packet classifier. The overall sub-system architecture considered in [36] for packet classification, per-flow queuing and scheduling is shown in Figure 12.19.

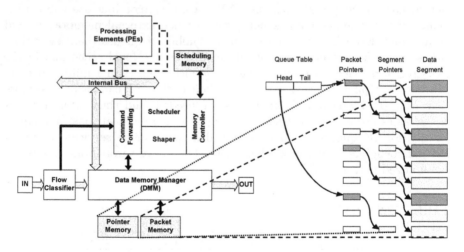

FIGURE 12.19: Packet buffer manager on a system-on-chip architecture.

The main function of the DMM is to store the incoming traffic to the data memory, retrieve parts of the stored packets and forward them to the internal processing elements (PEs) for protocol processing. The DMM is also responsible to forward the stored traffic to the output, based on a programmable traffic-shaping pattern. The specific design reported in [36] supports two incoming and two outgoing data paths at 2.5 Gbps line rate each; there is one for receiving traffic from the network (input), one for transmitting traffic to the network (output), and one bi-directional for receiving and sending traffic from/to the internal bus. It performs per flow queuing for up to 512K flows.

The DMM operates both at fixed length or variable length data items. It uses DRAM for data storage and SRAM for segment and packet pointers. Thus, all manipulations of data structures occur in parallel with data transfers, keeping DRAM accesses to a minimum. The architecture of the DMM is shown in Figure 12.20. It consists of five main blocks: the data queue manager (DQM), data memory controller (DMC), internal scheduler, segmentation block and reassembly block. Each block is designed in a pipeline fashion to exploit parallelism and increase performance. In order to achieve efficient memory management in hardware, the incoming data items are partitioned into fixed size segments of 64 bytes each. Then, the segmented packets are stored in the data memory, which is segment aligned. Segmentation and reassembly blocks perform this function. The internal scheduler forwards the incoming commands from the four ports to the DQM, giving different service priorities to each port. The data queue manager organizes the incoming packets into queues. It handles and updates the data structures, kept in the pointer memory. The data memory controller performs the low level read and writes to the data memory minimizing bank conflicts in order to maximize DRAM throughput as described below.

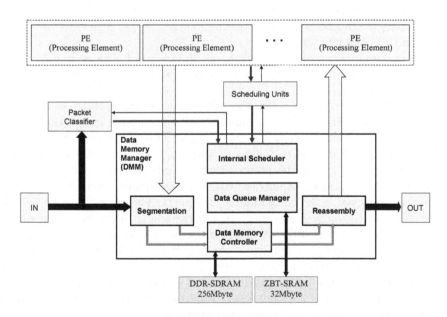

FIGURE 12.20: DMM architecture.

The DMM reported in [36] provides a set of commands in order to support the diverse protocol processing requirements of any device handling queues. Beyond the primitive commands of "enqueue" and "dequeue", the DMM features a large set of 18 commands to perform various manipulations on its data structures (a list of the commands is given in Table 12.3 in Section 12.5.2.2

along with the performance measured for the execution of these commands). Thus it can be incorporated in any embedded system that should handle queues.

DDR-DRAM has been chosen for the data memory because it provides adequate throughput and large storage space for the 512K supported queues, at a low cost as already discussed above. The DDR-SDRAM module used has a 64-pin data bus, which runs at 133 MHz clock frequency, providing 17.024 Gbps total throughput. The large number of the required pointer memory accesses requires a high throughput low latency pointer memory. SRAM has been selected as pointer memory, which provides the required performance. Typical SRAMs working at 133 MHz clock frequency provide 133M accesses per second or about 8.5 Gb/sec.

The data memory space is organized into fixed-size buffers (named segments), which is a usual technique in all memory management implementations. The length of segments is set to 64 bytes because this size minimizes fragmentation loss. For each segment, in the data memory, a segment pointer and a packet pointer are assigned. The addresses of the data segments and the corresponding pointers are aligned, as shown in Figure 12.19, in the sense that a data segment is indexed by the same address as its corresponding pointer. For example, the packet and segment pointers of the segment 0 are in the address 0 in the pointer memory.

The data queues are maintained as single-linked lists of segments that can be traversed from head to tail. Thus, head and tail pointers are stored per queue on a queue table. Head pointers point to the first segment of the head packet in the queue, while the tail pointer indicates the first segment of the tail packet. The DMM can handle traffic at variable length objects (i.e., packets) as well as at fixed-size data items. This is achieved by using two linked lists per flow: one per segment and one per packet. Each entry in the segment level list stores the pointer that indicates the next entry in the list. The maximum number of entries within a data queue equals the maximum number of segments the data memory supports. The packet pointer field has the valid bit set only in the entry that corresponds to the first segment of a packet. The packet pointer also indicates the address of the last segment of a packet. The number of entries of the packet lists is lower than the number of entries of the corresponding segment lists in a typical situation. However, in the worst case the maximum number of entries in the packet level lists is equal to the number of segment level lists, which equals the maximum number of the supported segments in the data memory.

Supporting two types of queues (packet, segment) requires two free lists, one per type. This results in double accesses for allocating and releasing pointers. The above flexible data structures, minimize memory accesses and can support the worst-case scenarios. The two types of linked lists are identical and aligned. In other words, there is only one linked list with two fields: segment and packet pointers. Segment pointers indicate the next segment in the list.

12.5.2.1   Optimization Techniques

The following subsections describe the optimization techniques used in the design to increase performance and reduce the cost of the system.

- **Free List Organization.** The DRAM provides high throughput and capacity at the cost of high latency and throughput limitations due to bank conflicts. A bank conflict occurs when successive accesses address the same bank, and in such case the second access must be delayed until the bank is available Bank conflicts reduce data memory throughput utilization. Hence, special care must be given to the buffer allocation and deallocation process. In [18] there is a proof of how, by using a single free list, a user can minimize the memory accesses during buffer releasing (i.e., delete or dequeue of a large size packet requires O(1) accesses to the pointer memory). However, this scheme increases the possibility of a bank conflict during an enqueue operation. On the other hand, using one free list per memory bank (total of eight banks in the current DRAM chips) minimizes or even avoids bank conflicts during enqueueing but increases the number of memory accesses during packet dequeueing-deletion to O(N). A trade-off of these two schemes, which minimizes the memory accesses and bank conflicts, is to use two free lists and allocate buffers for packet storing from the same free list. Additionally, the support of page-based addresses on the DRAM results in reduction up to 70 percent in the number of bank conflicts during writes and 46 percent during reads.

- **Memory Access Reordering.** The execution of an incoming operation, such as enqueue, dequeue, delete or append packet, sends read and write commands to the pointer memory to update the corresponding data structures. Successive accesses may be dependent. Due to access dependencies, the latency to execute an operation is increased. By reordering the accesses in an effective manner, the execution latency is minimized and thus the system performance increased. This reordering is performed for every operation and was measured to achieve a 30 percent reduction in the access latency.

- **Memory Access Arbitration.** Using the described free list organization, the write accesses to the data memory can be controlled to minimize the bank conflicts. Similar control cannot be performed to read accesses because they are random and unpredictable. Thus, a special memory access arbiter is used in the data memory controller block to shape the flow of read and write accesses to avoid bank conflicts. Memory accesses are classified in four FIFOs (one FIFO per port). The arbiter implements a round-robin policy. It selects an access only if it belongs to a non-busy bank. The information for bank availability is achieved by keeping the data memory access history (last three accesses). This

function can reduce bank conflicts by 23 percent. It also reduces the
hardware complexity of the DDR memory controller.

- **Internal Backpressure.** The data memory manager uses internal
  backpressure to delay incoming operations that correspond to blocked
  flows or blocked devices. The DMM keeps data FIFOs per output port.
  As soon as these FIFOs are about to overflow, alarm backpressure signals
  are asserted to suspend the flow of incoming operations related to this
  blocked datapath. Internal backpressure avoids overflows and data loss.
  This technique achieves DDM engine architecture reliability by using
  simple hardware.

12.5.2.2   Performance Evaluation of Hardware Queue Management Engine

Experiments on the DMM design were performed with the support of software
and micro-code specifically developed for an IP packet filtering application
executed on the embedded micro-engines of the PRO3 NPU presented in [37].

TABLE 12.3: Packet Command and Segment Command Pointer Manipulation
Latency

Packet Command	Segment Command	Clock Cycles (5 ns)	Pointer Memory Accesses r: Read; w: Write
Enqueue	Enqueue	10	4r4w
Read	Read	10	3r
Dequeue	Read_N	10	3r
Append	Dequeue_N	13	Min 5 (3r2w)
			Max 8 (3r5w)
Ignore	Overwrite	10	3r
Delete	Overwrite_Segment_length	7	2r1w
Ignore+Delete	Dequeue	13	Min 5 (3r2w)
			Max 8 (3r5w)
	Ignore	4	0
	Ignore+Overwrite_Segment_length	7	2r1w
	Overwrite_Segment_length+Append	11	6r4w
	Overwrite_Segment+Append	11	6r3w

In Table 12.3, the commands supported by the DMM engine are listed.
Note that the packet commands are internally translated into segment commands and only segment commands are executed at the low level controller.
Table 12.3 also shows the measured latency of these commands when executing the pointer manipulation functions. The actual data access at the data
memory can be done almost in parallel with the pointer handling. In particular, the data access can start after the first pointer memory access of each

command has been completed. This is achieved because the pointer memory accesses of each command have been scheduled so that the first one provides the data memory address. Hence, DMM can always handle a queue instruction within 65 ns. Since the data memory is accessed at about 50-60 ns (at the average case), and the major part of the queue handling is done in parallel with the data access, the above DMM engine introduces a minimum latency on the whole system. In other words, in terms of latency, you get the queue handling almost "for free", since the DMM latency is about the same as that of a typical (support only read and write) DRAM subsystem.

Table 12.4 depicts the performance results measured after stressing the DMM with real TCP traffic plugged to the NPU ingress interface (supporting one 2.5 Gbps ingress and one 2.5 Gbps egress interface). This table demonstrates the performance of the DMM in terms of both bandwidth and number of instructions serviced. It also presents the memory bandwidth required by our design to provide the performance specified.

TABLE 12.4: Performance of DMM

Number of Flows	AVG packet size (bytes)	MOperations/s serviced	Pointer Memory BW (Gb/s)	DMM BW (Gb/s)
2	100	8.22	4.53	7.60
2	90	10.08	4.45	9.72
2	128	11.26	4.40	9.20
4	128	10.05	3.70	8.40
4	128	9.44	3.80	9.20
Single	64	10.47	2.68	5.32
Single	64	13.70	3.74	7.04
Single	64	15.43	4.50	9.52
Single	50	13.43	4.42	6.88

Since the DMM in the case of the above 2.5 Gbps NPU should actually service each packet four times, the maximum aggregate throughput serviced by it is 10 Gb/sec. From the results of Table 12.4 it can easily be derived that the worst case is when there is only one incoming flow, which consists of very small packets. This worst case can still be served by the DMM engine operating at 200 Mhz while at the same time having a very large number of idle cycles (more than 25 percent even in the worst case). As described above, a simple DRAM can provide up to 17 Gb/sec of real bandwidth while the SRAM up to 8.5 Gb/sec. The maximum memory bandwidth utilization figures show that even in the worst case scenario the bandwidth required by the DRAM is up to about 14 Gb/sec (equal to the DMM bandwidth plus the measured 37 percent overhead due to bank conflicts and fragmentation) and that of the SRAM is 4.5 Gb/sec. As the internal hardware of the DMM in any of these cases is idle for more than 30 percent of the time, the specific DMM engine design could provide even a sustained bandwidth of 12 Gb/sec.

12.6 Scheduling Engines

Scheduling in general is the task of regulating the start and end times of events that contend for the same resource, which is shared in a time division multiplexing (TDM) fashion. Process scheduling is found as a major function of operating systems that control multi-threaded/multiprocessing computer systems ([31], [16], [5]. Packet scheduling is found in modern data networks as a means of guaranteeing the timely delivery of data with strict delay requirements, hence guaranteeing acceptable QoS to real-time applications and fair distribution of link resources among flows and users. Scheduling in a network processor environment is required either to resolve contention for processing resources in a fair manner (task scheduling), or to distribute in time the transmission of packets/cells (in a network medium) due to traffic management rules (traffic scheduling and/or shaping).

Although electronic technology advances rapidly, all of the NPU architectures discussed above are able to perform protocol processing at wire speed only on a long observation window, imposing buffering needs prior to processing. In the context of network processing described in this chapter, when applying complex processing at the processing elements (PEs), a long latency is inadvertently introduced. In order to efficiently utilize the processing capabilities of the node without causing QoS deterioration to packets from critical applications, an efficient queuing and scheduling mechanism (we will use the term *task scheduling* hereafter) for the regulation of the sequence of events related to the processing of the buffered packets is required. An additional implication stems from the multiprocessing architectures, which are most times employed to achieve the required performance that cannot be achieved by a single processing unit. This introduces an additional consideration in the scheduler design with respect to the maintenance of coherent protocol processing to cope with pipelined or parallel processing techniques, which are also very common.

In the outgoing path of the dataflow through the network processing elements, the transmission profile of the traffic leaving the nodes needs appropriate shaping, to achieve the expected delay and jitter levels. Since the internal scheduling and processing may have altered the temporal traffic properties (i.e., delaying some packets more than others, causing the so-called jitter or burstiness in the traffic profile), or because an application requirement to implement rate control for ingress traffic by a traffic scheduler or shaper appropriately adjusting the temporal profile of packet transmission (called hereafter *traffic scheduler*) is imposed.

12.6.1 Data Structures in Scheduling Architectures

In this section we will describe the basic building blocks of scheduling entities able to support both fixed and variable size packets, and to operate at high speeds, consuming few hardware resources. Such functional entities are frequently found as specialized micro-engines in several NPU architectures, or can be the basic functional elements of specialized NPUs designed to implement complex scheduling of packets across many thousands of packet flows and across many network ports.

The algorithmic complexity of proportionate time sharing solutions is based on per-packet time interval calculations and marking/time-stamping. Such algorithms are applicable only for scheduling tasks that have a predetermined completion time and increased complexity. Many studies have focused on analyzing the trade-offs between accurate implementation of algorithms theoretically shown to achieve fair scheduling among flows and simplified time representation and computations along with aggregate traffic handling to reduce memory requirements related to the handling of many thousands of queues. The simplest scheme for service differentiation is based on serving in simple FIFO order flows, classified based on the destination and/or priority of the corresponding traffic. This service discipline can be applied to traffic with different destinations (output ports or processing units), through the instatiation of multiple FIFOs, to avoid head of line (HOL) blocking.

A frequently employed technique that can reduce complexity and increase the throughput of the scheduler implementation with insignificant performance degradation is the grouping of flows with similar requirements into scheduling queues (SQ). Therefore, while a large number of actual data queues (DQs) can be managed by the queue manager, only a limited number of SQs need to be managed by the scheduler, greatly reducing memory requirements. In the simplest case such grouping can be used to implement a strict prioritized service, i.e., highest priority FIFOs are always serviced first until they become empty. This may also lead to starvation of lower priority queues. In order to avoid the starvation problem, queues need to be served in a cyclic fashion. In the simplest case flows within the same priority group are serviced in round-robin (RR) fashion as in [21]. A more general extension of the above approach results in a weighted round-robin service among NS flow groups (SQs) with proportional service (possibly extended in hierarchical hybrid schemes, e.g., implementing strict priorities between super groups): In this case the flows of the same priority are grouped in NS queues, which are served in a weighted round-robin manner, following an organization similar to that described in [42].

The rationale beyond grouping is to save and move implementation resources from detailed flow information to more elaborate resource allocation mechanisms and to improve overall performance applying the proper classification scheme. Assuming that an information entry is kept per schedulable entity, grouping a number of NF flows to a number NS of scheduling queues, a (NF-NS) reduction in storage requirements is achieved. The economy on memory resources regarding the number of pointers is significant since only 2* NS pointers for the management of the NS priority queues are required (and can be stored on-chip), plus NF next (flow) pointers. Flows are grouped to scheduling queues according to some classification rule which depends on the application/configuration. Although the mapping of flows to scheduling queues requires some information maintenance, it requires much less than the size of saved information. Apart from the memory space requirement reduction, reducing the number of schedulable entities facilitates a high decision rate in general, which proves to be mandatory for high-speed network applications.

12.6.2 Task Scheduling

In NPUs, the datapath through the system originates at the network interface, where packets are received and buffered in the input queue. Considering a parallel implementation of multiple processing elements (PEs) the role of the task scheduler is to assign a packet to each of the PEs whenever one of them becomes idle, after verifying that the packet/flow is eligible for service. This latter eligibility check of a packet from a specific flow before forwarding to a processor core (PE) is mandatory in order to maintain the so-called processor consistency. A multiprocessor is said to be processor consistent if the result of any execution is the same as if the operations of each individual processor appear in the sequential order specified by its program. To do this effectively, the scheduler can pick any of the packets in the selection buffer, cross-checking a state table indicating the availability of PEs as well as potential state-dependencies of a specific flow (e.g., packets from the same flow may not be forwarded to an idle processor if another packet is already under processing in one of the PEs in order to avoid state dependencies). A packet removed from the selection buffer for processing is replaced by the next packet from the input queue. Processed packets are placed into an output queue and sent to the outgoing link or the switch fabric of a router.

Hardware structures for the efficient support of load balancing of traffic in such multiprocessor SoCs in very high speed applications have appeared only recently. The assumptions and application requirements in these cases differ significantly from the processing requirements and programming models of high-speed network processing systems. NPUs represent a typical multiprocessing system where parallel processing calls for efficient internal resource management. However, the network processor architectures usually follow the run-to-completion mode, distributing processing tasks (which actually represent packet and protocol processing functions) on multiple embedded processing cores, rather than following complex thread parallelism. Load balancing has also been studied in [22]. The analysis included in [22] followed the assumption of multiple network processors operating in parallel with no communication and state sharing capabilities between them, as well as the requirement to minimize re-assignments of flows to different units for this reason as well as to avoid packet reordering. These assumptions are relaxed when the processing units are embedded on a single system-on-chip and access to shared memories as well as communication and state locking mechanisms are feasible. Current approaches for processor sharing in commercial NPUs are discussed below.

The IBM PowerNP NP4GS3 [3] includes eight dyadic protocol processing units (DPPUs) and each one contains two core language processors (CLPs). Sixteen threads at most can be active, even though each DPPU can support up to four processes. The dispatch event controller (DEC) schedules the dispatch of work to a thread, and is able to load balance threads on the available DPPUs and CLPs, while a completion unit detects their state and maintains frame order within communication flows. However, it can process 4.5 million packets per second layer 2 and layer 3 switching, while operating at 133 MHz.

The IXP 2800 network processor [19] embeds 16 programmable multi-threaded micro-engines that utilize super-pipeline technology that allows the forwarding of data items between neighboring micro-engines. Hence, the processing is based on a high-speed pipeline mechanism rather than on associating one micro-engine to the processing of a full packet (although this latter case is possible via its local dispatchers).

The Porthos network processor [32] uses 256 threads in 8 processing engines. In this case, in-order packet processing is controlled mainly by software, and it is assisted by a hardware mechanism which tags each packet with a sequence number. However, load balancing capability is limited and completely controlled by software.

12.6.2.1 Load Balancing

An example of such an on-chip core has been presented in [25]. The scheduler/load balancer presented in [25] is designed to allocate the processing resources to the different packet flows in a fair (i.e., weighted) manner according to pre-configured priorities/weights, whereas the load balancing mechanism supports efficient dispatching of the scheduled packets to the appropriate destination among a set of embedded PEs. The main datapath of the NPU in this case is the one examined in previous chapters and is shown in Figure 12.21.

This set of PEs shown in Figure 12.21 may be considered to be of similar capacity and characteristics (so effectively there is only one set) or may be differentiated to independent sets. In any case each flow is assigned to such a set. A load-balancing algorithm is essential to evenly distribute packets to all the processing modules. The main goal for the load balancing algorithm is that the workloads of all the processing elements are balanced and throughput degradation is prevented. Implementation of load balancing in this scheme is done in two steps. First, based on the results of the classification stage, packet flows undergoing the same processing (e.g., packets from similar interfaces using the same framing and protocol stacks and enabled services use a pre-defined set of dedicated queues) are distributed among several queues, which

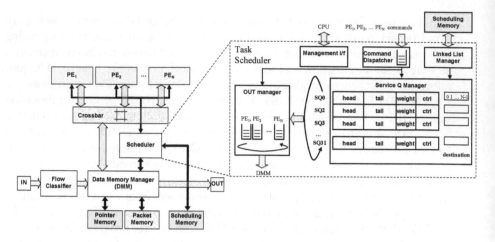

FIGURE 12.21: Details of internal task scheduler of NPU architecture described in [25].

represent the set of internal flows that are eligible for parallel processing. In the second step, the task scheduler based on the information regarding the traffic load of these queues (i.e., packet arrival events) selects and forwards packets based on an appropriate service policy. The application software that is executed by a PE on a given packet is implicitly defined by the flow/queue to which the packet belongs.

Pre-scheduled flows are load-balanced to the available PEs by means of a strict service mechanism. The scheme described in [25] is based on the implementation of an aging mechanism used in the core of a crossbar switch fabric (this on-chip interconnection architecture is usually called network-on-chip, NoC). The reference crossbar switch used is based on a traditional implementation of a shared-memory switch core; all target ports corresponding to a programmable processing core (PE) access this common resource with the aid of a simple arbitrating mechanism.

A block diagram of the load-balancing core described in [25] is shown in Figure 12.22. The main hardware data structures used (assuming 64 available on-chip PEs and 1M flows) are the following:

- The free list maintains the occupancy status of the rest tables; any PE may select any waiting flow residing in the flow memory.

- The FLOW memory records the flow identifier of a packet waiting to be processed.

- The DEST memory stores a mask denoting the PEs that can process this packet (i.e., can execute the required application code).

- The AGE memory stores the virtual time (in the form of a bit vector) based on the arrival time of this packet (it is called virtual, because it represents the relative order among the scheduled flows).

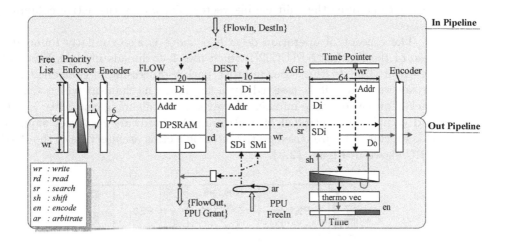

FIGURE 12.22: Load balancing core implementation [25].

Two pipeline operations are executed in parallel for incoming packets and packets that have completed processing at the PEs. The in pipeline is triggered if no PE service is in progress. It is responsible to store the flow identifier in an available slot provided by the free list, and mark the corresponding destination mask in the DEST table. The virtual time indicator is updated and stored in the AGE table, aligned with the flow identifier and the destination mask, which is used to guarantee service according to the arrival times of the requests. Finally, a filter/aggregation mask is updated to indicate the PEs needed to serve the waiting flows. After applying the filter mask to the PE availability vector indicator, the DEST table is searched to discover the flows scheduled to this PE. This is a "don't care" search, since the matching flows could be load-balanced to other PEs as well. In order to serve the oldest flow, the AGE table is searched in the next stage, based on the previous outcome. The outcome of this searching is: a) to read the AGE table and produce the winner flow to be served, and b) to shift all the younger flows to the right and automatically erase the age of the winner. The flow memory is read after encoding the previous outcome, and finally, the free list and DEST tables are updated accordingly. The basic structures used include multi-ported memories, priority enforcers, a content addressable memory (CAM) that allows ternary search operations (for the implementation of the DEST table). The most complex data structure is the AGE block. This is also a CAM, which differs in that it performs exact matches. Additionally, it has a separate read/write port and supports a special shift operation; it shifts each column vector to the right when indicated by a one in a supplied external mask. The circuit thermo vec performs a thermometer decoding. It transforms the winner vector produced by the priority enforcer to a sequence of ones from the located bit position until the left most significant bit. This is "ANDed" with the virtual time vector to produce the shift enable vector. Thus, only the active columns are shifted.

The concept of operation described above is also similarly found in the case of the Porthos NPU ([32]) which uses 256 threads in eight PEs (called tribes in [32]). In this case, in-order packet processing is controlled mainly by software, and it is assisted by a hardware mechanism which tags each packet with a sequence number. However, load balancing capability is limited and completely controlled by software. The hardware resources supporting this functionality are based on the interconnection architecture of the Porthos NPU shown in Figure 12.23.

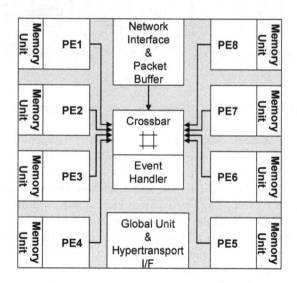

FIGURE 12.23: The Porthos NPU interconnection architecture [32].

The Porthos interconnect block consists of three modules: event handling, arbiter and crossbar (comprising 10 input and 8 output 64-bit wide ports supporting backpressure for busy destinations). The event module collects event information and activates a new task to process the event. It spawns a new packet processing task in one of the PEs via the interconnect logic based on external and timer (maskable) interrupts. There are two (configurable) methods to which an interrupt can be directed. In the first method, the interrupt is directed to any PE that is not empty. This is accomplished by the event module making requests to all eight destination PEs. When there is a grant to one PE, the event module stops making requests to the other PEs and starts the interrupt handling process. In the second method, the interrupt is directed to a particular PE.

The arbiter module performs arbitration between sources (PEs, network block, event handling module and transient buffers) and destinations. The arbiter needs to match the source to the destination in such a way as to maximize utilization of the interconnect, while also preventing starvation using a round-robin prioritizing scheme. This matching is performed in two stages. The first stage selects one non-busy source for a given non-busy destination. The second stage resolves cases where the same source was selected for multiple destinations. The arbitration scheme implemented in Porthos is "greedy", meaning it attempts to pick the requests that can proceed, skipping over sources and destinations that are busy. In other words, when a connection is set up between a source and a destination, the source and destination are locked out from later arbitration. With this scheme, there are cases when the arbiter starves certain contexts. It could happen that two repeated requests, with overlapping transaction times, can prevent other requests from being processed. To prevent this, the arbitration may operate in an alternative mode. In the non-greedy mode for each request that cannot proceed, there is a counter that keeps track of the number of times that request has been skipped. When the counter reaches a configurable threshold, the arbitration will not skip over this request, but rather wait at the request until the source and destination become available. If multiple requests reach this priority for the same source or destination, one-by-one they will be allowed to proceed in a strict round-robin fashion.

An architectural characteristic of the Porthos NPU is the support for the so-called task migration from one PE to another, i.e., a thread executing on a PE transferred to another. When migration occurs, a variable amount of context follows the thread. Specific thread migration instructions are supported specifying a register that contains the destination address and an immediate response that contains the amount of thread context to preserve. Thread migration is a key feature of Porthos, providing support to the overall loosely coupled processor/memory architecture. Since the memory map is not overlapped, every thread running in each PE has access to all memory. Thus, from the standpoint of software correctness, migration is not strictly required. However, the ability to move the context and the processing to an engine that is closer to the state that the processing is operating on allows a flexible topology to be implemented. A given packet may do all of its processing in a specific PE, or may follow a sequence of PEs (this decision potentially is made on a packet-by-packet basis). Since a deadlock can occur when the PEs in migration loops are all full, Porthos uses two transient buffers in the interconnect block to break such deadlocks, with each buffer capable of storing an entire migration (66 bits times maximum migration cycles). These buffers can be used to transfer a migration until the deadlock is resolved by means of atomic software operations at a cost of an additional delay.

12.6.3　Traffic Scheduling

The Internet and the associated TCP/IP protocols were initially designed to provide best-effort (BE) service to end users and do not make any service quality commitment. However, most multimedia applications are sensitive to available bandwidth and delay experienced in the network. To satisfy these requirements, two frameworks have been proposed by IETF: the integrated services (IntServ), and the differentiated services (DiffServ) [47], [26]. The IntServ model provides per-flow QoS guarantee and RSVP (resource reservation protocol) is suggested for resource allocation and admission control. However, the processing load is too heavy for backbone routers to maintain state of thousands of flows. DiffServ is designed to scale to large networks and gives a class-based solution to support relative QoS. The main idea of Diff-Serv is to minimize state and per-flow information in core routers by placing all packets in fairly broad classes at the edge of network. The key ideas of DiffServ are to: (a) classify traffic at the boundaries of a network, and (b) condition this traffic at the boundaries. Core devices perform differentiated aggregate treatment of these classes based on the classification performed by the edge devices. Since it is highly scalable and relatively simple, DiffServ may dominate the next generation Internet in the near future. Its implementation in the context of a network routing/switching node is shown in Figure 12.24.

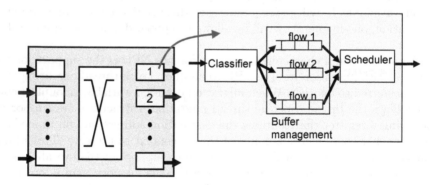

FIGURE 12.24: Scheduling in context of processing path of network routing/switching nodes.

In DiffServ, queues are used for a number of purposes. In essence, they are only places to store traffic until it is transmitted. However, when several queues are used simultaneously in a queueing system, they can also achieve effects beyond those for given traffic streams. They can be used to limit variation in delay or impose a maximum rate (shaping), to permit several streams to share a link in a semi-predictable fashion (load sharing) or move variation in delay from some streams to other streams. Queue scheduling schemes can be divided into two types: work-conserving and non-work-conserving. A policy is work-conserving if the server is never idle when packets are backlogged. Among work-conserving schemes, fair queueing is the most important category. WFQ (weighted fair queueing) [38], WF2Q, WF2Q+ [4] and all other GPS-based queueing algorithms belong to fair queueing. Another important type of work-conserving is the service curve scheme, such as SCED [8] and H-FSC [43]. The operation of these algorithms is schematically described in Figure 12.25.

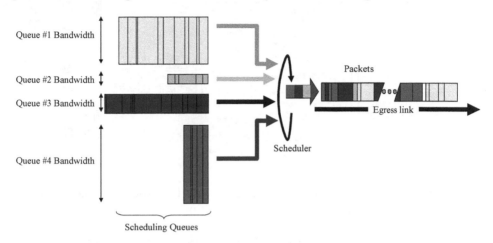

FIGURE 12.25: Weighted scheduling of flows/queues contending for same egress network port.

All these schemes present the traffic distortion [29] problem and traffic characterization at the entrance of the network would not be valid inside the network. And traffic can get more bursty in the worst case. In downstream switches, more buffer spaces are required to handle traffic burstiness and the receiver also needs more buffer space to remove jitter. Non-work-conserving schemes (also called shapers) are proposed in order to control traffic distortion inside a network. A policy is non-work-conserving if the server may be idle even when packets are backlogged. From the definition we can see that non-work-conserving schemes allow the output link to be idle even when there are packets waiting for service in order to maintain the traffic pattern. So bandwidth utilization ratio may be not be high in some cases.

The design of weighted schedulers can follow the generic architecture described above for task scheduling to implement multiple traffic management mechanisms in an efficient way. An extension of the NPU architecture that could exploit these traffic management extensions is shown in Figure 12.26(a). This architecture suits better the needs of multi-service network elements found in access and edge devices that act as traffic concentrators and protocol gateways. This architecture represents a gateway-on-chip paradigm exploiting the advances in VLSI technology and SoC design methodologies that enable the easy integration of multiple IP cores on complex designs. In cases like this the queuing and scheduling requirements are complicated. Apart from the high number of network flows and classes of service (CoS) that need to be supported, another hierarchy level is introduced that necessitates the extension of the scheduler architecture described above to support multiple virtual and physical output interfaces as shown in Figure 12.26(b).

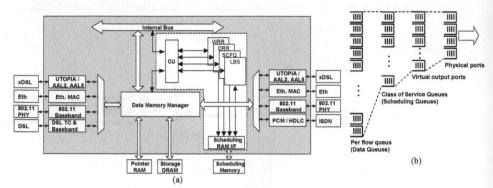

FIGURE 12.26: (a) Architecture extensions for programmable service disciplines. (b) Queuing requirements for multiple port support.

The generic scheduler architecture, as described in Figure 12.21, and following the organization presented in Figure 12.26 (a) and (b) which incorporates the internal to the NPU task scheduler inherently supports these hierarchical queuing requirements by means of independent scheduling per DQ, SQ and destination (port). Furthermore, the same module can implement different service disciplines (like WRR and DRR) in a programmable fashion with the same hardware resources. Thus, by proper organization of flows under SQs per CoS, efficient virtual and physical port scheduling can also be achieved as described in [35]. Implementation of more scheduling disciplines can also be achieved easily, by simply adding the service execution logic (finite state machine or FSM) as a co-processing engine, since the implementation area is small and operation and configuration is independent among them. Even a large number of schedulers could be integrated at low cost. Apart from the implementation of additional FSMs and potentially the associated on-chip memory (although insignificant) the only hardware extension required is the extension of the arbiter and memory controller modules to support a larger number of ports. The required throughput of the pointer memories used remains the same as long as the aggregate bandwidth of the incoming network interfaces is at most equal to the throughput offered by the DMM unit. The only limitation is related to the number of supported SQs, which represent one CoS queue each. Thus, the number of independently scheduled classes of service is directly proportional to the hardware resources that will be allocated for the implementation of the SQ memories and priority enforcers for fast searching in these memories, which can be extended to very high numbers of SQs as presented in [24]. In addition, functionality already present in the current scheduler implementation allows for deferring service of one SQ and manipulation of its parameters under software control. This feature offers itself for easy migration of one CoS from one scheduling discipline to another in this extended architecture.

With these extensions the NPU can efficiently support concurrent scheduling mechanisms for network traffic, crossing even dissimilar interfaces. Scheduling of variable length packet flows having as destinations packet interfaces (like Ethernet, packet-over-SONET etc.) can be scheduled by means of a packet scheduling algorithm like DRR or self-clocked fair queueing (SCFQ, [12]). The efficient implementation of packet fair queuing algorithms like SCFQ, according to the generic methodology presented in this section has also been discussed in [39]. Moreover, a novel feature of this architecture is its flexibility to implement hierarchical scheduling schemes only with pointer movement without necessitating data movement. Scheduling packets over multiple interfaces of the same type (e.g., multiple Ethernet interfaces) is easily achieved by assigning appropriate weights (that represent the relative share of a flow with respect to the aggregate capacity of the physical links) and different destinations (port) per flow. The only remaining hardware issue that requires attention is the handling of busy indication signals from the different physical ports to determine schedulable flows/SQs.

12.7 Conclusions

State-of-the-art telecommunication systems require modules with increased throughput in terms of packets processed per second and with advanced functionality extending to multiple layers of the protocol stack. High-speed data-path functions can be accelerated by hard wired implementations integrated as processing cores in multi-core embedded system architectures. This allows each core to be optimized either for processing intensive functions to alleviate bottlenecks in protocol processing or intelligent memory management techniques to sustain the throughput for data and control information storage and retrieval and exceed the performance of legacy SW-based implementations on generic microprocessor based architectures, which cannot scale to gigabit-per-second link rates.

The network processing units (NPUs) that we examined in this chapter in the strict sense are fully programmable chips like CPUs or DSPs but, instead of being optimized for the task of computing or digital signal processing, they have been optimized for the task of processing packets and cells. In this sense NPUs combine the flexibility of CPUs with the performance of ASICs, accelerating the development cycles of system vendors, forcing down cost, and creating opportunities for third-party embedded software developers.

NPUs in the broad sense encompass both dedicated and programmable solutions:

- Dedicated line-aggregation devices that combine several channels of high-level data link control support sometimes optimized for a specific access system such as DSL

- Intelligent memories, e.g., content-addressable memories that support efficient searching of complex multi-dimensional lookup-tables

- Application-specific ICs optimized for one specific protocol processing task, e.g., encryption

- Programmable processors optimized for one specific protocol processing task, e.g., header parsing

- Programmable processors optimized for several protocol processing tasks

The recent wave of network processors is aimed at packet parsing and header analysis. Two evolutions favor programmable implementations. First, the need to investigate and examine more header fields covering different layers of the OSI model, make an ASIC implementation increasingly complex. Secondly, flexibility is required to deal with emerging solutions for QoS and other services that are not yet standardized. The challenge for the programmable network processors lies in the scalability to core applications running at 10 Gbits/s and above (which is why general-purpose processors are not up to the job).

The following features of network processors have been taken into account to structure this case study:

- Target application domain (LAN, access, WAN, core/edge etc.).

- Target function (data link processing including error control, framing etc., classification, data stream transformation including encryption, compression etc., traffic management including buffer management, prioritization, scheduling, shaping, and higher layer protocol/FSM implementation)

- Architecture characteristics including:

  - Architecture based on instruction-set processor (ISP), programmable state machine (PSM), ASIC (non-programmable), intelligent memory (CAM)
  - Type of ISP (RISC, DSP)
  - Centralized or distributed architecture
  - Programmable or dedicated
  - DSP acceleration through extra instructions or co-processors

  − Presence of re-configurable hardware

- Software development environment (for programmable NPUs)

- Performance in terms of data rates

- Implementation: processing technology, clock speed, package, etc.

Review Questions and Answers

[Q_1] **What is the range of applications that are usually executed in network processing units?**
Refer to Section 12.1 of the text. Briefly, network processing functions can be summarized as follows:
- Implementation of physical ports, physical layer processing and traffic switching
- Framing
- Classification
- Modification
- Content/protocol processing
- Traffic engineering, scheduling and shaping algorithms

[Q_2] **What are the processing requirements and the bottlenecks that led to the emergence of specialized NPU architectures?**
Two major factors drive the need for NPUs: i) increasing network bit rate, and ii) more sophisticated protocols for implementing multi-service packet-switched networks. NPUs have to address the above communications system performance issues by coping with three major performance-related resources in a typical data communications system:

1. Processing cores (limited processing capability and frequency of operation of single, general purpose, processing units)
2. System bus(es) (limited throughput and scalability)
3. Memory (limited throughput and scalability of centralized architectures)

[Q_3] **What are the main differences in NPU architectures targeting access/metro networks compared to those targeting core networks?**
Due to the different application requirements there are the following differences:
- Overall throughput (access processors usually achieve throughputs in

the order of 1 Gbps, which is adequate for most access network technologies, whereas core networks may require an order of magnitude higher bandwidth i.e., 10-40 Gbps)

• Number of processing cores (single-chip IADs can integrate only a couple of general purpose CPUs, whereas high-end NPUs can integrate 4-64 processing cores)

• Multiplicity and dissimilarity of interfaces/ports (access processors frequently must support bridging between multiple networks of different technologies, whereas core processors are required to interface to high-speed line-cards and switching fabrics through a limited set of standardized interfaces)

• Architectural organization (access processors frequently require custom processing units since intelligent content processing, e.g., (de)encryption, (de)compression, transcoding, content inspection, etc. is usually pushed to the edge of the network, whereas core processors require ultimate throughput and traffic management which is addressed through massively parallel, pipelined and programmable FSM architectures with complicated memory management and scheduling units)

[Q_4] **Why is latency not very important when packet-processing tasks are executed on a network processor? What happens when such a task is stalled?**
Usually a network processor time-shares between many parallel tasks. Typically such tasks are independent, because they process packets from different flows. So, when a task is stalled (e.g., on a slow external memory access or a long-running coprocessor operation) the network processor switches to another waiting task. In this way, almost no processing cycles get wasted.

[Q_5] **Which instruction-level-parallel processor architecture is more area-efficient: superscalar, or VLIW? Why?**
Very long instruction word (VLIW) is more area-efficient than superscalar, because the latter includes a lot of logic dedicated to "discovering" at run time instructions that can be executed simultaneously. VLIW architectures include no such logic; instead they require that the compiler schedules instructions statically at compile time.

[Q_6] **What are the pros and cons of homogeneous and heterogeneous parallel architectures?**
By specializing each processing element to a narrowly defined processing task, heterogeneous architectures can achieve higher peak performance for each individual task. On the other hand, with such architectures, one has to worry about load balancing: the system architects need to choose the correct number and performance of each type of PE, a problem with no general solution, while the users must be careful to code

the applications in a way that balances the load between the different kinds of available PEs. With homogeneous architectures, the architect only needs to replicate a single type or PE as many times as silicon area allows, while the user can always take advantage of all available PEs. This of course comes at the cost of lower peak PE performance.

[Q.7] **Define multi-threading.**
A type of lighweight time-sharing mechanism. Threads are akin to processes in the common operating system sense, but hardware support allows very low (sometimes zero) overhead when switching between threads. So, it is possible to switch to a new thread even when the current thread will be stalled for a few clock cycles (e.g., an external memory access or an operation executed on a coprocessor). This allows the processor to take advantage of (almost) all processing cycles, by making progress on an alternate thread when the current one is stalled even briefly.

[Q.8] **Explain how the PRO3 processor overlaps processing with memory accesses.**
Refer to Section 12.3.2 of the text.

[Q.9] **Mention some types of custom instructions specific to network processing tasks.**
• Extraction of bit fields of arbitrary lengths and from arbitrary offsets within a word
• Insertion of bit fields of arbitrary lengths into arbitrary offsets within a word
• Parallel multi-way branches (or parallel comparisons, as in the IXP architecture)
• CRC/checksum calculation or hash function evaluation

[Q.10] **Define the problem of packet classification**
Refer to the introductory part of Section 12.4 of the text.

[Q.11] **Name a few applications of classification**
• Destination lookup for IP forwarding
• Access control list (ACL) filtering
• QoS policy enforcing
• Stateful packet inspection
• Traffic management (packet discard policies)
• Security-related decisions
• URL filtering
• URL-based switching
• Accounting and billing

[Q.12] **What are the pros and cons of CAM-based classification architectures?**

CAM-based lookups are the fastest and simplest ways to search a rule database. However, CAMs come at high cost and have high power dissipation. In addition, the capacity of a CAM device may enforce a hard upper bound on database size. (Strictly speaking, the same is true for algorithmic architectures, but since these usually rely on low-cost DRAM, it is easier to increase the memory capacity for large rule databases.)

[Q_13] How does iFlow's Address Processor exploit embedded DRAM technology?

First of all, it combines the database storage with all the necessary lookup and update logic into one device, thus reducing overall cost. Second, it takes advantage of a very wide internal memory interface to read out many nodes of the data structure and make that many comparisons in parallel, thus improving performance.

[Q_14] What are the main processing tasks of a queue management unit?

Refer to the introductory parts of Sections 12.5 and 12.5.2 of the text.

[Q_15] What are the criteria of selecting memory technology when designing queue management units for NPU?

Refer to Sections 12.5.1 and 12.5.2 of the text. Briefly, the memory technology of choice should provide:
• Adequate throughput depending on the data transaction (read/write, single/burst etc.) requirements
• Adequate space depending on the storage requirements
• Limited cost, board space and power consumption

[Q_16] What are the main bottlenecks in queue management applications and how are they addressed in NPU architectures?

Refer to Sections 12.5.1 and 12.5.2 of the text. Briefly, the main performance penalties are due to timing limitations related to successive memory operations depending on the memory technology. DRAM is an indicative case of memory technology which requires sophisticated controllers due to the limitations in the order of accesses, depending on its organization in banks, its requirements for pre-charging cycles, etc., to enhance its performance and better utilize its resources. Such controllers can enhance memory throughput through multiple techniques e.g., appropriate free list organization, appropriate scheduling of accesses requested by multiple sources enforcing reordering, arbitration, internal backpressure, etc.

[Q_17] What are the similarities and differences between task and traffic scheduling?

Both applications are related to resource management based on QoS criteria. In general scheduling refers to the task of ordering in time the execution of processes, which can either be processing tasks that require the exchange of data inside an NPU or transmission of data packets in a limited capacity physical link. In both cases the data on which the process is going to be executed are ordered in multiple queues served with an appropriate discipline that guarantees some performance criteria (delay, throughput, data loss etc.). Depending on the application, different requirements need to be met, in order delivery, rate-based flow limiting, etc. Task processing has three important differences in the way it should be implemented: i) the finish time of a processing task in contrast to packet transmission delays, which depend only on link capacity and packet length, may be unknown or hard to determine (e.g., due to the stochastic nature of branch executions that depend on the content of data which are not *a priori* known to the scheduler), ii) the availability of the resources varies dynamically and may have specific limitations due to dependencies in pipelined execution or atomic operations in parallel processing, etc., and iii) the optimization of throughput in task scheduling may require load balancing, i.e., distribution of tasks to any available resource whereas traffic scheduling needs to coordinate requests for access to the same predetermined resource (i.e., port/link).

[Q.18] **What are the main processing tasks of a traffic scheduling unit?**
Refer to Section 12.6 of the text. Briefly, traffic scheduling requires the implementation of an appropriate packet queuing scheme (a number of priority queues, possibly hierarchically organized) and the implementation of an appropriate arbitration scheme either in a deterministic manner or in the most complex case computing per packet information (finish times) and sorting appropriately among all packets awaiting service (e.g., DRR, SCFQ, WFQ-like algorithms etc.).

Bibliography

[1] BGP routing tables analysis reports. http://bgp.potaroo.net.

[2] Matthew Adiletta, Mark Rosenbluth, Debra Bernstein, Gilbert Wolrich, and Hugh Wilkinson. The next generation of Intel IXP network processors. *Intel Technology Journal*, 6(3):6–18, 2002.

[3] J.R. Allen, B.M. Bass, C. Basso, R.H. Boivie, J.L. Calvignac, G.T. Davis, L. Frelechoux, M. Heddes, A. Herkersdorf, A. Kind, J.F. Logan, M. Peyravian, M.A. Rinaldi, R.K. Sabhikhi, M.S. Siegel, and M. Waldvogel. IBM PowerNP network processor: hardware, software, and applications. *IBM Journal of Research and Development*, 47(2/3):177–193, 2003.

[4] Jon C. R. Bennett and Hui Zhang. Why WFQ is not good enough for integrated services networks. In *Proceedings of NOSSDAV '96*, 1996.

[5] Haiying Cai, Olivier Maquelin, Prasad Kakulavarapu, and Guang R. Gao. Design and evaluation of dynamic load balancing schemes under a fine-grain multithreaded execution model. Technical report, *Proceedings of the Multithreaded Execution Architecture and Compilation Workshop*, 1997.

[6] C-port Corp. C-5 network processor architecture guide, C5NPD0-AG/D, May 2001.

[7] Patrick Crowley, Mark A. Franklin, Haldun Hadimioglu, and Peter Z. Onufryk. *Network Processor Design: Issues and Practices*. Morgan Kaufmann, 2003.

[8] R. L. Cruz. SCED+: efficient management of quality of service guarantees. In *Proceedings of INFOCOM'98*, pages 625–642, 1998.

[9] EZchip. Network processor designs for next-generation networking equipment. http://www.ezchip.com/t_npu_whpaper.htm, 1999.

[10] EZchip. The role of memory in NPU system design. http://www.ezchip.com/t_memory_whpaper.htm, 2003.

[11] V. Fuller and T. Li. Classless inter-domain routing (CIDR): The internet address assignment and aggregation plan. RFC4632.

[12] Jahmalodin Golestani. A self-clocked fair queueing scheme for broadband applications. In *Proceedings of INFOCOM'94 13th Networking Coference for Global Communications*, volume 2, pages 636–646, 1994.

[13] Network Processing Forum Hardware Working Group. Look-aside (LA-1B) interface implementation agreement, August 4 2004.

[14] Pankaj Gupta and Nick McKeown. Classifying packets with hierarchical intelligent cuttings. *IEEE Micro*, 20(1):34–41, 2000.

[15] Pankaj Gupta and Nick McKeown. Algorithms for packet classification. *IEEE Network*, 15(2):24–32, 2001.

[16] A. El-Mahdy I. Watson, G. Wright. VLSI architecture using lightweight threads (VAULT): Choosing the instruction set architecture. Technical report, Workshop on Hardware Support for Objects and Microarchitectures for Java, in conjunction with ICCD'99, 1999.

[17] IBM. PowerNP NP4GS3.

[18] Motorola Inc. Q-5 traffic management coprocessor product brief, Q5TMC-PB, December 2003.

[19] Intel. Intel IXP2400, IXP2800 network processors.

[20] ISO/IEC JTC SC25 WG1 N912. Architecture of the residential gateway.

[21] Manolis Katevenis. Fast switching and fair control of congested flow in broadband networks. *IEEE Journal on Selected Areas in Communications*, 5:1315–1326, Oct. 1987.

[22] Manolis Katevenis, Sotirios Sidiropoulos, and Christos Courcoubetis. Weighted round robin cell multiplexing in a general-purpose ATM switch chip. *IEEE Journal on Selected Areas in Communications*, 9, 1991.

[23] Donald E. Knuth. *The Art of Computer Programming*, volume 3: Sorting and Searching. Addison-Wesley, 1973.

[24] George Kornaros, Theofanis Orphanoudakis, and Ioannis Papaefstathiou. Active flow identifiers for scalable, qos scheduling. In *Proceedings IEEE International Symposium on Circuits and Systems ISCAS'03*, 2003.

[25] George Kornaros, Theofanis Orphanoudakis, and Nicholas Zervos. An efficient implementation of fair load balancing over multi-CPU SoC architectures. In *Proceedings of Euromicro Symposium on Digital System Design Architectures, Methods and Tools*, 2003.

[26] K. R. Renjish Kumar, A. L. An, and Lillykutty Jacob. The differentiated services (diffserv) architecture, 2001.

[27] V. Kumar, T. Lakshman, and D. Stiliadis. Beyond best-effort: Router architectures for the differentiated services of tomorrow's internet. *IEEE Communications Magazine*, 36:152–164, 1998.

[28] Sridhar Lakshmanamurthy, Kin-Yip Liu, Yim Pun, Larry Huston, and Uday Naik. Network processor performance analysis methodology. *Intel Technology Journal*, 6, 2002.

[29] Wing-Cheong Lau and San-Qi Li. Traffic distortion and inter-source cross-correlation in high-speed integrated networks. *Computer Networks and ISDN Systems*, 29:811–830, 1997.

[30] Panos Lekkas. *Network Processors. Architectures, Protocols, and Platforms*. McGraw-Hill, 2004.

[31] Evangelos Markatos and Thomas Leblanc. Locality-based scheduling in shared-memory multiprocessors. Technical report, *Parallel Computing: Paradigms and Applications*, 1993.

[32] Steve Melvin, Mario Nemirovsky, Enric Musoll, Jeff Huynh, Rodolfo Milito, Hector Urdaneta, Koroush Saraf, and Myers Llp. A massively multithreaded packet processor. In *Proceedings of NP2, Held in conjunction with HPCA-9*, 2003.

[33] Aristides Nikologiannis and Manolis Katevenis. Efficient per-flow queueing in DRAM at OC-192 line rate using out-of-order execution techniques. In *Proceedings of ICC2001*, pages 2048–2052, 2001.

[34] Mike O'Connor and Christopher A. Gomez. The iFlow address processor. *IEEE Micro*, 21(2):16–23, 2001.

[35] Theofanis Orphanoudakis, George Kornaros, Ioannis Papaefstathiou, Hellen-Catherine Leligou, and Stylianos Perissakis. Scheduling components for multi-gigabit network SoCs. In *Proceedings SPIE International Symposium on Microtechnologies for the New Millennium, VLSI Circuits and Systems Conference, Canary Islands*, 2003.

[36] Ioannis Papaefstathiou, George Kornaros, Theofanis Orphanoudakis, Kchristoforos Kachris, and Jacob Mavroidis. Queue management in network processors. In *Design, Automation and Test in Europe (DATE2005)*, 2005.

[37] Ioannis Papaefstathiou, Stylianos Perissakis, Theofanis Orphanoudakis, Nikos Nikolaou, George Kornaros, Nicholas Zervos, George Konstantoulakis, Dionisios Pnevmatikatos, and Kyriakos Vlachos. PRO3: a hybrid NPU architecture. *IEEE Micro*, 24(5):20–33, 2004.

[38] Abhay K. Parekh and Robert G. Gallager. A generalized processor sharing approach to flow control in integrated services networks: The single-node case. *IEEE/ACM Transactions on Networking*, 1:344–357, 1993.

[39] Jennifer Rexford, Flavio Bonomi, Albert Greenberg, and Albert Wong. Scalable architectures for integrated traffic shaping and link scheduling in high-speed ATM switches. *IEEE Journal on Selected Areas in Communications*, 15:938–950, 1997.

[40] Tammo Spalink, Scott Karlin, Larry Peterson, and Yitzchak Gottlieb. Building a robust software-based router using network processors. In *Proceedings of the 18th ACM Symposium on Operating Systems Principles (SOSP)*, pages 216–229, 2001.

[41] V. Srinivasan, S. Suri, G. Varghese, and M. Waldvogel. Fast and scalable layer four switching. In *Proceedings of ACM Sigcomm*, pages 203–214, September 1998.

[42] Donpaul C. Stephens, Jon C. R. Bennett, and Hui Zhang. Implementing scheduling algorithms in high speed networks. *IEEE JSAC*, 17:1145–1158, 1999.

[43] Ian Stoica, Hui Zhang, and T.S.E Ng. A hierarchical fair service curve algorithm for link-sharing, real-time, and priority services. *IEEE/ACM Transactions on Networking*, 8(2):185–199, 2000.

[44] Sandy Teger and David J. Waks. End-user perspectives on home networking. *IEEE Communications Magazine*, 40:114–119, 2002.

[45] K. Vlachos, T. Orphanoudakis, Y. Papaefstathiou, N. Nikolaou, D. Pnevmatikatos, G. Konstantoulakis, J.A. Sanches-P, and N. Zervos. Design and performance evaluation of a programmable packet processing engine (ppe) suitable for high-speed network processors units. *Microprocessors and Microsystems*, 31(3):188–199, May 2007.

[46] David Whelihan and Herman Schmit. Memory optimization in single chip network switch fabrics. In *Design Automation Conference*, 2002.

[47] Xipeng Xiao and Lionel M. Ni. Internet QoS: A big picture. *IEEE Network*, 13:8–18, 1999.

[48] Wenjiang Zhou, Chuang Lin, Yin Li, and Zhangxi Tan. Queue management for qos provision build on network processor. In *Proceedings of the The Ninth IEEE Workshop on Future Trends of Distributed Computing Systems (FTDCS'03)*, page 219, 2003.

Index

2D mesh topology, 113

abstraction, 170
    behavioral, 170, 176, 182, 188
    data, 170, 187
    data register access, 186
    levels of, 166
    scan-to-functional hierarchy, 186
    state, 166
    structural, 169, 176, 184
    target, 184
    temporal, 169, 176, 179, 182, 189
abstraction levels, 271
ADL, 39, 40
application
    delay, 228
    layer, 312
    specific, 7
    specific routing algorithm, 112
    task graphs, 217
application programming interface, 271
application-specific
    memory-aware customization, 48
    NPU instruction set, 421
    on-chip interconnect techniques, 40
    processor techniques, 37, 44
architecture description languages, 40
ARM, 22, 23
ASIP, 42
asynchronicity, 158
automated test equipment, 158
autonomous power management, 339
autonomous power management
    Intel Centrino, 356
    multi-core, 351

board support package, BSP, 314
buffer allocation, 246, 249, 254, 258
built in self test, 8

C programming the picoArray, 375
C-5e processor, 413
cache miss ratio, 249
CASPER, 350
CASPER
    branch predictor, 354
    cpi calculations, 350
    current calculations, 350
    global power management unit, 355, 358
    load miss queue, 354
    local power management unit, 354, 358
    power calculations, 350
    power control registers, 354
    register file, 353
    trap logic unit, 351, 354
Cell architecture, 283
CellBE, 282
Chandy-Lamport
    snapshot algorithm, 182
classifier rules, 423
clock gating, 342
clocks
    asynchronous, 159
    multi-synchronous, 159
cloud computing, 26
CMT architecture simulator, 350
code size, 259, 261, 263
communication
    source-synchronous, 159
communication
    graph, 132
compilation techniques, 245

compiler optimizations, 9
compilers
    picoArray, 376
component-based OS, 326
connection, 170
content-addressable memory (CAM), 424
critical section, 276
cryptography, 11
CUDA, 284
CUSTARD, 53
cut edges, 213

data dependency, 248, 251, 252, 259
data locality, 249
DDR2, 24
deadlock, 278
debug
    communication-centric, 175
    computation-centric, 175
    example, 190
    instruments
        computation-specific, 178
        protocol-specific, 178
    intrusive control, 164
    intrusive observation, 164, 170
    methods
        logical, 171
        optical, 171
        physical, 171
    monitor, 178
        communication, 179
        computation, 179
    non-intrusive observation, 170
    pervasive, 173
    process, 166
    real-time trace, 170, 171
    relative, 173
    run/stop, 170, 172
debugging, 158
    picoArray, 381–388
defect tolerance, 18
degree priority routing algorithm, 146
design browser, 383
design for test, 8

diamond property, 294
divide and conquer, 8
DPM, 95
DSP, 15, 21, 23
DVFS, 343
DVS, 95
dynamic
    power management, 346
    voltage and frequency scaling, 5

earliest deadline first (EDF), 299
edge
    congestion, 213
    dilation, 213
    expansion, 213
embedded systems, 3, 8, 9, 11
emulation, 157
endochrony, 294
energy efficiency, 5
error, 164
    certain, 164
    constant, 164
    intermittent, 164
    lack of reproducibility, 168
    permanent, 164
    transient, 164
    uncertain, 164
Esterel, 287, 288
example
    femtocell, 392
    picoArray behavioral model, 380
    picoArray VHDL, 375
    Viterbi decoder, 389

fast pattern processor (FPP), 415
fault, 164
fault
    chains, 116
    rings, 116
    tolerance, 15
fault-tolerant routing algorithm, 112
femtocell, 392, 396
FileIO on the picoArray, 387
fork and join model, 273
formal verification, 157

functional programming language (FPL), 427
functional validation, 158
fusion, 246, 249, 251

GALS, 13, 159
Garnet, 89
gated clocking, 5
general purpose OS, GPOS, 317
genetic algorithm, 14
GPGPU, 283
graceful degradation, 15
graph partitioning, 211
green systems, 4

handshake, 176
    sequence, 159
    valid-accept, 159, 181
hardening
    benefits, 391
    in picoArray products, 388
    Viterbi decoder, 389
hardware abstractions layer, 312
Hardware Architecture Description Languages, 55
hardware/software co-design, 54
heterogeneous multi-core, 310
heterogeneous multiprocessor, 8
HiBRID, 15
hierarchical task graph, 324
hierarchy, 8
HMC-SoC, 310
homogeneous multiprocessor, 8, 19

iFlow address processor, 427
iFlow classifier, 428
iFlow packet processor (iPP), 413
initiator, 159
instrumentation
    computation-specific, 180
    protocol-specific, 181
integrated access device (IAD), 406
Integrated Circuit Debug Environment, 184, 187, 190
intellectual property, 156
interconnect, 3, 12

debug control, 178, 180, 182, 183
debug data, 178, 183
event distribution, 178, 181, 182
interface description language, 327
interprocessor communication, 13
IP access points, 114
IP core, 5
IP reuse, 7, 72
IP state, 159
IPC, 19
irregular mesh topology, 112
IXP 2800 processor, 413

JouleTrack, 81

leakage reduction, 74
lexicographic dependency, 251, 252
lexicographic order, 247
LISATek platform, 57
livelock, 278
load balancing, 417, 445
logic connection, 189
logic synthesis, 6
loop shifting, 251
low power design, 73, 74
LUNA, 89
LUSTRE, 287

manufacturability, 73
manufacturing test, 157, 172, 176, 184, 187
mapping, 210
mapping function, 132
medical, 26, 27
medical imaging, 23
memory architecture, 11
memory architecture
    for network processors , 431
memory bandwidth, 9, 433
memory optimization
    application-level, 244
    techniques, 247, 248, 433
memory space, 249, 262
message, 176
    request, 163
    response, 163

message passing interface (MPI), 279, 281
METIS, 208
MIMD, 8
model
  micro-threading , 297
  process based threading , 297
models of computation, 324
modulo operator elimination, 259
Molen, 53
monitor, 178, 278
Moore's law, 13, 156
Mpeg4, 227
MPOC, 19
MPSoC, 8, 244–246, 250
MRICDF, 301
multimedia, 27
multimedia applications, 244, 247
multiprocessor, 8, 26
multiprocessor systems-on-chips, 244
multiway branch acceleration, 422
mutex, 277

NAUTY, 208
NEATO, 208
network processing unit (NPU), 401
network-on-chip, 13, 112, 170, 202
NoC, 13
  simulation models, 223
  topology models, 218
NOCEXplore, 90
node
  congestion, 213
  expansion, 213
non-determinism, 158, 162
non-pre-emptive scheduling, 210
Nostrum, 89
NPU multi-threading, 418

object code (OC), 289
observability
  internal, 158
  lack of, 167
observation
  intrusive, 170

non-intrusive, 170, 171
odd-even turn model, 118
OIP avoidance pre-routing algorithm, 112
OMAP, 5, 22
OMAP platform, 10
OMNeT++ simulation framework, 216
on-chip bus picoArray, 373
on-chip variability, 7
OpenMP, 280
operating system layer, 312
optical proximity correction, 7
optimization, 256
ORINOCO, 81
oversized IP, 112

packet data unit (PDU), 401, 408
parallel processing, 256
parallelism, 13
partitioning, 252, 256
physical design, 8
picoArray
  behavioral simulation models, 379
  compiler, 376
  debugging, 381–388
  hardware functional accelerators, 374
  inter-picoArray interface, 374
  interconnect, 373
  overview, 371
  PC101, 370
  PC102, 370, 375, 389–391
  PC20x, 370, 375, 388, 389, 391, 392
  simulation, 378–381
  tools flow, 375
PIRATE, 90
PKtool, 82
  Bluetooth, 86
  DCT, 86
  power models, 83
  power states, 85
  SATD, 86
place and switch on picoArray, 381

platform, 10, 23
platform template, 156
point-to-point topology, 202
polling, 163
polyhedral model, 247
POSIX threads, 272
power
    dissipation, 4, 6, 23
    gating, 5, 343
    management, 5, 24
    management rtos, 347
    tools, 203
power/ground noise, 341
PowerNP processor, 416
PowerOpt, 81
pre-emptive scheduling, 210
predicated execution, 422
PRO3 processor, 413
probabilistic methods, 76
probe effect, 164, 182
probes
    definition, 387
    on the picoArray, 387
    with FileIO, 388
process
    reduction, 158
    refinement, 157
processing time, 256, 259
processor
    consistency, 444
    customization, 39
program transformation, 248
programming model, 270
protocol
    communication, 159, 180
Pthreads, 275

quality of service, 180
queue management, 431, 434, 435

race condition, 276
RASIP, 52
rate monotonic algorithm (RMA), 299
real time

constraints, 341
system, 14
systems, 339
trace, 170
reconfigurable instruction set processors, 52
refinement, 167
replay
    deterministic, 172
    guided, 176
    instant, 172
    partially deterministic, 176
reproducibility
    lack of, 168
residential gateway (RG), 404
RISC NPU, 410
root cause, 158
    deduction of, 168
round trip time, 231
routing algorithm
    negative-first, 118
    north-last, 118
    west-first, 118
routing switch processor (RSP), 415
RTL power estimation, 77
run, 158, 162, 166
run-time monitoring, 340

safety-critical applications, 339
scan chains, 172, 184, 187
scheduling, 210, 345
scheduling
    shaping, 442, 451
    task scheduling, 442, 444
    traffic scheduling, 442, 450
scope, 165, 166, 168, 170
SCOTCH partitioning tool, 214
SDFG multi-threading model, 298
security, 11
semaphore, 278
short circuit current, 75
SIGNAL, 287
SimplePower, 81
simulation, 157
simultaneous multi-threading, 300

smart caching, 344
smart reflex, 5
software defined radio, 393
software organization, 311
software transactional memory (STM),
    279
source routing for deviation points
    (SRDP), 146
Spidergon STNoC, 219
sporadic processing, 339
state
    consistent global, 160
    erroneous, 164
    global, 160
    sampling, 159
static code analysis
    picoArray, 383
step
    barrier, 190, 192
    single, 190
StreamIt, 285
supercomputer, 3
switching activity, 75
synchro-tokens, 173
synchronization
    barrier, 169, 190
    semaphore, 163, 169
synchronous data flow graphs, 298
synchronous programming languages,
    286
synchrony hypothesis, 286
synthetic traffic models, 206
system level
    design, 73
    power estimation, 77
    tools, 203
system-on-chip, 3, 6, 8, 156, 244, 401

target, 159
task graph embedding, 221
task-optimized cores (TOPs), 415
TAXYS, 300
Tcl scripting picoArray tool chain,
    385
temperature, 7

temperature monitoring, 340
temporal multi-threading, 300
temporary array, 249, 254
ternary content-addressable memory
    (TCAM), 424
test
    control block, 186
    generation, 8
test access port, 176, 184
    controller, 183, 185
    test clock, 185
    test data input, 185
    test data output, 185
test-point register, 183
testability, 73
TGFF, 209
thread building blocks, 280
tiling, 246, 249, 253
time triggered architectures, 291
time-to-market, 156
timing analysis, 8
TLM, 73, 81
topology graph, 132
trace, 162
traffic management, 402, 442, 452
transaction, 163, 176, 179
    element, 176
transactional
    block, 278
    memory, 278
trie, 425
turn models, 116
turns-tables (TT), 146

unimodular transformation, 256, 260

verification, 8
VHDL programming picoArray, 375
video, 5, 9, 12, 13, 15
VIPER, 16
virtual channels, 116
VLIW, 19
voltage, 7
VxWorks, 347

warp processing, 53

Wattch, 81
weak endochrony, 294
WiMAX, 388
windows threads, 272
Wirth's law, 2
wrapper, 11

Xpipes, 89
Xtensa processor, 60
XY-deviation tables (XYDT), 146

yield, 18

9 780367 384302